T0133728

Introduction to

Software Engineering

Second Edition

Chapman & Hall/CRC Innovations in Software Engineering and Software Development

AIMS AND SCOPE

This series covers all aspects of software engineering and software development. Books in the series will be innovative reference books, research monographs, and textbooks at the undergraduate and graduate level. Coverage will include traditional subject matter, cutting-edge research, and current industry practice, such as agile software development methods and service-oriented architectures. We also welcome proposals for books that capture the latest results on the domains and conditions in which practices are most effective.

PUBLISHED TITLES

Building Enterprise Systems with ODP: An Introduction to Open Distributed Processing
Peter F. Linington, Zoran Milosevic, Akira Tanaka, and Antonio Vallecillo

Computer Games and Software Engineering
Kendra M. L. Cooper and Walt Scacchi

Evidence-Based Software Engineering and Systematic Reviews
Barbara Ann Kitchenham, David Budgen, and Pearl Brereton

Fundamentals of Dependable Computing for Software Engineers
John Knight

Introduction to Combinatorial Testing
D. Richard Kuhn, Raghu N. Kacker, and Yu Lei

Introduction to Software Engineering, Second Edition
Ronald J. Leach

Software Designers in Action: A Human-Centric Look at Design Work
André van der Hoek and Marian Petre

Software Development: An Open Source Approach
Allen Tucker, Ralph Morelli, and Chamindra de Silva

Software Engineering: The Current Practice
Václav Rajlich

Software Essentials: Design and Construction
Adair Dingle

Software Metrics: A Rigorous and Practical Approach, Third Edition
Norman Fenton and James Bieman

Software Test Attacks to Break Mobile and Embedded Devices
Jon Duncan Hagar

CHAPMAN & HALL/CRC INNOVATIONS IN SOFTWARE ENGINEERING AND SOFTWARE DEVELOPMENT

Introduction to Software Engineering

Second Edition

Ronald J. Leach

Howard University
Washington, DC, USA

CRC Press is an imprint of the
Taylor & Francis Group an **informa** business
A CHAPMAN & HALL BOOK

CRC Press
Taylor & Francis Group
6000 Broken Sound Parkway NW, Suite 300
Boca Raton, FL 33487-2742

Printed on acid-free paper
Version Date: 20150916

International Standard Book Number-13: 978-1-4987-0527-1 (Hardback)

Library of Congress Cataloging-in-Publication Data

Leach, Ronald J., author.
Introduction to software engineering / Ronald J. Leach. -- Second edition.
pages cm. -- (Chapman & Hall/CRC innovations in software engineering and software development series)
Includes bibliographical references and index.
ISBN 978-1-4987-0527-1
1. Software engineering. I. Title.

QA76.758.L33 2016
005.1--dc23 2015030294

Visit the Taylor & Francis Web site at
http://www.taylorandfrancis.com

and the CRC Press Web site at
http://www.crcpress.com

Contents

Preface to the Second Edition

Since the first edition of this book was published, there have been enormous changes to the computer software industry. The inventions of the smartphone and the tablet have revolutionized our daily lives, and have transformed many industries. Online data is ubiquitous. The pace of deployment of ever more advanced digital capabilities of modern automobiles is breathtaking. Cloud computing has the potential to both greatly increase productivity and greatly reduce the maintenance and upgrade costs for large organizations. A student studying software engineering might not even recall a time when these devices and software systems had not been ubiquitous.

An experienced software professional probably would observe that there are problems in this paradise, and that most of these problems had been around in one form or another for many years, regardless of the advances in technology. Nearly all these problems are in the area of software. Here are some examples.

The first problem is in the education of computer science students. In the mid-1990s, my university was partnering with a major research university on educational issues in computer science. An unnamed faculty member at our partner university was quoted as saying the following about the education of his students, although the same comment could have been applied to almost any college or university at that time: "We teach our students to write 300-line programs from scratch in a dead language." This is a powerful statement and had a major influence on the curriculum in my department. That influence continues to this very day.

Unfortunately, this statement is still generally applicable to the way that software development is still taught, at least in the United States. The only difference is that the dead language, which at the time was Pascal, generally is not taught anywhere.

The rest of the statement is true, because there is little systematic effort to reuse the large body of existing code for applications and subsystems. Students are encouraged to use any available software development toolkit (SDK) and to use an application programming interface (API) in much of their software development. If they have been introduced to cloud computing in their coursework, they probably have used interfaces such as HTTP, SOAP, REST, or JSON. They generally are not encouraged to reuse code written by others in any systematic way. This is somewhat surprising, because so much source code is available on the Internet, and because the effective, systematic reuse of large-scale source code components to make prototypes that morph into complete working projects is at the heart of the rapid development process known as "agile programming" or "agile development."

Lack of student exposure to, and interaction with, very large systems has other ramifications. Students often ignore the possibilities in working with such long-lived systems that provide essential services and will continue to do so for the foreseeable future.

This book will encourage reuse of existing software components in a systematic way as much as possible, with the goal of providing guidance into the way that software is currently developed and is likely to be developed in the foreseeable future, regardless of the software development life cycles that the reader may encounter, including the classical waterfall model, rapid prototyping model, spiral model, open source model, agile method, or any other technique that the reader is likely to encounter during his or her career.

Here is another example. The operating systems of smartphones and tablets are not identical to the ones most commonly used on many companies' computer offerings. This often causes interoperability problems. For example, most smartphones running some form of the Android operating system do not have quite the same version of Linux as their basis, which leads to some software incompatibilities. The Apple iOS operating system for iPhones does not have precisely the same type of file system organization as Apple computers, which, themselves, use an underlying operating system called Apple Darwin that is also based on Linux.

The relative insecurity of many smartphones and tablets causes major issues for organizations that wish to limit access to their critical data. Many banking applications (apps) deployed on smartphones to be interoperable with bank databases have had major security breaches. Even highly touted fingerprint readers have been compromised.

Most software sold in online app stores is developed without very much effort given to documentation. The general feeling is that there appears to be little concern originally intended in the app's postdeployment effort known as "software maintenance" that is so costly for traditional systems. Generally speaking, an app developer is expected to maintain the software by correcting any errors, and at no cost to the purchaser of an app. This unpaid effort can be costly to developers in terms of time and effort, but is almost always necessary in order to continue to achieve good reviews, even if the initial reviews were good.

Online data can be extremely useful, but many people, and many organizations as well, wish to limit this access. As the second edition of this book is being written in 2015, much of the news is filled with stories about the release of government data that is meant to have been confidential. The evident lack of appropriate data security is evidence of the incorrect design of databases and poor deployment of database access controls, which are, again, evidence of a software engineering problem, primarily in computer security.

This problem of data security is not unique to government systems; many private companies have had their confidential intellectual property compromised. This is another example of a problem in computer security.

Of course, everyone is well aware of the almost constant stream of information about consumer confidential information being accidently released, or even stolen, from many major companies. This is yet another example of a problem in computer security.

If you have younger siblings or friends, you may be aware of the "common application" software often used for the college admissions process. In its initial rollout in

2013, this web-based software often crashed during use, leaving applications in various states of completion, with applicants unsure if they should restart the application process. Clearly, the common application software was not efficient, reliable, or even usable at that time.

There are also technical problems at the healthcare.gov website with the enrollment in insurance exchanges that are a major part of the Affordable Care Act. The problems include slow responses, broken and incorrect links, and internal inconsistencies within the website. (As with the privately developed common application software, many of these technical implementation problems are highly likely to be fixed long before the time this book appears in print.)

The constant upgrading of automobile software is also a concern. Many problems with the entertainment and communications software have been reported in the literature. I have a personal example: my favorite car. After an upgrade of the satellite radio software at the provider's site, the software on the car kept contacting the satellite for programming data even while the car was turned off. This depleted the battery and required two tows, two replacement batteries, and two days of work by the dealer's top mechanic until the car was fixed. This is a problem in software configuration management, because the newest version of the satellite provider's system did not interface correctly with existing satellite-enabled radios. It is also a problem in software design, because there was no safety mechanism, such as a time-out, to protect the battery. Deployment of software to multiple locations is a constant issue in modern software that uses networks and a variety of end-user devices. (You might see this particular car, a convertible with a vanity license plate. If you do, please smile and buy another copy of this book. Thank you.)

The problem that my car had is only a small portion of the kinds of problems that might occur as we move toward an "Internet of things." The term refers to the way that many people think that nearly every digital device will be connected and accessible by the Internet. In this sense, far more than standard computer equipment and smartphones can be accessible from the Internet. You are probably aware of concerns about accessing power plants and the electric grid via the Internet, and the security concerns about a potential vulnerability due to security breaches. Indeed, almost every device you use in everyday life, including human-driven cars, driverless cars, elevators, thermostats, medical devices, appliances, and electronic locks, just to name a few, will be accessible from the Internet. Many companies, including Google and many automobile manufacturers, have projects to develop autonomous vehicles that will largely replace human drivers. (The first known use of the term *Internet of things* was by Kevin Ashton in the June 2009 issue of the *RFID Journal*.)

Here is a specific example with which I am very familiar: the use of the Internet for monitoring elevators. Here is the context. Nearly all modern elevators use microprocessors to control operation of the elevator cars and to monitor the elevator's "health and safety." Because efficient operation of an elevator often uses a more complex data structure than a stack or queue for scheduling, an elevator simulation has been a favorite topic of discussion in data structures books for many years. The first such elevator scheduling simulation I ever saw was in volume 1 of Donald Knuth's *The Art of Computer Programming* (Knuth, 1973).

The microprocessors often relate safety status information from all the elevator company's installed elevators back to a database in a single communications center. In most elevator companies, this database is linked to the emergency service of a local fire department in the event of system failure in the form of a car stuck between floors or a failure of some of the redundant mechanical and electrical systems. In some elevator companies' systems, the database can be examined for determination of which problems recur, so that they can be fixed either on site or by reprogramming microprocessors, thereby helping reduce expensive service calls. All of this activity takes place in an environment in which the Internet communications and database technologies are constantly changing, whereas the typical microprocessor is created specifically for a particular generation of elevators and might be difficult to replace or even reprogram. Some additional information on the issue of elevator monitoring and fault determination can be found in my paper "Experiences Analyzing Faults in a Hybrid Distributed System with Access Only to Sanitized Data" (Leach, 2010).

Cloud computing, a very hot topic now, is, in some sense, a very old idea that goes back to the early 1960s. In the 1960s, most computing was done on mainframes, to which a collection of terminals was attached. Software was installed and updated on the mainframe, and no changes were needed on the terminals. All control resided in the hands of the mainframe operators. If the mainframe went down, or the software crashed, remote users on terminals were unable to accomplish anything. Such remote users may have had no knowledge of where the mainframe was located, although they needed to understand the mainframe's hardware and software architecture to program it efficiently.

When the personal computer revolution began, control of access and software devolved to the person who controlled the personal computer on his or her desk. This created problems for system administrators, because they had to support large numbers of users, many of whom were not particularly knowledgeable about computers.

The movement to cloud computing allows system administrators to keep all the applications and data in what, to users, appears to be a single place, "the cloud," that may, in fact, be a collection of servers and networks stored in one or more data centers in a location that may not be known to the remote users.

Some of the issues in cloud computing involve scalability; that is, in having computer systems function well even if they are forced to respond to far more users, far more complex programs that implement far more complex algorithms, and far larger data sets than they were originally designed or originally intended for. A 2013 article by Sean Hull in the *Communications of the ACM* (Hull, 2013) describes some of the issues in scalability.

There are often performance issues with cloud-based software. Proper data design and security prevent remote users from having unlimited access, but poor data design and poor security make the potential access dangers of placing everything on a cloud much more serious.

Yet another concern with cloud-based systems is the ownership and management of the cloud. Two articles in the May 7, 2015, issue of the *New York Times* illustrate the point. Nick Wingfield writes in "Zynga Is Trimming Its Staff and Its Game Ambitions" about reducing the number of games produced, eliminating sports games, and moving the data center (where a player's behavior is analyzed) to a cloud service such as Amazon. A second article, "Europe Adds E-Commerce to Antitrust Inquiries," by Mark Scott, describes some European concerns about the broad reach of several large companies such as Amazon, Google, and Apple. Consider the problems that might arise if a company placed a major part of its business on a cloud platform that is being discontinued or modified due to legal or other pressures.

Some software development processes currently in use were not common when the first edition of this book was produced. Perhaps the most important new process is known as "agile programming." Agile programming is an attempt to improve the efficiency of the software development process by having software professionals experienced in a particular application domain develop software using reusable components such as APIs and subsystems whose performance and capability are well known to the developers.

There is a belief that the interaction between software and hardware in modern network-centric systems should be viewed as Software as a Service (SaaS). This notion helps with abstraction, but places serious loads on both system performance and the treatment of unexpected software interactions not described in APIs.

The term *SaaS* is often used in the context of cloud-based computing, where the term *cloud* is synonymous for remote storage of data and applications, where the actual location of the data or application used is probably not known to the user. Two of the most commonly used cloud storage services are Amazon Cloud Services and IBM Cloud Services. SaaS is often used to monetize the use of software services on a pay-as-you-go basis.

Cloud computing often includes the concept of virtualization, in which a software application executes on a remote machine that runs a virtual image of an operating system together with a set of all necessary cooperating applications. The idea is that maintenance of applications in virtual environments is much easier to control than maintaining the same applications on multiple physical computers. Virtualization is an old idea, in the sense that there have been emulators for, say, allowing Apple Mac applications to run on a computer running Microsoft Windows, allowing Windows operations to run on modern Apple computers, or run Sun Solaris applications within a Windows environment, without having the hassle of a dual-boot computer running multiple operating systems, one at a time.

Ideally, applications running in different virtual environments are sealed off from other applications. However, many questions have been raised about the security of applications running in virtualized environments, especially in cases where two different virtualized environments are executing on the same server.

Here is another example of a software problem that the use of cloud computing probably cannot solve. On May 6, 2015, I was asked to be an expert witness in a lawsuit involving a company that produced software with poor performance and security flaws. There was some question about how close the company's software was to an existing, legacy system. (This opportunity became available to me one day before the two aforementioned articles in the *New York Times*.)

It should be clear that the area of software engineering is evolving quickly, but that many of the problems we now face are older ones in new guises. The goal of this book is to provide you with the fundamentals in order to be able to have a satisfying professional career as a software engineer regardless of what changes occur in the future, even if many of these changes cannot be predicted or are even disruptive in nature. Ideally, you will understand all the software development methodologies and processes discussed in this book at a reasonably sophisticated level, and will be proficient in at least one of them.

Do not think, however, that all software engineering issues occur with new systems. The Ada programming language, although not as popular as it was in the late 1980s and early 1990s, is still used for thirty-two air traffic control systems, twenty-six commercial airplane systems, thirteen railroad transportation systems (including the English Channel Tunnel), twenty scientific space vehicles, eight commercial satellites, two nuclear power plants, one automobile robot assembly plant, eight banking systems, and, of course, many military applications. Ada's strong support for advanced software engineering principles is a primary issue in these safety-critical areas.

The second edition follows the same organization as the first edition and is comprised of nine primary chapters. Many chapters will include one or more sections that provide typical views of software managers on the relevant technical activities. The purpose of the "managerial view" is to provide you with additional perspective on software engineering, not to prepare you to be a manager in the immediate future. In most organizations, a considerable amount of varied project experience is essential before a person is given a managerial position at any level of responsibility.

Chapter 1 contains a brief introduction to software engineering. The goals of software engineering are discussed, as are the typical responsibilities of team members. Both the classical waterfall and iterative software development models such as rapid prototyping and the spiral approach are discussed. New to this chapter is extensive material on both open source and agile software development methods.

Chapter 2 contains a brief overview of project management. The intention is to present the minimum amount of information necessary to educate the student about the many types of software development environments that are likely to be encountered in the professional work force. The chapter has been rewritten to emphasize coding practices, such as reducing coupling between modules, instead of presenting long examples of actual source code as exemplars. An overview of cost and scheduling information is also given here. We also begin our discussion of a case study of agile software development. The case study will be continued throughout the book.

Chapters 3 through 8 are devoted to requirements; design; coding; testing and integration; delivery, installation, and documentation; and maintenance. Although these six chapters are presented in the order in which they would occur in the classical waterfall software development model, the material works well with other software development approaches.

We present a unique approach in Chapter 3, which is devoted to the requirements process, with heavy emphasis on the movement from preliminary, informal requirements to more complete ones that describe a system in detail and can be used as the basis for a test plan. There is a hypothetical dialogue that illustrates how requirements might be elicited from a potential customer. The goal of the chapter is to indicate the importance of requirements understanding, even if there is no known customer, such as may be the case for development of apps for mobile devices. The large software project that we will discuss in each of the remaining chapters in this book is introduced in this chapter. Issues of interoperability and user interfaces are discussed here as part of the requirements process. The requirements traceability matrix developed here is used throughout the book. The chapter also continues the case study on agile programming through the development of the requirements for an actual system.

Chapter 4 is devoted to software design. Both object-oriented and procedurally oriented design are discussed, and several commonly used design representations are given. We consider matching preexisting software patterns to the software's requirements as part of the design process. Software reuse is also emphasized here. Large-scale, reusable software components are at the heart of the continuing case study on agile software development.

The topic of coding is discussed in Chapter 5, where we examine software implementation techniques. This chapter emphasizes source code file organization, naming conventions, and other coding standards. Since the reader is presumed to have had considerable experience writing source code, this chapter is brief. The role of coding, especially coding standards, in agile development is included.

In Chapter 6, we discuss testing and integration in detail. Both "white-box" and "black-box" testing are discussed, as are testing methods for object-oriented programs. The "big-bang," "bottom-up," and "top-down" methods of software integration are discussed here. Since software integration is at the heart of many agile development projects, we describe this in our continuing case study.

Chapter 7 is quite short and is devoted primarily to delivery and installation. It also includes a discussion of internal and external documentation and a brief discussion of online help systems. Due to the unique nature of agile software development processes, the continuing case study is very brief in this chapter on delivery and installation.

Chapter 8 describes something that is foreign to most beginning software engineering students: software maintenance. It is discussed in detail because maintenance accounts for a large percentage of the funds spent on software projects and because many students will begin their professional careers as software maintainers. As was the case with Chapter 7, the continuing case study is very brief in this chapter on evolution and software maintenance.

Chapter 9 lists a set of open research problems in software engineering and provides some suggestions to help you read the existing software engineering literature.

There are three appendices: one on software patents, one on command-line arguments, and one on flowcharts.

Any book is the result of a team effort. The efforts of Senior Acquisitions Editor Randi Cohen, at Chapman & Hall/CRC Press; Richard LeBlanc, editor of the Innovations in Software Engineering and Software Development Series; many anonymous reviewers; students who read through the manuscript; and the Taylor & Francis "book team," especially Jay Margolis and Adel Rosario, are gratefully acknowledged.

Preface to the First Edition

SOFTWARE ENGINEERING LIES AT the heart of the computer revolution. Software engineering may best be described as the application of engineering techniques to develop and maintain software that runs properly and is constructed in an efficient manner.

Like most areas of computer science, software engineering has been influenced by the Internet and the ready availability of web browsers. You are certainly aware that the Internet has been instrumental in creating new job opportunities in the areas of web page design and server software development using tools such as the Common Gateway Interface (CGI) and Java. Many organizations use both the Internet and networks called "intranets," which are used only within the organization.

However, the Internet is hardly without problems. Poorly designed web pages, links that become outdated because there is no systematic plan for keeping them current, servers that crash under the load from multiple users, applications that use excessive resources, web pages and frame layouts that are unattractive when viewed on 15-inch monitors instead of the developers' 21-inch systems, and computer security failures are familiar issues to any heavy Internet user.

Less evident is the problem of separation of program functionality into servers and clients. A changing technology with relatively few standards, many of which are at least slightly inconsistent, causes additional problems. As you will see, these are essentially problems in software engineering, which we will address in this book. Software engineering is necessary for writing programs for single user machines connected to a network, multi-user machines that are either standalone or connected to networks, and to networks themselves.

Software is used in automobiles, airplanes, and many home appliances. As the boundaries between the telecommunications, entertainment, and computer industries continue to blur in multimedia and networking, the need for software will only increase, at least for the foreseeable future. Software engineering is essential if there is to be any hope of meeting the increasing needs for complex, high-quality software at affordable prices.

Almost equally important is the more subtle influence of the Internet on the process of software development. Because of the complexity of modern software systems, nearly all software development is done by teams. As you will see later is this book, networks can be used to aid in the development process.

This book is intended for juniors and seniors majoring in computer science. At some institutions, the book may also be used as a text for a graduate level course intended for those students who did not take an undergraduate course. A student reading this book will have taken several programming courses, including one that uses a modern programming language such as C, C++, Java, Ada, or Swift. At a minimum, the student will have had a follow-up course in data structures. Ideally, the student will have some experience with software projects larger than a few hundred lines of code, either as part of an internship or in formal classes.

A book intended for an audience of undergraduates must be short enough for students to read it. It must show the student that there is a major difference between the reality of most industrial software engineering projects and the small, individually written programs that they have written in most of their computer science classes. I believe that projects are an important part of any software engineering course, other than those taught to experienced professionals or advanced graduate students. Therefore, the use of appropriate team projects will be considered as essential to reinforce the material presented in the book.

Like most of us, students learn best by doing. However, their learning is enhanced if a systematic framework is provided, along with many examples of actual software industry practice where appropriate. Examples of this practical approach are present in most chapters of this book. The goal is to take students from an educational situation in which they have written relatively small programs, mostly from scratch and by themselves, and move them toward an understanding of how software systems that are several orders of magnitude more complex are developed.

Whenever it is pedagogically sound, we emphasize some approaches to software development that are used currently in industry and government. Thus, we discuss the Internet as a medium for publishing requirements and design documents, thereby making project coordination easier. This is done because many companies are either currently using, or intend to use, either the Internet or a local "intranet" for communication and coordination of software projects. This trend will certainly continue. The impact of the Java language as a "software backplane" unifying software development across multiple platforms cannot be ignored. Active-X and CORBA (Common Object Request Broker Architecture) are also heavily used in large-scale software projects with components written in more than one programming language.

Object-oriented technology (OOT) presents a special problem for the author of any introductory book on software engineering. Certainly the availability of good class libraries is having an impact on the first software engineering projects, as evidenced by the incorporation of a panel on this topic at CSC'96. The Java language is generally considered to be more object-oriented than C++ because C++ has support for both procedural and object-oriented programming, and Java enforces the object-oriented paradigm. There are more books on Java programming than on any other programming language. There are four alternative approaches to object-oriented technology in a software engineering book: ignore OOT; use only OOT; use both, thus having two parallel tracks within the same book; or use a hybrid. Each approach has proponents and opponents.

This book uses a hybrid approach, describing object-oriented design and functional decomposition as alternative approaches in a systematic way. The emphasis is on showing that the goals of software engineering; namely, to write programs that are:

- Efficient
- Reliable
- Usable
- Modifiable
- Portable
- Testable
- Reusable
- Maintainable
- Interoperable with other software
- Correct

This book contains a simple but non-trivial running example that is available in electronic form. The software example is composed of components in the C and C++ languages that will be integrated with certain commercial software applications. Multiple languages are used to reflect the reality of modern software development. The example illustrates the need for proper design and implementation of software components as part of the solution to a problem. There is emphasis on all the steps needed in software engineering: requirements; design, including design tradeoffs; testing; installation; and maintenance. Coding, with the exception of coding standards, is touched on only briefly. The book also covers computation and proper use of some software metrics such as lines of code or the McCabe cyclomatic complexity. It stresses the use of software tools whenever possible, and also includes a brief discussion of software reuse. The software associated with the book is available for testing and experimentation. There are several spreadsheets for project schedule and metrics.

The book is organized into nine chapters. Many chapters will include one or more sections that provide typical views of software managers on the relevant technical activities. The purpose of the "managerial view" is to provide you with additional perspective on software engineering, not to prepare you to be a manager in the immediate future. In most organizations, a considerable amount of varied project experience in essential before a person is given a managerial position at any level of responsibility.

Chapter 1 contains a brief introduction to software engineering. The goals of software engineering are discussed, as are the typical responsibilities of team members. Both the classical waterfall and iterative software development models such as rapid prototyping and the spiral approach are discussed.

Chapter 2 contains a brief overview of project management. The intention is to present the minimum amount of information necessary to educate the student about the type of software development environment that is likely to be encountered in the professional work force. An overview of cost and scheduling information is also given here.

We present a unique approach in Chapter 3, which is devoted to the requirements process, with heavy emphasis on the movement from preliminary, informal requirements to more complete ones that describe a system in detail and can be used as the basis for a test plan. We present a hypothetical dialogue that illustrates how requirements might be elicited from a potential customer. The large software project that we will discuss in each of the remaining chapters in this book is introduced in this chapter. Issues of interoperability and user interfaces are discussed here as part of the requirements process. The requirements traceability matrix developed here is used throughout the book.

Chapter 4 is devoted to software design. Both object-oriented and procedurally oriented design are discussed and several commonly used design representations are given. We consider matching preexisting software patterns as part of the design process. Software reuse is also emphasized here.

The topic of coding is discussed in Chapter 5, where we examine software implementation techniques. This chapter emphasizes source code file organization, naming conventions, and other coding standards. Since the reader is presumed to have had considerable experience writing source code, this chapter is brief.

In Chapter 6, we discuss testing and integration in detail. Both "white-box" and "black-box" testing are discussed, as are testing methods for object-oriented programs. The "big-bang," "bottom-up," and "top-down" methods of software integration are discussed here.

Chapter 7 is quite short and is devoted primarily to delivery and installation. It also includes a discussion of internal and external documentation and a brief discussion of online help systems.

Chapter 8 describes something that is foreign to most beginning software engineering students: software maintenance. It is discussed in detail because maintenance accounts for a large percentage of the funds spent on software projects and because many students will begin their professional careers as software maintainers.

Chapter 9 lists a set of open research problems in software engineering and provides some suggestions to help you read the existing software engineering literature.

Some of the design documents, software, and spreadsheets described in this book are available for you to download from the website http://imappl.org/~rjl/Software-Engineering.

Any book is the result of a team effort. The efforts of the editor, Dawn Mesa at CRC Press, Taylor & Francis, many anonymous reviewers, students who read through the manuscript, and the CRC Press/Taylor & Francis "book team," especially Helena Redshaw and Schuyler Meder, are gratefully acknowledged.

To the Instructor and the Reader

The subject of software engineering is a vast one and is far too large to be covered in detail in any single volume of reasonable size. I have chosen to provide coverage of what I believe are the basics, at a level of detail that is appropriate in an age where much of the critical information is available for free on the World Wide Web. Nearly all topics discussed herein include some examples, often historical ones, if there is sufficient information on both the successes of software projects using the ideas behind these topics and some of the pitfalls that occurred using these ideas. State-of-the-art information is provided when appropriate to those software engineering trends that can be expected to be important for the foreseeable future. The choices made in this book are strongly motivated by the likelihood that you will have many types of positions in your (I hope, long and distinguished) careers in many different technology environments.

The book contains a discussion of a major software project that is intended to provide a basis for any term project that the instructor may provide. I believe that a first course in software engineering is greatly improved by students formed into teams to complete a software project making use of many of the ideas in this book. There is also a case study of an actual complex project that was created using an agile development process. Of course, any team software development within an academic course is artificial, in the sense that the pressures typically involved in industry are not generally present within a single class project.

Note that this section is addressed to both instructors and readers. This is because I think that students in their software teams and instructors must all work together to ensure that the learning process is as efficient as possible. Often the instructor will act as both a "customer" and a "manager's manager," who interfaces primarily with the software team's leader. In some cases, the team projects will have external customers, either at the college or university, or with a representative of some potential employer. In most other situations, the instructor fills both roles. The situation in these class projects provides an idealized model of what a student may encounter in industry.

The student is expected to be able to write programs (at least small ones) competently before enrolling in a software engineering course. They should be open to the idea that there are many things they need to learn in order to help create the much larger software systems that affect every part of modern life. Design techniques are illustrated, but students are expected to be at least familiar with the design representations they may need before entering the course. There is no attempt to provide extensive tutorials on dataflow diagrams and UML, for example. Even a student who expects to be an independent entrepreneur will benefit from the discussions in this book.

The instructor is expected to determine which software development life cycle will be chosen for primary discussion in the course: classical waterfall model, rapid prototyping model, spiral model, open source model, agile method, or a similar life cycle. He or she may select one particular software development life cycle for use by every team in the class, allow the students to select their own particular software development life cycle, or even have the entire class work on the same project, each group using a different software development life cycle and comparing the results.

The instructor may have additional goals, depending on the institution's curriculum: assessing and improving the student's performance in the many deliverable products that are in the form of technical reports or presentations. (It is my experience talking to hundreds of employers during my nine years as a department chair, that most employers place a high value on excellence in written and oral communications.)

In an age when an enormous amount of source code is freely posted and there is so much sharing of ideas among students in multiple universities, there is little benefit to demanding that students do not copy code from other sources. Indeed, they should be encouraged to reuse code, as long as they attribute their sources for source code components, and have a systematic way of determining what the software component actually does, how it is documented, and provide some estimate of the component's quality (perhaps by looking at the code structure).

CHAPTER 1

Introduction

1.1 THE NEED FOR SOFTWARE ENGINEERING

The first commonly known use of the term *software engineering* was in 1968 as a title for a NATO conference on software engineering. An article by A. J. S. Rayl in a 2008 NASA magazine commemorating NASA's fiftieth anniversary states that NASA scientist Margaret Hamilton had coined the term earlier (Rayl, 2008).

The nature of computer software has changed considerably in the last forty-five or so years, with accelerated changes in the last fifteen to twenty. Software engineering has matured to the extent that a common software engineering body of knowledge, known by the acronym SWEBOK, has been developed. See the article "Software Engineering Body of Knowledge" (SWEBOK, 2013) for details. Even with this fine collection of knowledge of the discipline of software engineering, the continuing rapid changes in the field make it essential for students and practitioners to understand the basic concepts of the subject, and to understand when certain technologies and methodologies are appropriate—and when they are not. Providing this foundational knowledge is the goal of this book. You will need this foundational knowledge to be able to adapt to the inevitable changes in the way that software will be developed, deployed, and used in the future.

We begin with a brief history. In the late 1970s and early 1980s, personal computers were just beginning to be available at reasonable cost. There were many computer magazines available at newsstands and bookstores; these magazines were filled with articles describing how to determine the contents of specific memory locations used by computer operating systems. Other articles described algorithms and their implementation in some dialect of the BASIC programming language. High school students sometimes made more money by programming computers for a few months than their parents made in a year. Media coverage suggested that the possibilities for a talented, solitary programmer were unlimited. It seemed likely that the computerization of society and the fundamental changes caused by this computerization were driven by the actions of a large number of independently operating programmers.

However, another trend was occurring, largely hidden from public view. Software was growing greatly in size and becoming extremely complex. The evolution of word processing software is a good illustration.

In the late 1970s, software such as Microsoft Word and WordStar ran successfully on small personal computers with as little as 64 kilobytes of user memory. The early versions of WordStar allowed the user to insert and delete text at will, to cut and paste blocks of text, use italics and boldface to set off text, change character size, and select from a limited set of fonts. A spelling checker was available. A small number of commands were allowed, and the user was expected to know the options available with each command. Lists of commands were available on plastic or cardboard templates that were placed over the keyboard to remind users of what keystroke combinations were needed for typical operations. The cardboard or plastic templates were generally sold separately from the software.

Microsoft Word and WordStar have evolved over time, as has most of their competition, including Apple's Pages word processing software. Nearly every modern word processing system includes all the functionality of the original word processors. In addition, modern word processing software usually has the following features:

- There is a graphical user interface that uses a mouse or other pointing device. (On tablets and smartphones, this interface is based on the touch and movement of one or more fingers.)
- There is a set of file formats in which a document can be opened.
- There is a set of file formats in which a document can be saved.
- There is a set of conversion routines to allow files to be transferred to and from different applications.
- There is a large set of allowable fonts.
- The software has the capability to cut and paste graphics.
- The software has the capability to insert, and perhaps edit, tables imported from a spreadsheet.
- The software has the capability to insert, and perhaps edit, other nontextual material.
- There are facilities for producing word counts and other statistics about the document.
- There are optional facilities for checking grammar.
- The software has the capability for automatic creation of tables of contents and indices.
- The software has the capability to format output for different paper sizes.
- The software has the capability to print envelopes.
- The software has the capability to compose and format pages well enough to be considered for "desktop publishing." Many inexpensive printers are capable of producing high-quality output.
- Automatic backups are made of documents, allowing recovery of a file if an unexpected error in the word processing software or a system crash occurs.

Because of the proliferation of printers, a word processing system must contain a large number of printer drivers. Most word processing systems often include an online help facility.

The added complexity does not come free, however. The late 1990s versions of most word processors required nine 1.44 MB floppy disks for its installation and the executable file itself was larger than four megabytes. Now such software is sold on CDs or DVDs, as downloads, or as a preinstalled option whose cost is included in the price of new computers.

Some are sold both as stand-alone systems and as part of "office suites" that are integrated with other applications, such as spreadsheets, presentation graphics software, and database management software. Microsoft Word is generally sold in such suites. Apple iWorks has recently been made available for free, and many open source packages that include many of the necessary capabilities are available also. Before Apple made the decision to make iWorks free, its three primary components, Pages, Numbers, and Keynote, were sold both separately and as an integrated suite.

The need to support many printers and to allow optional features to be installed increases word processing systems' complexity. Most printer drivers are available as downloads.

The latest versions of most word processing systems consist of many thousands of files. New releases of the word processing software must occur at frequent intervals. If there are no releases for a year or so, then many users who desire additional features may turn to a competitor's product.

Even if a single individual understood all the source code and related data files needed for a new release of the word processing software, there would not be enough time to make the necessary changes in a timely manner. Thus, the competitive nature of the market for word processing software and the complexity of the products themselves essentially force the employment of software development teams. This is typical of modern software development—it is generally done by teams rather than by individuals. The members of these teams are often referred to as "software engineers." Software engineers may work by themselves on particular projects, but the majority of them are likely to spend most of their careers working as part of software development teams. The teams themselves will change over time due to completion of old projects, the start of new projects, and other changes in individuals' careers and rapid technological changes.

The issues involved in the word processing software systems discussed in this section are typical of the problems faced by software engineers. The requirement of being able to cut and paste graphics and tables is essentially forced by the marketplace. This, in turn, requires that the cut-and-paste portion of the word processing software must interface with graphics and spreadsheet applications.

The interface can be created for each pair of possible interoperable applications (word processor and graphics package, word processor and spreadsheet, Internet browser, etc.). Alternately, there can be a single standard interface between each application (word processor, graphics package, spreadsheet, etc.) and the operating system, or some other common software. Figures 1.1 and 1.2 illustrate the two approaches. Note that Figure 1.1 indicates a system that has a central core that communicates between the various applications and the operating system.

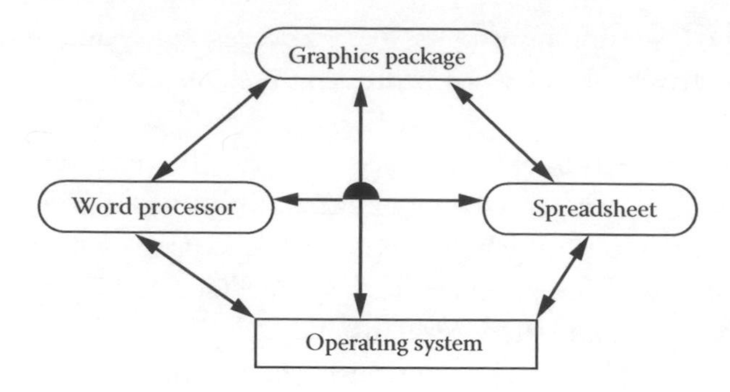

FIGURE 1.1 A complex interconnection system.

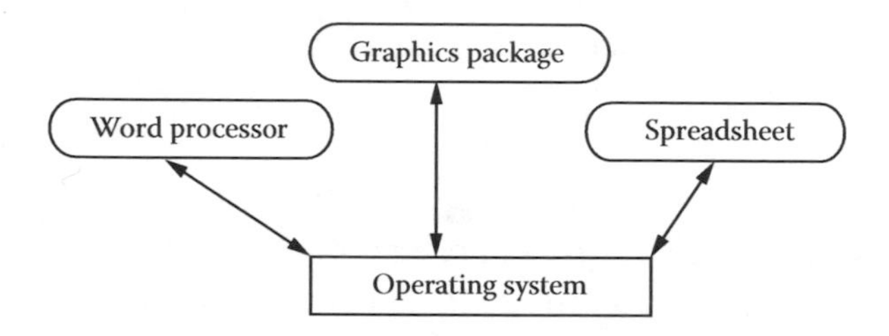

FIGURE 1.2 A conceptually simpler interconnection system.

The design illustrated in Figure 1.2 is conceptually simpler, with the major complications of device drivers, standards, and interfaces hidden in the interfaces. In either case, the word processing software must adhere to a previously specified interface. This is typical of the software industry; software that does not interface to some existing software is rare.

This simple model illustrates the need for a systematic approach to the issue of system complexity. Clearly, the complexity of a software system is affected by the complexity of its design.

There are also quality issues involved with software. Suppose that, in its rush to release software quickly and beat its competitors to market, a company releases a product that contains serious flaws. As a hypothetical example, suppose that a word processor removes all formatting information from a file, including margins, fonts, and styles, whenever the sequence of commands Save, Check spelling, Insert page break, and Save is entered. It is unlikely that any customer using this inadequately tested version of the word processing software would ever use this product again, even if it were free.

Sometimes the decision to add new features to a product may be based on technological factors such as the Internet or the cloud. At the time that this book is being written, several companies that produce word processing software are developing new applications that are network based and require subscriptions instead of an upfront cost. Several options are possible, regardless of whether data is stored either locally or remotely:

1. Have all new software reside on the user's local computer, as is presently the case for word processors for personal computers
2. Have a remote application be invoked over a network whenever the user selects a previously stored local document

3. Have the core of the application reside on the local computer, with specialized features invoked from a remote server only if needed

4. Have both the document and the remote application reside on the remote server

The advantage of the first alternative is that there is no change in the company's strategy or basic system design. The risk is a lack of ability to perform advanced operations such as having a document that can be easily shared by several users who are widely scattered, or that recovery might be compromised. There is also risk of the popular perception that the software is not up to date in its performance.

Using the second alternative means that the distribution costs are essentially reduced to zero and that there is a steady revenue stream automatically obtained by electronically billing users. The software must become more complex when issues such as security of data, security of billing information, and performance response become very important.

The third alternative has some of the best features of the first two. For example, distribution costs are reduced and the minimal core of the software, which resides on a local computer, can be smaller and simpler. Unfortunately, this alternative also shares the common disadvantages of the first two in terms of technology and complexity.

The fourth alternative is the natural extension of the second and third alternatives. There are some performance drawbacks to this approach, as any user of the Internet is aware.

Performance issues are critical because of potential delays in having, say, formatting changes appear quickly on a user's screen if remote storage is used.

There are obvious security issues if a user's critical or confidential data is stored on a remote server beyond the control of a user or the user's organization. The article by Richard Stallman of the Free Software Foundation (Stallman, 2010) is well worth reading on this subject.

Regardless of the choice made, it is clear that most word processing software will become more complex in the future. Word processing programs, including the one used to write this book, must interface with multiple applications to share printers and utility functions. They must also share data with other applications.

The problems just described for producers of word processing software are similar to those of software development organizations producing software for a single client or a small set of potential clients. For example, software used to control the movements of multiple trains sharing the same track system for one geographical area of a country's railway system must be able to interface with software used for similar purposes in another geographical area.

This is not an abstract problem. The coordination of independently developed local systems can be a problem. In Australia, the lack of coordination of even the size of railroad tracks in the different states and territories caused major problems in the development of an Australian national railroad system until national standards were developed and adhered to. There recently have been similar software problems with standardization of railroad speed monitoring in the United States. (Fortunately, in the United States, track sizes have been standardized for many years.)

Similar requirements hold for the computer software used to control aircraft traffic, the software used within the airplanes themselves, the central core of a nuclear power plant, and chemical processes in an oil company, or to monitor the dozens of medicines delivered to a patient in a hospital.

Such systems have an even higher degree of testing and analysis than does word processing software, because human life and safety are at stake. The term *safety-critical* is used to describe such systems. Obviously safety-critical software systems require more care than, for example, computer games.

Note, however, that many computer games are very complex and place many demands on a computer to achieve the level of realism and speed that many users now demand. This is the reason that many games are written for powerful game machines such as those found in the Microsoft Xbox and Nintendo 3DS families than for personal computers. In another direction, some relatively simple games written for smartphones, such as *Angry Birds* with its rudimentary graphics, have been wildly successful.

What will happen to the computer gaming industry? I mentioned the changes planned by game-producing company Zynga in the "Preface to the Second Edition" of this book. These were done in reaction to current and expected market trends.

At a 2008 talk to an overflow crowd at the Engineering Society of Baltimore located in the historic Mount Vernon area of the city, Sid Meier of Firaxis Games said something that amazed members of the audience not familiar with technical details of his career: He programmed in C! It was basically Objective C, but he felt he got the most control of the game-playing rules and logic doing this, leaving development of video, sound, music, and interaction with the object-oriented game engine to others. This division of labor largely continues at Firaxis even now. The point is, many older technologies continue in use for a long time, and are even critically important in what we might consider to be state-of-the-art systems.

Let us summarize what we have discussed so far. We talked about the explosive growth of personal computers in the 1980s and 1990s. The growth has continued to the present day, with computing now done on smartphones, tablets, and other devices. As indicated, many of the initial versions of some of the earlier software products have evolved into very large systems that require more effort than one individual can hope to devote to any project. What is different now?

A moment's thought might make you think that standards such as Hypertext Markup Language (HTML) and the Java programming language with its application programming interfaces have changed everything. There are more recent developments. Application frameworks and cloud computing have been major players, along with highly mobile devices such as smartphones and tablets with their specialized user interfaces.

There are also inexpensive, high-quality application development frameworks and software development kits. There are many sixteen-year-olds who are making a large amount of money as web page designers, although this is relatively less frequent in 2014, since many elementary school students in the United States are taught how to design a web page before they reach their teens. One of the job skills most in demand now is "web master"—a job title that did not even exist in 1993. A casual reading of online job listings might lead you to believe that we have gone back to the more freewheeling days of the 1980s.

Certainly, the effects of technology have been enormous. It is also true that nearly anyone who is so inclined can learn enough HTML in a few minutes to put together a flashy web page.

However, the problem is not so simple. Even the most casual user of the Internet has noticed major problems in system performance. Delays make waiting for glitzy pictures and online animations very unappealing if they slow down access to the information or services that the user desired. They are completely unacceptable if they cannot display properly on many portable devices with small screen "real estate."

Proper design of websites is not always a trivial exercise. As part of instruction in user interface design, a student of mine was asked to examine the main websites of a number of local universities to obtain the answer to a few simple questions. The number of selections (made by clicking a mouse button) ranged from five to eleven for these simple operations. Even more interaction was necessary in some cases because of the need to scroll through online documents that were more than one screen long, or even worse, more than one screen wide. Efficiency of design is often a virtue.

Several issues may not be transparent to the casual user of the Internet. Perhaps the most troublesome is the lack of systematic configuration management, with servers moving, software and data being reorganized dynamically, and clients not being informed. Who has not been annoyed by the following famous message that appears so often when attempting to connect to an interesting website?

```
ERROR 404: File not found.
```

It is obvious what happened to the information that the user wanted, at least if there was no typing error. As we will see later in this book, maintenance of websites is often a form of "configuration management," which is the systematic treatment of software and related artifacts that change over time as a system evolves.

There are also major issues in ensuring the security of data on servers and preventing unwanted server interaction with the user's client computer. Finally, designing the decomposition of large systems into client and server subsystems is a nontrivial matter, with considerable consequences if the design is poor.

It is clear that software engineering is necessary to have modern software development done in an efficient manner. These new technologies have refocused software engineering to include the effects of market forces on software development. As we will see, these new technologies are amenable to good software engineering practice.

We will consider project size and its relationship to software teams in more detail in the next section.

1.2 ARE SOFTWARE TEAMS REALLY NECESSARY?

You might plan to be a developer of stand-alone applications for, say, smartphones, and wonder if all this formality is necessary. Here are some of the realities. There are well over one million apps for iPhones and iPads on the App Store and as a result, it can be hard to have people locate your app. You may have heard about the programmer who developed an app that sold 17,000 units in one day and quit his job to be a full-time app developer. Perhaps you heard about Nick D'Aloisio, a British teenager who sold an application named *Summly* to Yahoo! The price was $30 million, which is a substantial amount of money for someone still in high school. (See the article at http://articles.latimes.com/2013/mar/26/business/la-fi-teen-millionaire-20130326.)

Some anecdotal information appears to suggest that teams of software engineers are not necessary for some systems that are larger than the typical smartphone app. For example, the initial version of the MS-DOS operating system was largely developed by two people, Bill Gates and Paul Allen of Microsoft. The Apple DOS disk operating system for the hugely successful Apple II family of computers was created by two people, Steve Jobs and Steve Wozniak, with Wozniak doing most of the development. The UNIX operating system was originally developed by Dennis Ritchie and Kenneth Thompson (Ritchie and Thompson, 1978). The first popular spreadsheet program, VisiCalc, was largely written by one person, Dan Bricklin. The list goes on and on.

However, times have changed considerably. The original PC-DOS operating system ran on a machine with 64K of random access, read-write memory. It now runs only as an application in a terminal window whose icon is found in the accessories folder on my PC. The Apple DOS operating system used even less memory; it ran on the Apple II machine with as little as 16K memory. UNIX was originally developed for the PDP-11 series of 16-bit minicomputers. Versions of VisiCalc ran on both the Apple II and the 8086-based IBM PC. The executable size of their successor programs is much, much larger.

Indeed, PC-DOS has long been supplanted to a great extent by several variants of Microsoft Windows such as Windows 8 and 10. Apple DOS was supplanted by Mac OS, which has been supplanted by OS X, Yosemite, and the like. UNIX still exists, but is much larger, with the UNIX kernel, which is the portion of the operating system that always remains in memory, being far larger than the entire memory space of the PDP-11 series computers for which it was originally written. The size of the UNIX kernel led Linus Torvalds to develop a smaller kernel, which has grown into the highly popular Linux operating system.

VisiCalc no longer exists as a commercial product, but two of its successors, Excel and Numbers, are several orders of magnitude larger in terms of the size of their executable files. The current version of Excel consists of millions of lines of source code written in C and related languages. See the book and article by Michael Cusumano and Richard Selby (Cusumano and Selby, 1995, 1997) for more information on Excel.

The additional effort to develop systems of this level of complexity does not come cheap. In his important book *The Mythical Man-Month*, Fred Brooks gives the rule of thumb that a software project that was created by one or two persons requires an additional eight or nine times the original development effort to change it from something that can be used only by its originators to something useful to others (Brooks, 1975). Brooks's rule of thumb appears to be valid today as well. It was interesting to have Brooks's book quoted by Hari Sreenivasan when discussing problems with the deployment of the healthcare.gov website for enrollment in the insurance exchanges available under the Affordable Care Act (*PBS NewsHour*, October 23, 2013.)

It is important to understand the distinction between initial prototype versions of software, which are often by very small groups of people, and commercial software, which requires much larger organizational structures. Today, any kind of commercial software requires a rock-solid user interface, which is usually graphical, at least on desktop devices, laptops, and larger, full-screen tablets. User interfaces for smartphones may be much simpler in order to be useful on these devices' smaller screens. Heads-up and other wearable devices, such as Google Glass, have their own interface issues, which, unfortunately, are far beyond the scope of this book.

Testing is essential in modern software development, and a technical support organization to answer customers' questions is necessary. Documentation must be clear, complete, and easy to use for both the first-time user who is learning the software and the experienced user who wishes to improve his or her productivity by using the advanced features of the package. Thus, the sheer pressure of time and system size almost always requires multiple individuals in a software project. If development of a software product that fills a need requires six months, multiplication by Brooks's conservative factor of eight means that the software will take at least four years to be ready for release, assuming that the time needed and the effort needed in terms of person-hours scale up in the same way. Of course, the product may be irrelevant in four years, because the rest of the software industry is not standing still. (This is one reason that agile software processes have become more popular.)

Most successful projects that have given birth to software companies have gotten much larger in terms of the number of people that they employ. Clearly, even software that was originally developed by one or two entrepreneurs is now developed by teams. Even the wildly successful companies and organizations that make Internet browsers, the initial versions of which were largely due to the efforts of a small team, are now employers of many software engineers in order to increase the number of available features, to ensure portability to a number of environments, to improve usability by a wide class of users, and even to provide sufficient documentation.

How large should such a software team be? The most important factor seems to be the size of the project, with capability of the team members, schedule, and methodology also important. We will discuss estimation of software project size in Chapter 2. For now, just remember one thing: your software is likely to be developed within a team environment.

Modern teams may be highly distributed, and some major portions of a team project may be totally outsourced. Outsourcing brings its own set of complications, as the experience of two of my relatives, one with a major project in electronic healthcare records systems and the other with the interface from a website to a database, shows. A 2013 article by Lisa Kaczmarczyk has an interesting discussion of some of the cultural issues with coordination of outsourced software development (Kaczmarczyk, 2013). Just keep in mind that the problem of coordination with entities performing this outsourced work can be highly complex, and that you can learn a lot by observing how senior personnel or management in your organization handle the problems with outsourced software that will inevitably occur.

1.3 GOALS OF SOFTWARE ENGINEERING

Clearly organizations involved with producing software have a strong interest in making sure that the software is developed according to accepted industry practice, with good quality control, adherence to standards, and in an efficient and timely manner. For some organizations, it is literally a matter of life and death, both for the organization and for potential users of the software. *Software engineering* is the term used to describe software development that follows these principles.

Specifically, the term *software engineering* refers to a systematic procedure that is used in the context of a generally accepted set of goals for the analysis, design, implementation, testing, and maintenance of software. The software produced should be efficient, reliable, usable, modifiable, portable, testable, reusable, maintainable, interoperable, and correct. These terms refer both to systems and to their components. Many of the terms are self-explanatory; however, we include their definitions for completeness. You should refer to the recently withdrawn, but still available and useful, IEEE standard glossary of computer terms (IEEE, 1990) or to the "Software Engineering Body of Knowledge" (SWEBOK, 2013) for related definitions.

Efficiency—The software is produced in the expected time and within the limits of the available resources. The software that is produced runs within the time expected for various computations to be completed.

Reliability—The software performs as expected. In multiuser systems, the system performs its functions even with other load on the system.

Usability—The software can be used properly. This generally refers to the ease of use of the user interface but also concerns the applicability of the software to both the computer's operating system and utility functions and the application environment.

Modifiability—The software can be easily changed if the requirements of the system change.

Portability—The software system can be ported to other computers or systems without major rewriting of the software. Software that needs only to be recompiled in order to have a properly working system on the new machine is considered to be very portable.

Testability—The software can be easily tested. This generally means that the software is written in a modular manner.

Reusability—Some or all of the software can be used again in other projects. This means that the software is modular, that each individual software module has a well-defined interface, and that each individual module has a clearly defined outcome from its execution. This often means that there is a substantial level of abstraction and generality in the modules that will be reused most often.

Maintainability—The software can be easily understood and changed over time if problems occur. This term is often used to describe the lifetime of long-lived systems such as the air traffic control system that must operate for decades.

Interoperability—The software system can interact properly with other systems. This can apply to software on a single, stand-alone computer or to software that is used on a network.

Correctness—The program produces the correct output.

These goals, while noble, do not help with the development of software that meets these goals. This book will discuss systematic processes and techniques that aid in the efficient development of high-quality software. The software systems that we will use as the basis of our discussion in this book are generally much too large to be developed by a single person. Keep in mind that we are much more interested in the process of developing software, such as modern word processors that will be used by many people, than in writing small programs in perhaps obsolete languages that will be used only by their creators.

1.4 TYPICAL SOFTWARE ENGINEERING TASKS

There are several tasks that are part of every software engineering project:

- Analysis of the problem
- Determination of requirements
- Design of the software
- Coding of the software solution
- Testing and integration of the code
- Installation and delivery of the software
- Documentation
- Maintenance
- Quality assurance
- Training
- Resource estimation
- Project management

We will not describe either the analysis or training activities in this book in any detail. Analysis of a problem is very often undertaken by experts in the particular area of application, although, as we shall see later, the analysis can often benefit from input from software engineers and a variety of potential users. Think of the problem that Apple's software designers faced when Apple decided to create its iOS operating system for smartphones using as many existing software components from the OS X operating system as possible.

Apple replaced a Linux-based file system where *any* running process with proper permissions can access *any* file to one where files are in the province of the particular application that created them. The changes in the user interface were also profound, using the "capacitive touch" where a swipe of a single finger and a multifinger touch are both part of the user interface.

Unfortunately, a discussion of software training is beyond the scope of this book.

We now briefly introduce the other activities necessary in software development. Most of these activities will be described in detail in a separate chapter. You should note that these tasks are generally not performed in a vacuum. Instead, they are performed as part of an organized sequence of activities that is known as the "software life cycle" of the organization. Several different approaches to the sequencing of software development activities will be described in the next section about software life cycles. For now, just assume that every software development organization has its own individual software life cycle in which the order of these activities is specified.

The requirements phase of an organization's software life cycle involves precisely determining what the functionality of the system will be. If there is a single customer, or set of customers, who is known in advance, then the requirements process will require considerable discussion between the customer and the requirements specialists on the software team. This scenario would apply to the process of developing software to control the flight of an airplane.

If there is no immediately identifiable customer but a potential set of individual customers, then other methods such as market analysis and preference testing might be used. This approach might be appropriate for the development of an app for a smartphone or tablet, or a simple software package for the Internet.

The process of developing software requirements is so important that we will devote all of Chapter 3 to this subject.

The design phase involves taking the requirements and devising a plan and a representation to allow the requirements to be translated into source code. A software designer must have considerable experience in design methodology and in estimating the trade-offs in the selection of alternative designs. The designer must know the characteristics of available software systems, such as databases, operating systems, graphical user interfaces, and utility programs that can aid in the eventual process of coding. Software design will be discussed in Chapter 4.

Coding activity is most familiar to students and needs not be discussed in any detail at this point. We note, however, that many decisions about coding in object-oriented or procedural languages might be deferred until this point in the software life cycle, or they might have been made at the design or even the requirements phase. Since coding standards are often neglected in many first- and second-year programming courses, and they are critical in both large, industrial-type systems and for small apps that are evaluated for quality in an app store before they are allowed to be placed on sale, coding standards will be discussed in some detail. Software coding will be covered in Chapter 5.

A beginning software engineer is very likely to be assigned to either software testing or software maintenance, at least in a large company. Software testing is an activity that often begins after the software has been created but well before it is judged ready to be released to its customer. It is often included with software integration, which is the process of combining separate software modules into larger software systems. Software integration also often requires integration of preexisting software systems in order to make large ones. Software testing and integration will be discussed in Chapter 6.

Documentation includes much more than simply commenting the source code. It involves rationales for requirements and design, help files, user manuals, training manuals, technical guides such as programming reference manuals, and installation manuals. Even internal documentation of source code is much more elaborate than is apparent to a beginning programmer. It is not unusual for a source code file to have twice as many lines of documentation and comments as the number of lines that contain actual executable code.

After the software is designed, coded, tested, and documented, it must be delivered and installed. Software documentation, delivery, and installation will be discussed in Chapter 7.

The term *software maintenance* is used to describe those activities that are done after the software has been released. Maintenance includes correcting errors found in the software; moving the software to new environments, computers, and operating systems; and enhancing the software to increase functionality. For many software systems, maintenance is the most expensive and time-consuming task in the software development life cycle. We have already discussed maintenance of stand-alone apps. Software maintenance will be discussed in more detail in Chapter 8.

Of course, all these activities must be coordinated. Project management is perhaps the most critical activity in software engineering. Unfortunately, it is difficult to truly understand the details of project management activities without the experience of actually working on a large-scale software development project. Therefore, we will be content to present an overview of project management activities in Chapter 2. Each of the later chapters will also include a section presenting a typical managerial viewpoint on the technical activities discussed in the chapter.

Quality assurance, or QA, is concerned with making certain that both the software product that is produced and the process by which the software is developed meet the organization's standards for quality. The QA team is often responsible for setting the quality standards. In many organizations, the QA team is separate from the rest of the software development team. QA will be discussed throughout the book rather than being relegated to a separate chapter. We chose this approach to emphasize that quality cannot simply be added near the end of a software project's development. Instead, the quality must be a result of the engineering process used to create the software.

1.5 SOFTWARE LIFE CYCLES

In the previous section, we discussed the different activities that are typically part of the systematic process of software engineering. As we saw, there are many such activities. Knowing that these activities will occur during a software product's lifetime does not tell you anything about the timing in which these activities occur. The activities can occur in a rigid sequence, with each activity completed before the next one begins, or several of the activities can occur simultaneously. In some software development organizations, most of these activities are performed only once. In other organizations, several of the activities may be iterated several times, with, for example, the requirements being finalized only after several designs and implementations of source code have been done. The timing of these activities and the choice between iterative and noniterative methods are often described by what are known as software development models.

We will briefly discuss six basic models of software development life cycles in this introductory section:

- Classical waterfall model
- Rapid prototyping model
- Spiral model
- Market-driven model
- Open source development model
- Agile development model

(Only the first four life cycle models listed were described in the first edition of this book. There is a seventh software development life cycle, called hardware and software concurrent development, or codevelopment, that is beyond the scope of this book.)

Each of these software life cycle models will be discussed in a separate section. For simplicity, we will leave documentation out of our discussion of the various models of software development. Keep in mind that the models discussed describe a "pure process"; there are likely to be many variations on the processes described here in almost any actual software development project.

1.5.1 Classical Waterfall Model

One of the most common models of the software development process is called the classical waterfall model and is illustrated in Figure 1.3. This model is used almost exclusively in situations where there is a customer for whom the software is being created. The model is essentially never applied in situations where there is no known customer, as would be the case of software designed for the commercial market to multiple retail buyers.

In the simplest classical waterfall model, the different activities in the software life cycle are considered to occur sequentially. The only exception is that two activities that are adjacent in Figure 1.3 are allowed to communicate via feedback. This artificiality is a major reason that this model is no longer used extensively in industry. A second reason is that technology changes can make the original set of requirements obsolete. Nonetheless, the model can be informative as an illustration of a software development life cycle model.

A life cycle model such as the one presented in Figure 1.3 illustrates the sequence of many of the major software development activities, at least at a high level. It is clear, for example, that specification of a system's requirements will occur before testing of individual modules. It is also clear from this abstraction of the software development life cycle

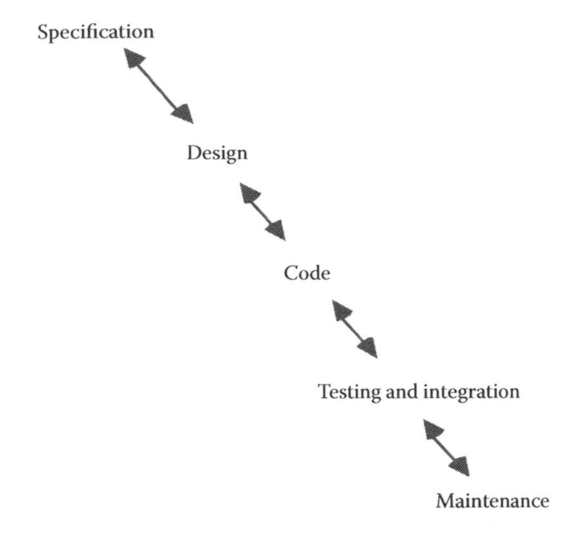

FIGURE 1.3 A simplified illustration of the classical waterfall life cycle model.

that there is feedback only between certain pairs of activities: requirements specification and software design; design and implementation of source code; coding and testing; and testing and maintenance.

What is not clear from this illustration is the relationship between the different activities and the time when one activity ends and the next one begins. This stylized description of the classical waterfall model is usually augmented by a sequence of milestones in which each two adjacent phases of the life cycle interface. A simple example of a milestone chart for a software development process that is based on the classical waterfall model is given in Figure 1.4.

Often, there are several opportunities for the software system's designers to interact with the software's requirements specification. The requirements for the system are typically presented in a series of requirements reviews, which might be known by names such as the preliminary requirements review, an intermediate requirements review, and the

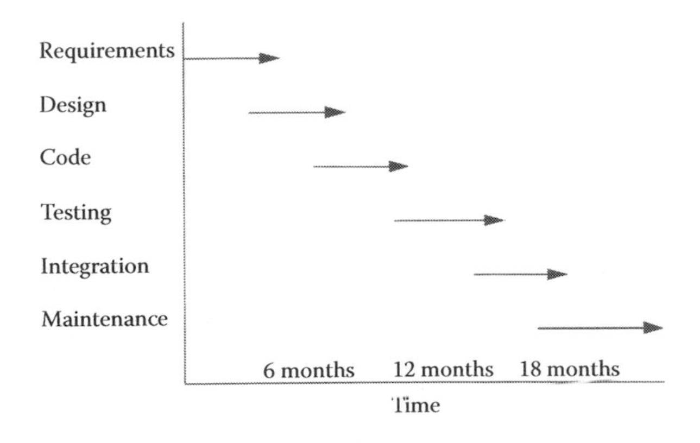

FIGURE 1.4 An example of a milestone chart for a classical waterfall software development process.

final (or critical) requirements review. In nearly all organizations that use the classical waterfall model of software development, the action of having the requirements presented at the critical requirements review accepted by the designers and customer will mark the "official" end of the requirements phase.

There is one major consequence of a decision to use the classical waterfall model of software development. Any work by the software's designers during the period between the preliminary and (accepted) final, or critical, requirements reviews may have to be redone if it is no longer consistent with any changes to the requirements. This forces the software development activities to essentially follow the sequence that was illustrated in Figure 1.3.

Note that there is nothing special about the boundary between the requirements specification and design phases of the classical waterfall model. The same holds true for the design and implementation phases, with system design being approved in a preliminary design review, an intermediate design review, and the final (or critical) design review. The pattern applies to the other interfaces between life cycle phases as well.

The classical waterfall model is very appealing for use in those software development projects in which the system requirements can be determined within a reasonable period and, in addition, these requirements are not likely to change very much during the development period. Of course, there are many situations in which these assumptions are not valid.

1.5.2 Rapid Prototyping Model

Due to the artificiality of the classical waterfall model and the long lead time between setting requirements and delivering a product, many organizations follow the rapid prototyping or spiral models of software development. These models are iterative and require the creation of one or more prototypes as part of the software development process. The model can be applied whether or not there is a known customer. The rapid prototyping is illustrated in Figure 1.5. (The spiral model will be discussed in Section 1.5.3.)

The primary assumption of the rapid prototyping model of software development is that the complete requirements specification of the software is not known before the software

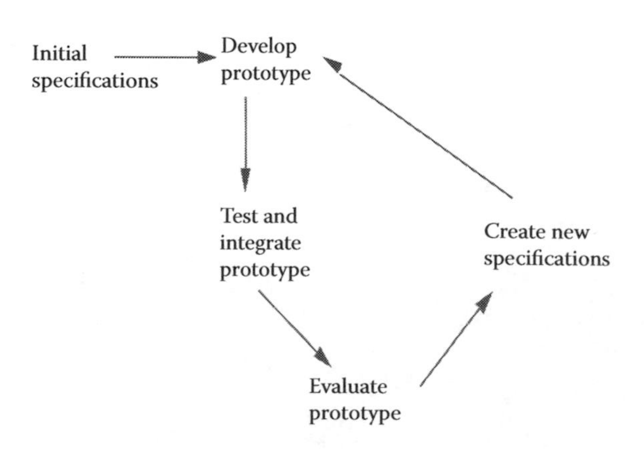

FIGURE 1.5 The rapid prototyping model of the software development life cycle.

is designed and implemented. Instead, the software is developed incrementally, with the specification developing along with the software itself. The central point of the rapid prototyping method is that the final software project produced is considered as the most acceptable of a sequence of software prototypes.

The rapid prototyping method will now be explained in more detail, illustrating the basic principles by the use of a system whose development history may be familiar to you: the initial development of the Microsoft Internet Explorer network browser. (The same issues apply to the development of the Mozilla Firefox, Apple Safari, and Google Chrome browsers, to a large extent.) Let us consider the environment in which Microsoft found itself. For strategic reasons, the company decided to enter the field of Internet browsers. The success of Mosaic and Netscape, among others, meant that the development path was constrained to some degree. There had to be support for HTML and most existing web pages had to be readable by the new browser. The user interface had to have many features with the same functionality as those of Netscape and Mosaic. However, the user interfaces had to be substantially different from those systems in order to avoid copyright issues.

Some of the new system could be specified in advance. For example, the new system had to include a parser for HTML documents. In addition, the user interface had to be mouse driven.

However, the system could not be specified fully in advance. The user interface would have to be carefully tested for usability. Several iterations of the software were expected before the user interface could be tested to the satisfaction of Microsoft. The users testing the software would be asked about the collection of features available with each version of Internet Explorer. The user interface not only had to work correctly, but the software had to have a set of features that was of sufficient perceived value to make other potential users change from their existing Internet browsers if they already used one or purchase Microsoft Internet Explorer if they had never used such a product before.

Thus, the process of development for this software had to be iterative. The software was developed as a series of prototypes, with the specifications of the individual prototypes changing in response to feedback from users. Many of the iterations reflected small changes, whereas others responded to disruptive changes such as the mobile browsers that work on smaller screens on smartphones and tablets with their own operating systems and the need to interoperate with many Adobe products, for example

Now let us consider an update to this browser. There have been changing standards for HTML, including frames, cascading style sheets, a preference for the HTML directive <p> instead of <div>, and a much more semantically powerful version, HTML 6.

Here the browser competition is between IE and several browsers that became popular after IE became dominant, such as Mozilla's Firefox, Apple's Safari, and Google's Chrome. Of course, there are many software packages and entire suites of tools for website development, far too many to mention. Anyone creating complex, graphics-intensive websites is well aware of the subtle differences between how these browsers display some types of information.

Not surprisingly, there are many web page coding styles that once were popular but have become obsolete. Following is an example of the effect of coding styles, given in the context of a software engineering project that illustrates the continued use of multiple prototypes. The project shows some of the steps needed for a prototype-based software project. You will recognize many of these steps that will appear in one form or another throughout this book.

As part of my volunteer activities for a nonprofit cultural organization located in Maryland, I had to reconstruct a complex website that had been taken over by criminal intruders who had turned it into a SPAM relay site. As a response to the SPAMming, the hosting company disallowed access to the website.

The problem was caused primarily by a combination of a weak password set by the website's initial creator and a plaintext file in a third-party application that did not protect the security of an essential mySQL database. I also thought that the website's hosting company had poor security, so I was happy to move the website to a new hosting company. The problem was exacerbated by the fact that the software originally designed to create the website was no longer available and, therefore, could not be placed on my personal computer that I used to recreate the website. (This unavailability of the software that was originally used to create a particular software artifact is all too common in the software engineering community, due to changes in companies and organizations that produce applications, operating systems, compilers, development environments, and so on.)

Here are the steps that had to be followed:

1. Move the HTML and other files from the organization's website at the original hosting company to my home computer.
2. Identify the directory structure used to organize the files that comprised the website.
3. Examine the HTML of the main pages in each directory to determine their function. This meant looking for files that were named something like index.htm or index.html.
4. Develop a naming standard for file names, and rename files as necessary to make sure that links worked in a consistent manner.
5. Keep a log of all changes, with the results of the changes. This was invaluable, since I could only work intermittently on this problem when I had free time for this volunteer activity.
6. Test the links in the pages to see how the directory structure was reflected in the portion of the website that was on my home computer. Fix them if possible, editing the raw HTML.
7. While the previous steps were occurring, evaluate new website creation tools to determine if it would be better to develop the website using a tool or raw HTML. Cost was a major issue for this cash-strapped organization. This evaluation of available tools is part of a systems engineering effort that we will meet again in Chapter 3.
8. Begin porting files to the hosting environment at the new hosting company. (I had already decided that hosting this service on my home computer was out of the question for security reasons and my own contract with my Internet service provider.)

9. Make the initial renewed website minimally functional, with only static web pages with working links.
10. Test the website iteratively as it is being created, making sure that all links work directly.
11. Remove unnecessary programming, such as JavaScript programming on otherwise simple pages for which there was little benefit to the programming. This was done for security purposes.
12. Release this portion of the website to the members and the entire outside world.
13. Replace the database software using a new version of mySQL, together with the updated version of our application software. This was all done using improved security access. Place all of this inside a newly created password-protected area of the website used exclusively for our members.
14. Give access to a select few individuals to test the security features throughout the reengineering process. Using the results of their tests, I would fix all problems as necessary.
15. Release a final version to the user community.

We now consider each of these steps in turn.

Fortunately, I was given permission from the website's original Internet hosting service to access the account long enough to move all the website's files onto my home computer. I was able to use a very fast freeware software utility, *Filezilla*, which used the FTP file transfer protocol for concurrent file transfer, which was much faster than using a batch transfer of each directory and each subdirectory.

Of course, the original website creator did not have a complete backup of all the files. The files were named without adherence to any consistent file naming standard, so there was a long learning curve to see just what each file contained, and what hyperlinks went with which file.

To make matter worse, Netscape *Communicator*, the software tool used to create the website, no longer worked on the new hosting company's system. In fact, Netscape as a company no longer existed.

Experimentation with the software tools available convinced me that I had to recreate most of the website's pages from scratch, editing the raw HTML. Clearly the only sensible approach was an iterative one, with the detailed requirements and the implementation evolving together.

I did a simple calculation of the likely amount of data that would be stored on the new website, and what was the likely amount of data that would be transferred each month. Armed with this information, I chose a particular package from the hosting company. (Since the cost was small, I decided to absorb the hosting cost myself to avoid getting permission from a committee for each expenditure. It was easier for me to get a letter at the end of the year thanking me for my expenses on behalf of the organization than to ask for approval of each small expenditure. Armed with this letter, I could take the expense as a charitable contribution.) I then uploaded all the directories to the new web hosting company's site.

A portion of the original HTML created using the obsolete and unavailable website design software is shown next. It really looks ugly. I note that a few years ago, an undergraduate student named Howard Sueing, who is now a highly successful computer consultant, was working on a HTML translator that flagged oddities in HTML 4 that would cause problems if executed in HTML 5, and he described this code as "immature." It was immature, as you can see from the following HTML code fragment whose sole purpose is to insert a clickable tab labeled "Md-MAP" into a fixed position on a particular web page.

```
beginSTMB("auto","0","0","vertically","","0","0","2","3","#ff0000","","ti
   led",
"#000000","1","solid","0","Wipe right",
"15","0","0", "0","0","0", "0","2",
"#7f7f7f","false","#000000","#000000",
"#000000","simple");
appendSTMI("false","Md-MAP ","center","middle","","","-1",
   "-1","0","normal","#fbc33a","#ff0000","","1","-1","-1","blank.gif",
   "blank.gif","-1","-1","0","","mdmap01.html","_self","Arial","9pt","#0000
   00","normal","normal","none","Arial","9pt","#ffffff","normal","normal",
   "none","2","solid","#000000","#000000","#000000","#000000","#000000
   ","#000000","#000000","#000000","mdmap01.html","","","tiled","tiled");
...
endSTMB();
```

Note that there is little separation of objects and attributes, which is a common theme of object-oriented programming. What you do not see is that I had to manually enter carriage returns throughout the code, because these HTML statements had been produced without any breaks, so that what is shown in this book as two blocks that consisted of multiple lines of code was available only as two very long lines. Clearly, one of the first steps of the process of changing the code was to insert these hard carriage returns and to check that each insertion left the functionality of the relevant web page unchanged. (The three dots near the end of the code fragment reflect that many lines of HTML code have been omitted for reasons of clarity.)

Once I fixed the line breaks, so that I could intelligently read the code, I looked at the file structure and developed some naming conventions for file names. I then meticulously loaded the web pages one at a time, checking all hyperlinks and changing file names to the standard naming conventions as I went along. I created a simple spreadsheet indicating all the changes I made and whether they solved the problems I found.

I made the site operable as soon as possible, concentrating on the static web pages. I removed items such as pages that used JavaScript, but that did not, in my opinion, have enough functionality to justify the security problems in older versions of JavaScript. After this minimal website was up and running, and working properly, I could concentrate on the more dynamic and troublesome capabilities that I needed to include.

The last two steps were installation of a database system that was built on top of a mySQL database and the creation of a protected area for access by organization members only. Discussion of the database and security issues encountered would take us too far from our discussion of software engineering.

It was surprising to find that the primary problem with our members accessing the members-only area of the website was that their passwords often did not work because of the need to clear their browser's data cache. Many of the browsers available on user's computers had different ways of treating cached data and I had to respond to each user separately. It was clear to me that I hated doing customer service, but I gained an increased appreciation of the people who work in technical support at help desks. By this time, the project, which had been a combination of reengineering of the old website, reusing much of the code, and developing multiple prototypes (each updating of the website), had morphed into maintenance of the website.

Look over the description of the recreation of the website described earlier from the perspective of an iterative software engineering project. It is clear that several things were necessary for an efficient iterative software development life cycle. There must be an initial step to begin the process. In the rapid prototyping model described in Figure 1.5, an initial set of specifications is needed to start the prototyping process. In our example of reengineering a website, the specifications were to reproduce the functionality of the original website as much as possible.

There must also be an iterative cycle in order to update and revise prototypes. This is clear from the diagram in Figure 1.5. This iteration was the basic approach used to reengineer the aforementioned website.

Finally, there must be a way to exit from the iteration cycle. This step is indicated in Figure 1.5 by the "evaluate prototype" step. The basis for the evaluation is the perceived completeness, correctness, functionality, and usability of the prototype. This evaluation varies from organization to organization. In many cases, the final prototype is delivered as the final product; in others, the final prototype serves as the basis for the final delivered product. This was the case for the website reengineering project.

Was the project successful? The recreated website was up and running, it had the same functionality as the original website, the design of the individual webpages was made more consistent, and some security flaws were fixed. The recreated website was easy to modify. This project was certainly a success.

A milestone chart can also be developed for the rapid prototyping method of software development. A sample milestone chart for the rapid prototyping process model is illustrated in Figure 1.6. (The time periods shown are pure fiction.) Compare this chart to the milestone chart shown in Figure 1.4 for the classical waterfall software development process.

1.5.3 Spiral Model

Another iterative software development life cycle is the spiral model developed by Barry Boehm (Boehm, 1988). The spiral model differs from the rapid prototyping model primarily in the explicit emphasis on the assessment of software risk for each prototype during the evaluation period. The term *risk* is used to describe the potential for disaster in the

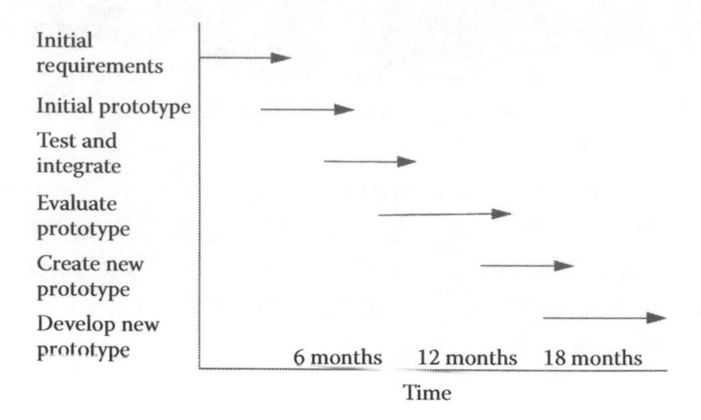

FIGURE 1.6 An example of a milestone chart for a rapid prototyping software development process.

software project. Note that this model is almost never applied to projects for which there is no known customer.

Clearly there are several levels of disasters, and different organizations and projects may view identical situations differently. Examples of commonly occurring disastrous situations in software development projects include the following:

- The software development team is not able to produce the software within the allotted budget and schedule.
- The software development team is able to produce the software and is either over budget or schedule but within an allowable overrun (this may not always be disastrous).
- The software development team is not able to produce the software within anything resembling the allotted budget.
- After considerable expenditure of time and resources, the software development team has determined that the software cannot be built to meet the presumed requirements at any cost.

Planning is also part of the iterative process used in the spiral development model.

The spiral model is illustrated in Figure 1.7.

As we saw before with the classical waterfall and rapid prototyping software development models, a milestone chart can be used for the spiral development process. This type of milestone chart is illustrated in Figure 1.8. Compare the illustration in Figure 1.8 to the milestone charts illustrated in Figures 1.4 and 1.6. For simplicity, only a small portion of a milestone chart is included.

Both the rapid prototyping and spiral models are classified as iterative because there are several instances of intermediate software that can be evaluated before a final product is produced.

It is easy to see the difference between iterative approaches and the classical waterfall model; the waterfall model has no provision for iteration and interaction with a potential

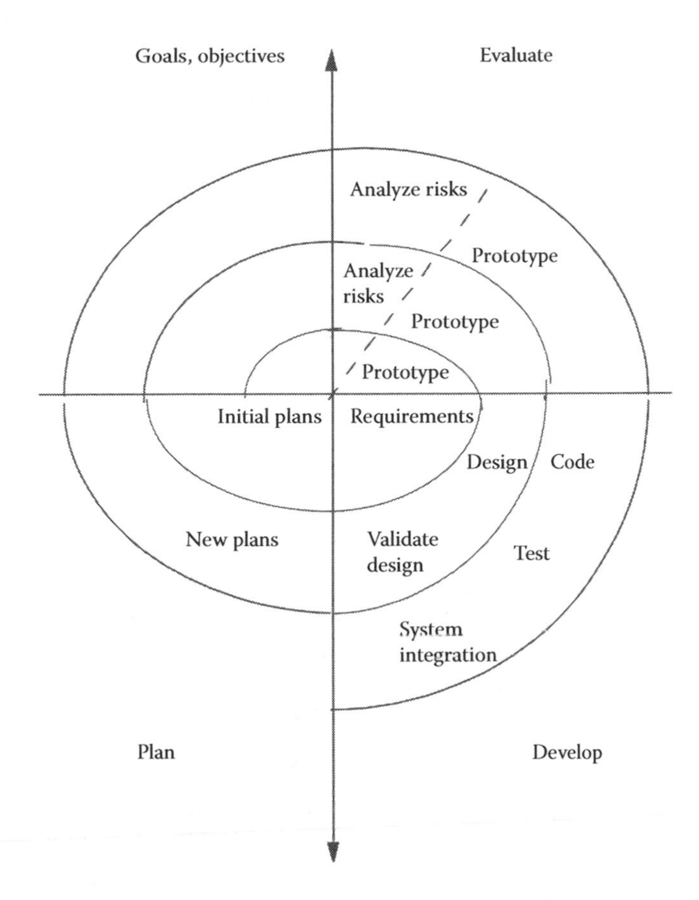

FIGURE 1.7 The spiral model of the software development life cycle.

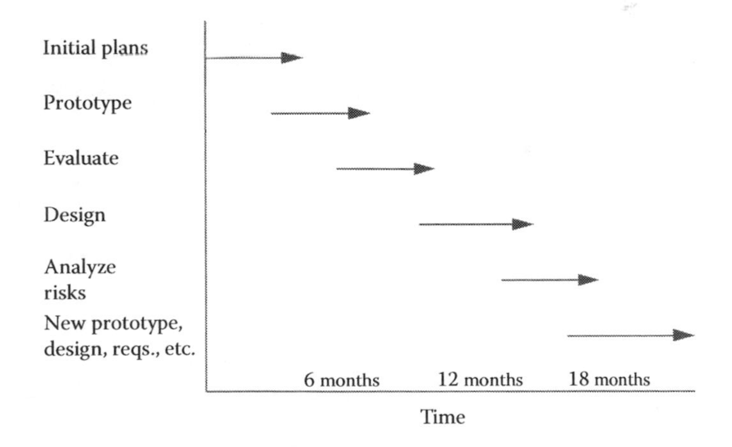

FIGURE 1.8 An example of a milestone chart for a spiral software development process.

customer after the requirements are complete. Instead, the customer must wait until final product delivery by the development organization.

Many, but not all, organizations that use the classical waterfall model allow the customer to participate in reviews of designs. Even if the customer is involved in evaluation of detailed designs, this is often the last formal interaction between the customer and the developer, and the customer is often uneasy until the final delivery (and often even more uneasy after delivery). It is no wonder that iterative software development approaches are popular with customers.

1.5.4 Market-Driven Model of Software Development

It should be clear that none of the models of the software development life cycle previously described in this book are directly applicable to the modern development of software for the general consumer market. The models previously discussed assumed that there was time for relatively complete initial requirements analysis (as in the classical waterfall model) or for an iterative analysis (as in the rapid prototyping and spiral models with their risk assessment and discarding of unsatisfactory prototypes). These models do not, for example, address the realities of the development of software for personal computers and various mobile devices. Here, the primary pressure driving the development process is getting a product to market with sufficient quality to satisfy consumers and enough desirable new features to maintain, and even increase, market share.

This is a relatively new approach to software development and no precise, commonly accepted models of this type of software development process have been advanced as yet, at least not in the general literature. The reality is that the marketing arm of the organization often drives the process by demanding that new features be added, even as the product nears its target release date. Thus there is no concept of a "requirements freeze," which was a common, unwritten assumption of all the previous models at various points.

We indicate the issues with this type of market-driven "concurrent engineering" in the stylized milestone chart illustrated in Figure 1.9. We have reduced the time frame from the previous milestone charts to reflect the reality that many products have several releases each year.

We emphasize that even small applications written by a single programmer have an expected maintenance cycle, leading to new releases. Suppose, for example, that you have written an app for a mobile phone that has sold a reasonably large number of downloads. If an error is found or if the underlying technology changes, the app must be updated. If

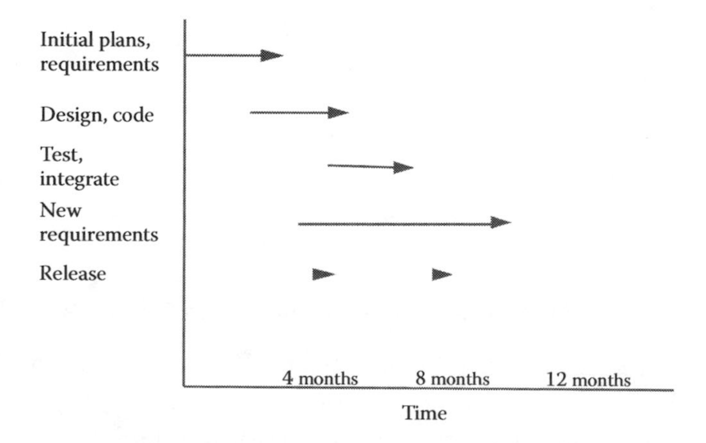

FIGURE 1.9 An example of a milestone chart for a market-based, concurrently engineered software development process.

the developer does not do so or is slow in making the changes, bad reviews result, and sales will plummet.

Even the most casual observer of the early twenty-first century will note that many of the most successful software companies were headed by visionaries who saw an opportunity for creating something that would be useful to huge numbers of people. I hope that some of the readers of this book have these insights and develop wildly successful software accordingly.

Fortunately, the software that anyone would now consider creating for a market of unknown customers is much easier to distribute than in the past. It will almost certainly be delivered by download, relieving the software's creator from the drudgery of creating a large number of copies of the software on CD or DVD, and arranging with stores for shelf space and maintenance of an inventory.

1.5.5 Open Source Model of Software Development

Open source software development is based on the fundamental belief that the intellectual content of software should be available to anyone for free. Hence, the source code for any project should be publicly available for anyone to make changes to it. As we will discuss later in this section, "free" does not mean "no cost," but refers to the lack of proprietary ownership so that the software can be easily modified for any new purpose if necessary.

Making source code available is the easy part. The difficulty is in getting various source code components developed by a wide variety of often geographically separated people to be organized, classified, maintainable, and of sufficient quality to be of use to themselves and others. How is this done?

In order to answer this question we will separate our discussion of open source projects into two primary categories: (1) projects such as new computer languages, compilers, software libraries, and utility functions that are created for the common good; and (2) reuse of the open source versions of these computer languages, compilers, software libraries, and utility functions to create a new software system for a specific purpose.

Here is a brief overview of how many open source projects involving computer languages, compilers, software libraries, and utility functions appear to be organized. The primary model for this discussion of open source software development is the work done by the Free Software Foundation, although there also are aspects of wiki-based software development in this discussion.

An authoritative person or group decides that a particular type of software application or utility should be created. Initial requirements are set and a relatively small number of software modules are developed. They are made publicly available to the open source community by means of publication on a website, possibly one separate from the one that contains relatively complete projects.

The idea is that there is a community of developers who wish to make contributions to the project. These developers work on one or more relevant source code modules and test the modules' performance to their own satisfaction, then submit the modules for approval by the authoritative person or group. If the modules seem satisfactory, they are uploaded to a website where anyone can download them, read the source code, and test the code. Anyone can make comments about the code and anyone can change it. The only rule is that the changes must be made available to anyone, as long as the source code is always made available, without charge.

It is the use of the achievements and creations of this widespread community with its free access that is the basis for open source software development.

How do members of the large group of open source programmers make money if they are not allowed to charge for their software creations? There are two common ways. The most senior and authoritative people often have substantial contracts to develop systems, with the only proviso that the source code for the software produced by open source organization bc madc available publicly free of charge. These large contracts are a major part of the funding that keeps the Free Software Foundation's continued operation possible, in addition to donations from volunteers and members of the various user communities.

In addition to the aforementioned funding, members of the open source community often make money by doing software consulting, using their detailed knowledge of the internal structure of many important software products that were developed as open source systems. Being a developer of some of the major modules used in a system is rather strong evidence of just how the system works and what issues are involved in the system's installation, configuration, and operation.

Note that many of the programmers who volunteer their time and effort to create open source modules for various projects are often employed by companies that develop large software systems. This appears to be a way that these volunteers feel that they can be more creative, since they can decide which projects are most interesting to them, rather than some of the projects to which they have been assigned. There is anecdotal evidence to support this view. I heard a statement to this effect while interacting with colleagues at an academic software engineering conference. When Bill Gates visited Howard University in 2006 while I was department chair, I asked him if he was aware of this, and he agreed that it was a likely scenario. (Howard University had sent more students to work at Microsoft in the 2004–2005 academic year than any other college or university in the United States, and Gates came to see us as part of a six-university tour that was intended to increase enrollment in computer science. Microsoft is not typically thought of as a company known for development of open source software.)

We now consider the second category of open source development: the reuse of the open source versions of these computer languages, compilers, software libraries, and utility functions in order to create a new software system for a specific purpose.

We have not attempted to give a precise definition of open source software development because of the variations in the way that it is used in actual practice. Note that open source does not necessarily mean free.

A July 28, 2014, article by Quentin Hardy in the *New York Times* describes how a company, Big Switch Networks, is moving from a development model that was entirely open source to one that includes a parallel development of completely open source software together with one in which many of the more commercially desirable features are proprietary, and therefore, not free. This approach has some things in common with the approach of several resellers of Linux software, especially Red Hat (now Fedora), Ubuntu, and Debian, which provide many things for free but sell consulting and some other services. In particular, some companies prefer to pay for the maintenance of open source software that they use rather than depend on volunteers.

For another perspective on the distinction between open source software and cost-free software, see the philosophy stated on the Free Software Foundation website at the URL https://www.gnu.org/philosophy/free-software-for-freedom.html. We will not discuss this issue of the precise definition of open source software any further in this book.

1.5.6 Agile Programming Model of Software Development

The roots of the concept of agile development can be traced back to at least as far as the "Skunk Works" efforts to engineer projects at the company now known as Lockheed Martin. (A *Wikipedia* article accessed May 5, 2015, states that the term *Skunk Works* "is an official alias for Lockheed Martin's Advanced Development Programs [ADP], formerly called Lockheed Advanced Development Projects.") Skunk Works came to the fore by the early 1940s, during World War II.

The term is still used to describe both highly autonomous teams and their projects, which are usually highly advanced. Many of the Skunk Works projects made use of existing technologies, a forerunner for component-based software development, with the efficient use of existing software components that can be as large as entire subsystems.

The principles of what is now known as agile programming are best described by the "Agile Manifesto," which can be found at http://agilemanifesto.org/principles.html and whose basic principles are listed here for convenience.

PRINCIPLES BEHIND THE AGILE MANIFESTO

We follow these principles:

Our highest priority is to satisfy the customer through early and continuous delivery of valuable software.
Welcome changing requirements, even late in development. Agile processes harness change for the customer's competitive advantage.
Deliver working software frequently, from a couple of weeks to a couple of months, with a preference to the shorter timescale.
Business people and developers must work together daily throughout the project.
Build projects around motivated individuals. Give them the environment and support they need, and trust them to get the job done.
The most efficient and effective method of conveying information to and within a development team is face-to-face conversation.
Working software is the primary measure of progress.
Agile processes promote sustainable development. The sponsors, developers, and users should be able to maintain a constant pace indefinitely.
Continuous attention to technical excellence and good design enhances agility.
Simplicity—the art of maximizing the amount of work not done—is essential.
The best architectures, requirements, and designs emerge from self-organizing teams.
At regular intervals, the team reflects on how to become more effective, then tunes and adjusts its behavior accordingly.

Some of these principles may seem obvious.

- Agile development is a form of prototyping.
- It depends on a large existing body of high-quality, working components that are especially well-understood by the development team.
- Due to the availability of toolkits, this model also works well for software in the form of apps that are intended for mobile devices. A primary example of this is the availability of free frameworks for the development of computer games.

The term *scrum* is often used in the context of agile development. Originally, the term was frequently used to describe the meetings at the end of each day to coordinate the ideas and accomplishments of the major agile stakeholders. The term was motivated by the sport of rugby. More often, the term *scrum* is used to describe the products of the important organization scrum.org. Often, the term is capitalized, and that is the pattern we will follow henceforth.

We will return to agile software development many times throughout this book.

1.5.7 Common Features of All Models of Software Development

Many organizations have their own written variants of these software development models, and recognize that any models are stylized approximations to actual software development practice in any case. Even a large defense contractor working on a multimillion dollar software project for the government might have to develop some prototypes in an iterative manner, due to disruptive changes in technology. A more market-driven life cycle might have many more things happening at the same time. We will not pursue these different nuances in life cycle models any further in this book.

There are many variants of these six life cycle models, but these are the most commonly used. Note that most of the activities in each of these life cycle models are similar. The primary differences are in the timing of activities and the expectation that some portions of the initial prototypes will be discarded if iterative approaches, such as the rapid prototyping or spiral development models, are used.

The point about timing of life cycle activities cannot be overemphasized. The various life cycle models are, at best, stylized descriptions of a process. Most organizations have their own set of procedures to be followed during software development. Of course, you should follow the guidelines set by your employer. Just be aware that these are the same sets of essential activities that must always occur in software development, regardless of the order in which they occur.

1.5.8 Software Evolution and the Decision to Buy versus Build versus Reuse versus Reengineer

Almost all software products, other than relatively small stand-alone programs written by a single programmer for his or her own use, will change over time. For example, in May 2014 the version of the iOS operating system for my iPhone was version 7.1.1, the operating system for my MacBook Air was 10.9.2, and the version of Word was 14.4.1.

My desktop PC was running a version of Windows 7 (Windows 7 Professional) and also can run an emulator for Windows XP that I needed for several years in order to run a particular version of publication software. (I no longer use the Windows XP emulator, because Microsoft has ended support for this very old operating system.) My desktop Mac uses the same operating system version as my MacBook Air.

The same situation held true for smaller devices. The app 2Screens Remote that allows my iPhone to control Apple Keynote presentations that had been loaded to the 2Screens app on my iPad was 1.5.2. I received notification that version 2.6.1 of the Fly Delta app is available and that I should upgrade to this latest release via the App Store.

You should think of all this software as evolving over time, whether with error fixes, increased features, or the need to have, say, applications software work with a new operating system. The term *software maintenance* is often used to describe this evolution. Clearly, this evolution of the software requires an effort in, at least, design, coding, testing, and integration. This effort obviously involves time, personnel, and other resources. With the typical movement of software engineers from one position to another, either inside or, most commonly, outside companies, there are real decisions to be made about maintaining this evolving software.

For long-lived systems, it is estimated that more than 80 percent of total costs during the software's useful life is during the maintenance phase after the initial delivery. Dan Galorath, the creator of the SEER software cost estimation tool, uses the figure 75 percent in his models. See the SEER website http://www.galorath.com/index.php/software_maintenance_cost. One reason these costs are so high is that it takes a long time for software maintenance engineers to learn the detailed structure of source code that is new to them.

If the expected costs to maintain the software for the rest of its expected life are too high, it may be more sensible to buy an existing software package to do many of the same things rather than incur high maintenance costs. Other alternatives may be to reengineer the software so that existing off-the-shelf software components can be used, or to have even greater percentages of software reused from other projects.

1.6 DIFFERENT VIEWS OF SOFTWARE ENGINEERING ACTIVITIES

As we have seen earlier in this chapter, most modern software development projects require teams. There are two obvious ways to organize teams: ensure that each person is a specialist with unique skills; have each person be a generalist, who is able to perform most, if not all, team responsibilities; or some combination of the two. Each of the organizational methods has unique advantages and disadvantages.

A team with specialists is likely to have individual tasks done more efficiently than if they were to be done by a generalist. Some technologies are so complex and are changing so rapidly that only a specialist can keep up with them.

On the other hand, a generalist is more able to fill in if there is a short-term emergency such as one of the team members being very sick or having a family emergency. Generalists often see connections between apparently unrelated subjects and can aid in identifying patterns of software development problems that have been solved before.

In any event, different people, both on and off particular software development teams, have different perspectives on the organization's software, the efficiency of the process of developing the software, and the particular project that they are working on.

A software project manager is interested in how the software development process is working, and whether the system will be produced on time and within budget. Decisions about the viability of the software may be made at a higher level, and many projects that are on time and under budget are canceled because of new releases by the competition, by corporate managers making some of the merged companies' systems redundant, or even by disruptive technologies.

The manager is especially concerned with the group he or she heads. It also is natural for a lower-level manager to be somewhat concerned with the work done by other groups on related projects. After all, an experienced manager is well aware that software projects can be canceled if they are expected to be delivered late or over budget. Higher-level management often has to make unpopular decisions based on whatever information is available. Thus, lack of performance of a team working on one subsystem can affect a project adversely, even if the other subsystems are likely to be produced on time by the teams responsible for their development.

Therefore, a manager will often require his or her team to produce measurements on the team's progress. These measurements could include the number of lines of code written, tested, or reused from other projects; the number and types of errors found in systems; and the number and size of software modules integrated into larger systems.

Software engineers often hate to fill out forms, because they feel that the time needed to do this takes away from their essential activities. It is hoped that understanding a manager's perspective will make this activity more palatable, even for the most harassed programmer.

1.7 SOFTWARE ENGINEERING AS AN ENGINEERING DISCIPLINE

The goal of this section is to convince you that software engineering is an engineering discipline. We will do this by presenting a collection of anecdotes relating to current software development practice. We will describe three classes of software application environments: HTML and other programming for the Internet, applications programs for personal computers, and applications that involve health and safety. In each case, we will illustrate the need for a disciplined approach to software engineering.

The current rush to create both mobile and traditional websites for the Internet, and the shortage of trained personnel to create them, means that jobs are readily available for people with minimal technical knowledge. However, at least for the case of traditional, nonmobile websites, this unusual situation is unlikely to continue far into the future, for several reasons.

Experimentation with leading edge technology may be appealing to many software engineers but will be less so to an organization's chief software officer if the future of the organization depends upon this technology working properly. This later situation is often called "bleeding edge technology." Management is much more likely to support small experimental projects than make major changes to its primary business practice, just to keep up with "leading edge technology."

You should note that when the rush to create the initial web pages is over, the web pages must be tested for accuracy of content, correctness of links, correctness of other programming using the CGI and related approaches, performance under load, appropriateness for both mobile devices and ones with larger screens, general usability, and configuration management. (Configuration management is a very general issue in software engineering. For now, think of it as the systematic approach used to make sure that changes in things such as Internet browser software standards do not make the pages unusable, at the same time making sure that the pages are still accessible by older versions of browsers.)

Here is an example of this problem: I recruited a research team to work on web page design. The team consisted of four freshmen majoring in computer science at Howard University. (The students were chosen because of the clever way they tore off portions of a cardboard pizza box when we ran out of paper plates at the departmental student–faculty party at the beginning of the academic year.) The goal of the project was to develop a website for an online archive of examples of source code with known logical software errors to be used by researchers in software testing to validate their theories and testing tools.

After three months of designing web pages, testing alternative designs, and writing HTML and CGI and Perl scripts, the web page was essentially produced in its final form. The remainder of the effort was devoted to the students learning software testing theory, porting examples written in one programming language to other standard languages, database design, and writing driver programs for procedures that were to be included in the software testing archive. Unfortunately, this website is no longer active.

In short, first-year college students did the web page work in a few months and then turned to standard software engineering activities. Writing web pages for small projects is too easy to serve as the basis for a career without considerable design training. The computer science content involved with the use of HTML appears to be very small. The computer science content is involved with the programming of servers, clients, and scripts in Java and other languages. Since HTML is a language, enhancing the language and developing HTML translators is also a computer science effort. It is in this programming portion that the software engineering activity occurs.

The experience of web design is reminiscent of the situation when personal computers became popular. One negative aspect of the tremendous publicity given to the early successful entrepreneurs in the software industry is the common perception among nonprofessionals that software engineering is primarily an art that is best left to a single talented programmer instead of going back to the origins of computers. The earliest computer programmers were scientists who wrote their own simple algorithms and implemented them on the computers available to them. The earliest computers were programmed by connecting wires in a wiring frame, although this arduous process was soon replaced by writing programs in the binary and octal code that was the machine language of the computer. Machine language programming was quickly replaced by assembly language programming where a small set of mnemonics replaced the numerical codes of machine language. All this encouraged the so-called ace programmer.

Finally, the first attempt at what we now call software engineering emerged in the development of subroutines, or blocks of code that perform one computational task and can be reused again and again. (Admiral Grace Murray Hopper is generally credited with having written the first reusable subroutine.) Higher-level languages were then developed, with more expressive power and with extensive support for software developments. Indeed, every modern software development environment includes compilers, editors, debuggers, and extensive libraries.

The most modern software development environments include tools to build graphical user interfaces and special software to assist with the software development process itself. The software even has a special name, CASE tools, where the acronym stands for computer-aided software engineering.

Even with the improvement in software development environments and the increasing complexity of modern software, the image of the lone programmer persists. Little harm is done if the lone programmer develops software only for his or her use. The lone programmer also is useful when developing small prototype systems.

For larger software projects, an engineering discipline is necessary. This means that the entire software development process is guided by well-understood, agreed-upon principles, and these principles are based on documented experience. The software development process must:

- Allow for accurate estimates of the time and resources needed for completion of the project
- Have well-understood, quantifiable guidelines for all decisions made
- Have identifiable milestones
- Make efficient use of personnel and other resources
- Be able to deliver the products on time
- Be amenable to quality control

The resulting software product must be of high quality and meet the other goals of software engineering. Above all, as an engineering discipline, both the efficiency of the process and the quality of the product must be measurable.

The issue of quality brings us to the discussion of the other two software application environments that we will consider in this chapter: applications programs for personal computers, and applications that involve health and safety.

There is a tremendous amount of competitive pressure in the area of software application development for personal computers and portable devices. Many software development organizations attempt to release new versions of major applications every six months. For some applications, the interval between releases may be as short as four months. This rapid software development life cycle requires a degree of "concurrent engineering," in which several different life cycle activities are performed simultaneously. The driving factor in this process is the need to provide new features in products.

This efficient production of software comes at a price, however. There is no way that the quality, as measured by such things as ease of installation, interoperability with other applications, or robustness, can be as high as it would be if there were a longer period for testing and design reviews, among other things. Contrary to some views, such software is tested. The decision to release a software product is based on a careful assessment of the number and severity of errors remaining in the software and the relative gain in market share because of new features. Companies such as Microsoft use the technique of software reliability to base decisions on formal data analysis.

It is worthwhile at this point to briefly discuss the software development process used at Microsoft (Cusumano and Selby, 1995, 1997) for many large software products. Most new software development at Microsoft has three phases:

1. In the planning phase, a vision statement is developed for the software based on customer analysis. The specifications are written, together with a list of features. Finally, a project plan is set.

2. In the development phase, development is performed together with testing and evolution of the specifications. This is an iterative process.

3. In the third and final phase, called "stabilization," comprehensive testing occurs both within Microsoft and with selected "beta sites," for testing by external users.

The later portions of this process are often called "sync and stabilize" to reflect that they allow individual software modules to be placed into the overall product. Most systems under development have a daily "build," in which a clean compilation is done. This allows the synchronization of individual modules, regardless of where the modules were created, which is critical in view of the global nature of software development. On a personal note, it was always interesting to see the number of daily builds in large projects such as various releases of the Windows operating system.

The stabilization occurs when the number of changes appearing in the software and the number of errors has been reduced below a level that is considered acceptable. The number of detected software errors for the entire software system is computed each day. A typical, but hypothetical, graph of the daily number of software errors for a month is indicated in Figure 1.10.

Since the number of errors remaining in the software illustrated in Figure 1.10 appears to have stabilized below 7, the software development in this hypothetical example would probably be suspended and the software released, at least if the remaining errors were considered to be small and to be of minimal importance. The implicit assumption is that the software will be sufficiently defect-free once the number of remaining known software errors is below a predefined level and the statistical history suggests that this number will remain low. This assumption is based on the subject known as software reliability, which we will discuss in some detail when we study testing and quality control in Chapter 6.

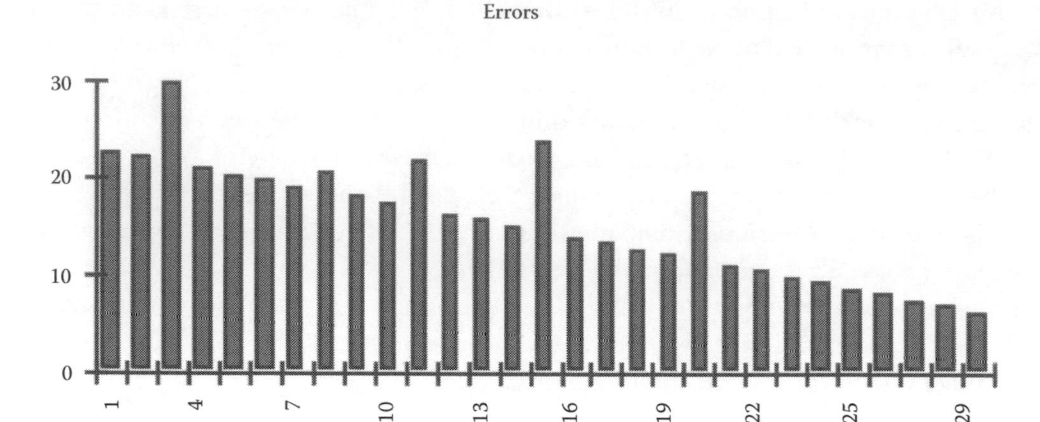

FIGURE 1.10 A typical graph of the daily number of software errors for a month.

You should be aware that each organization has its own standards for acceptable quality of the products it releases. However, no company could stay in business if the marketplace generally considered its products to be of extremely low quality.

Note that these systems are developed over a period of time, however short. Software such as Windows NT, which had a much longer development period, had over a thousand separate compilations (builds) before the latest version was released. Also, most Microsoft products now require large development teams. For example, the development team for Windows 95 had over 200 members (Cusumano and Selby, 1997). Newer versions of operating systems have had similar sizes.

It is clear that Microsoft has a software development process. This process has been tailored to the company's unique corporate culture, understanding of the marketplace, and its financial resources. Satya Nadella, who became Microsoft CEO in early 2014, has changed company practice to have a common software development environment across all platforms, aiming for greater efficiency of development and encouraging reuse.

Applications that involve health and safety generally have a more rigid software development process and a higher standard for logical software errors. The reasons are obvious: potential loss of life in a medical monitoring system, considerable hardship if airplanes are grounded because of an error in the air traffic control system, or major disruption to financial markets if database software used for the U.S. Social Security system fails. Considerable financial loss can occur if, for example, a hotel reservation system or a package tracking system has major failures. People could lose jobs. Clearly, there is an even higher need for a formal software development process. Systems in these areas are often much larger than in the areas previously discussed and, thus, in much more need of a careful software engineering process. The process must be carefully documented to meet the approval of agencies, such as the U.S. Food and Drug Administration for computer software that controls medical products.

We have illustrated that each type of software development area—Internet, personal computer applications, and safety-critical applications—involves some degree of software engineering, although the processes may be different in different areas. There is one other advantage to treating software engineering as an engineering discipline: avoidance of future risk. Risk analysis is a major part of any engineering discipline.

If risk analysis had been applied to software development when programmers were using two digits instead of four to represent dates, the software development industry would not have faced the "Year 2000 problem," also known as the "Y2K problem," whose cost had been expected to be billions of dollars. (This problem occurred because programmers were under pressure to save space in their code and data sets, and because they did not expect their software to be used twenty or more years after it was developed.) The risk analysis would have cost money, but the resources applied to solve the problem could have been used in much more profitable endeavors with sufficient planning.

It is appropriate at this point to mention a major controversy in the software engineering community: certification of software professionals. In the United States, certification is most likely to take place at the state-government level.

The arguments for certification are that too much software is either produced inefficiently or else is of low quality. The need to ensure the correctness of safety-critical software that affects human life is adding to the pressure in favor of certification. Most engineering disciplines require practicing professionals, or at least senior professionals in each organization, to pass examinations and be licensed, leading to the title "Professional Engineer." For example, a mechanical or civil engineer with this certification is expected to know if a particular type and size of bolt assembly component used in construction has sufficient strength for a specified application.

The primary argument against certification of software professionals is the perception of the lack of a common core of engineering knowledge that is relevant to software development. Many opponents of certification of software professionals believe that the field is too new, with few software engineering practices being based on well-defined principles. Certifying software components for reuse in applications other than the one for which it was created is a major concern.

The Association for Computing Machinery (ACM) has officially opposed licensing for software engineers, whereas the IEEE has supported it. I will not take a position on this issue, although I expect a consensus to arise in the next few years. You should note, however, that the number of undergraduate computer science programs accredited by ABET is increasing rapidly. ABET was originally an acronym for the Accreditation Board for Engineering Technology, but now only the term ABET is in common use. ABET-accredited computer science programs are specifically evaluated by the Computing Science Commission (CAC), a part of ABET, under the auspices of CSAB, which was formerly known as the Computer Science Accrediting Board. A recent article by Phillip A. Laplante (2014) in *Communications of the ACM* provides an interesting perspective on licensing of software professionals (Laplante, 2014).

1.8 SOME TECHNIQUES OF SOFTWARE ENGINEERING

Efficiency is one of the goals of software engineering, both in the efficiency of the development process and the run-time efficiency of the resulting software system. We will not discuss run-time efficiency of the software in this book, other than to note that it is often studied under the topics "analysis of algorithms" or "performance evaluation." Developing software efficiently often involves systematic efforts to reuse existing software, as we will see later in this section.

The other goals of software engineering that were presented in Section 1.3 are also important. However, focusing on efficiency of both the software product and the process used to develop the software can clarify the engineering perspective of the discipline known as "software engineering."

Following are five major techniques that are typically part of good software engineering practice and which may be familiar to you:

1. There must be a systematic treatment of software reuse.
2. There must be a systematic use of metrics to assess and improve both software quality and process efficiency.
3. There must be a systematic use of CASE tools that are appropriate for the project's development environment.
4. There must be a systematic use of valid software cost estimation models to assess the resources needed for a project and to allocate these resources efficiently.
5. There must be a systematic use of reviews and inspections to ensure that both the project's software and the process used to develop that software produce a high-quality solution to the problem that the software is supposed to solve.

Each of these techniques is briefly discussed next in a separate section. Note that these techniques are intended to support the efficient performance of the software engineering tasks listed in Section 1.4. Even if you are a lone programmer working on a small project, you need to work as efficiently as possible, so these guidelines apply to you, too.

1.8.1 Reuse

Software reuse refers to the development of software systems that use previously written component parts that have been tested and have well-defined, standard interfaces. Software reuse is a primary key to major improvements in software productivity. The productivity of programmers has remained amazingly constant over the last thirty-plus years, with the average programmer producing somewhere in the range of eight to twelve correct, fully documented and tested lines of code per day (Brooks, 1975). At first glance, this number seems appallingly low, but you should realize that this takes into account all meetings, design reviews, documentation checks, code walkthroughs, system tests, training classes, quality control, and so on. Keep in mind that although productivity in terms of lines of code has changed little, productivity in terms of problems solved has increased greatly due to the power of APIs and the ability to leverage efforts by reusing code and other artifacts as much as possible.

Software reuse has several different meanings in the software engineering community. Different individuals have viewpoints that depend upon their responsibilities. For example, a high-level software manager might view software reuse as a technique for improving the overall productivity and quality of his or her organization. As such, the focus would be on costs and benefits of organizational reuse plans and on schemes for implementing company-wide schemes.

Software developers who use reusable code written by others probably view software reuse as the efficient use of a collection of available assets. For these developers, who are also consumers of existing software, software reuse is considered a positive goal since it can improve productivity. A project manager for such development would probably view reuse as useful if the appropriate reused software were easy to obtain and were of high quality. The manager would have a different view if he or she were forced to use poorly tested code that caused many problems in system integration and maintenance because of its lack of modularity or adherence to standards.

On the other hand, developers who are producers of reusable code for use for their own projects and for reuse by others might view reuse as a drain on their limited resources. This is especially true if they are required to provide additional quality in their products or to collect and analyze additional metrics, all just to make source code more reusable by others.

A reuse librarian, who is responsible for operating and maintaining a library of reusable components, would have a different view. Each new reuse library component would have to be subjected to configuration management. (Configuration management is the systematic storage of both a newly edited document as well as the ability to revert back to any previous version. It is an essential part of every common software engineering process and is included in nearly all tools that support software engineering.) Configuration management is necessary if a file is being changed over a long period of time even if it is just being edited by a single person, and even more so if the file is being used and edited by several people. Some degree of cataloging would be necessary for future access. Software reuse makes the job of a reuse librarian necessary.

In general, the term *software reuse* refers to a situation in which some software is used in more than one project. Here "software" is defined loosely as one or more items that are considered part of an organization's standard software engineering process that produces some product. Thus, "software" could refer either to source code or to other products of the software life cycle, such as requirements, designs, test plans, test suites, or documentation. The term *software artifact* is frequently used in this context.

In informal terms, reuse can be defined as using what already exists to achieve what is desired. Reuse can be achieved with no special language, paradigm, library, operating system, or techniques. It has been practiced for many years in many different contexts. In the vast majority of projects, much of the necessary software has been already developed, even if not in-house.

Reusability is widely believed to be a key to improving software development productivity and quality. By reusing high-quality software components, software developers can simplify the product and make it more reliable. Frequently, fewer total subsystems are used and less time is spent on organizing the subsystems.

If a software system is large enough, programmers often work on it during the day and wait until the next day to check the correctness of their code by running it. Many systems are so large that they are compiled only once a day, to reduce computer load and to provide consistency.

The major improvement in computer productivity is due to the improvement in programming languages. A single line of code in a modern programming language—say, C++ or Java—may express the same computational information that requires many lines of assembly language. A spreadsheet language, such as Microsoft Excel or Numbers, is even more powerful than most modern, general-purpose programming languages for special applications. Table 1.1 shows the effects of the choice of programming languages on the relative number of statements needed to perform a typical computation. Some related data can be found in Capers Jones's book (Jones, 1993).

Table 1.2 presents another view of how the expressive power of programming languages can be compared: the cost of a delivered line of code.

Note that there is a much more efficient way to improve software productivity: reuse code. The efficient reuse of source code (and other software engineering items such as

TABLE 1.1 Comparison of the Expressive Quality of Several Program Languages

Language	Lines of Code
Assembly	320
Ada	71
C	150
Smalltalk	21
COBOL	106
Spreadsheet Languages	6

Source: Reifer, D. J., *Crosstalk: The Journal of Defense Software Engineering*, vol. 9, no. 7, 28–30, July 1996.
Note: The term "spreadsheet language" refers to commonly available, higher-level programming applications such as spreadsheets or databases.

TABLE 1.2 Comparison of the Dollar Cost of Delivered Source Line of Code for Several Program Languages in Several Different Application Domains

Application Domain	Ada83	C	C++	3GL	Domain Norm
Commercial command and control	50	40	35	50	45
Military command and control	75	75	70	100	80
Commercial products	35	25	30	40	40
Commercial telecommunications	55	40	45	50	50
Military telecommunications	60	50	50	90	75

Source: Reifer, D. J., *Crosstalk: The Journal of Defense Software Engineering*, vol. 9, no. 7, 28–30, July 1996.

requirements, designs, test cases, documentation, and so on) can greatly improve the way in which software is produced.

At first glance, it might be surprising to a beginning software engineer that software reuse is relevant in an age of network-centric computing with many applications running in the cloud and many other applications running on mobile devices. However, a moment's thought leads to the observation that a cloud computing environment, with its hiding of details of large-scale subsystems, is an excellent place for reused software.

1.8.2 Metrics

Any engineering discipline has a quantifiable basis. Software engineering is no exception. Here, the relevant measurements include the size of a software system, the quality of a system, the system's performance, and its cost. The most commonly used measurement of the size of a system is the number of lines of source code, because this measurement is the easiest to explain to nonprogrammers. We will meet other measurements of software system size in this book, including measurements of the size of a set of requirements.

There are two measurements of software system quality that are in common use. The first is the number of defects, or deviations from the software's specifications. The quality of a system is often measured as the total number of defects, or the "defect ratio," which is the number of defects per thousand lines of code. The terms *fault* and *failure* are sometimes used in the software engineering literature. Unfortunately, the terms *defect*, *error*, *fault*, and *failure* do not always mean the same thing to different people. The term *bug* has been in common use since the time when Grace Murray Hopper recounted a story of an error in the Mark II computer traced to a moth trapped in a relay. The "Software Engineering Body of Knowledge" uses the term *bug* (SWEBOK, 2013). However, we will follow the recommendations of the IEEE (1988) and use the following terminology: A software *fault* is a deviation, however small, from the requirements of a system; and a software *failure* is an inability of a system to perform its essential duties.

A second measurement of software system quality is the number of faults per thousand hours of operation. This measurement may be more meaningful than the number of faults per thousand lines of code in practice.

One other much less common measurement is an assessment of the quality of the user interface of a software system. Although the quality of a user interface is hard to measure, good user interfaces are often the keys to the success or failure of a software application in a crowded market. We will discuss user interfaces extensively in Chapter 4 as part of our study of software design.

The term *metrics* is commonly used in the software engineering literature to describe those aspects of software that we wish to measure. There is a huge number of metrics that are currently in use by software organizations. Many of the metrics are collected, but the resulting values are not frequently used as a guide to improving an organization's software development process or toward improving the quality of its products. The use of metrics is absolutely essential for a systematic process of software engineering. Without metrics, there is no way to evaluate the status of software development, assess the quality of the result, or track the cost of development.

Metrics can be collected and analyzed at almost any phase of a software life cycle, regardless of the software development life cycle used. For example, a project could measure the number of requirements the software system is intended to satisfy, or, more likely, the number of critical requirements could be measured. During testing, the number of errors found could be measured, with the goal of seeing how this number changed over time. If more errors appear to grow near the scheduled date of release, there is clearly a problem. Ideally, other metrics would have pointed out these problems earlier in the development cycle.

Our view is that metrics should be collected systematically, but sparingly. Our approach is to use the goals, questions, metrics (GQM) paradigm of Basili and Rombach (Basili and Rombach, 1988). The GQM paradigm consists of three things: goals of the process and product, questions we wish to ask about the process and product, and methods of measuring the answers to these questions. The GQM paradigm suggests a systematic answer to the question "which metrics should we collect?"

Typical goals include:

- Quantify software-related costs
- Characterize software quality
- Characterize the languages used
- Characterize software volatility (volatility is the number of changes to the software component per unit time)

There are clearly many questions that can be asked about progress toward these goals. Typical questions include the following (we will only list two questions per goal):

- What are the costs per project? (This question is used for costs.)
- What are the costs for each life cycle activity? (This question is used for costs.)
- How many software defects are there? (This question is used for quality.)
- Is any portion of the software more defect-prone than others? (This question is used for quality.)
- What programming languages are used? (This question is used for languages.)
- What object-oriented programming language features are used? (This question is used for languages.)
- How many changes are made to requirements? (This question is used for volatility.)
- How many changes are made to source code during development? (This question is used for volatility.)

The clarity of these questions makes the choice of metrics easy in many cases. We note that there are several hidden issues that make metrics data collection complicated.

For example, if there is no tracking mechanism to determine the source of a software error, then it will be difficult to determine if some portion of the software is more defect-prone than others. However, if you believe that this question must be answered in order to meet your stated goals, then you must either collect the data (which will certainly cost money, time, and other resources) or else change your goals for information gathering.

There are a few essentials for collection and analysis of metrics data in support of a systematic process of software reuse:

- Metrics should be collected on the same basis as is typical for the organization, with extensions to be able to record and analyze reuse productivity and cost data.
- Predictive models should use the reuse data, and the observed resource and quality metrics must be compared with the ones that were estimated.
- Metrics that measure quality of the product, such as errors per 1,000 source lines of code, perceived readability of source code, and simplicity of control flow, should be computed for each module.
- Metrics that measure the process, such as resources expended, percentage of cost savings, and customer satisfaction, should be computed for each module and used as part of an assessment of reuse effectiveness.

1.8.3 Computer Aided Software Engineering (CASE)

There are many vendors of CASE tools, far too many to discuss here in any detail. There are even more prototype CASE tools currently under development by academic and other research institutions. CASE tools can be complex to use and require a considerable amount of effort to develop. Commercial CASE tools are often expensive and require extensive training. I used one commercial product that cost $3,600, just for a single seat license at my university as part of a research project, with an annual software maintenance cost of nearly $500 per year. A second commercial product that I use often is much cheaper. (Many CASE tool companies give great discounts to academic institutions, thereby encouraging the development of a knowledgeable workforce and obtaining a tax deduction.) Why are these products so popular?

There are several reasons for this popularity. The most important is that software development is extremely expensive for most organizations and anything that improves efficiency and quality of the software process at reasonable cost is welcome. For many organizations, the cost of software development is so high that nearly any tool that improves productivity, however slight, is worth its price.

Organizations often wish to provide development environments that are considered cutting edge if they are to be competitive in attracting and retaining good personnel. An examination of the larger display ads for organizations hiring software engineers indicates the emphasis on providing good development environments.

Other reasons for the popularity of CASE tools include the need for a common, consistent view of artifacts at several different phases of the organization's software life cycle. Indeed, the requirements, design, and source code implementation of a project can be much more consistent if they are all stored in a common repository, which checks for such consistency. For the most part, this common repository of several different types of software artifacts is present only in expensive, high-end CASE tools.

There is a wide range of CASE tools. Some simple tools merely aid in the production of good quality diagrams to describe the flow of control of a software system or the flow of data through the software during its execution. Examples of control flow diagrams and data flow diagrams are shown in Figures 1.11 and 1.12, respectively.

The earliest popular graphical designs were called "flowcharts" and were control-flow oriented. The term *control flow* is a method of describing a system by means of the major blocks of code that control its operation. The nodes of a control flow graph are represented by boxes whose shape and orientation provides additional information about the program. For example, a rectangular box with horizontal and vertical sides means that a computational process occurs at this step in the program. Such boxes are often called "action boxes." A diamond-shaped box, with its sides at 45-degree angles with respect to the horizontal direction, is known as a "decision box." A decision box represents a branch in the control flow of a program. Other symbols are used to represent commonly occurring situations in program behavior.

Control flow diagrams indicate the structure of a program's control at the expense of ignoring the movement and transformation of data. This is not surprising, since control

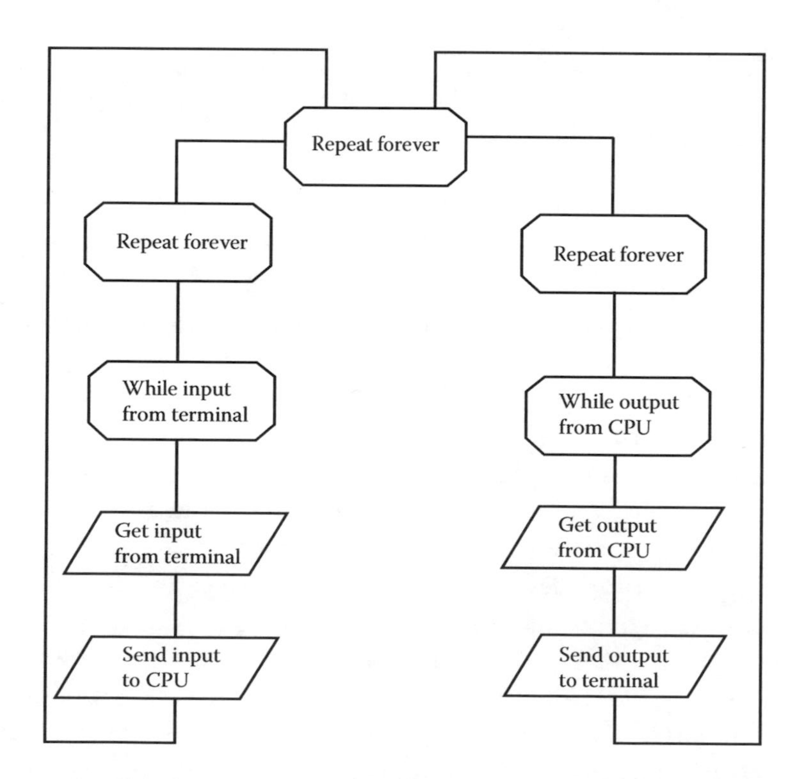

FIGURE 1.11 An example of a control flow diagram.

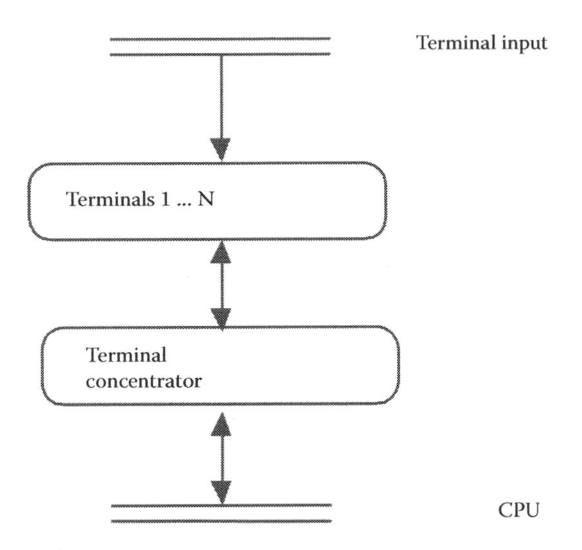

FIGURE 1.12 An example of a high-level data flow diagram with a data source and a data sink.

flow diagrams were developed initially for use in describing scientific programming applications. Programs whose primary function was to manage data, such as payroll applications, were often represented in a different way, using "data flow diagrams."

There is an often-voiced opinion that flowcharts are completely obsolete in modern software. This opinion is not completely correct. A United States patent named "Automatically Enabling Private Browsing of a Web Page, and Applications Thereof," invented by Michael David Smith and assigned to Google, uses diagrams that are related to flowcharts. One of the diagrams that comprises the application submitted to obtain this patent, number US 8,935,798 B1 and issued on January 13, 2015, is included in Appendix A.

Data flow representations of systems were developed somewhat later than control flow descriptions. The books by Yourdon (Yourdon, 1988) and DeMarco (DeMarco, 1979) are probably the most accessible basic sources for information on data flow design. Most software engineering books contain examples of the use of data flow diagrams in the design of software systems.

Since different data can move along different paths in the program, it is traditional for data flow design descriptions to include the name of the data along the arrows indicating the direction of data movement.

Data flow designs also depend on particular notations to represent different aspects of a system. Here, the arrows indicate a data movement. There are different notations used for different types of data treatment. For example, a node of the graph that represents a transformation of input data into output data according to some set of rules might be represented by a rectangular box in a data flow diagram.

A source of an input data stream, such as interactive terminal input, would be represented by another notation, indicating that it is a "data source." On the other hand, a repository from which data can never be recalled, such as a terminal screen, is described by another symbol, indicating that this is a "data sink."

You might ask at this point how these notations are used to describe larger systems. Certainly the systems you will create as practicing software engineers are much too large to be described on a single page. A flowchart will usually connect to another flowchart by having the action or decision box appear both on the initial page where it occurs and on the additional page where the box is needed.

Typical data flow descriptions of systems use several diagrams at different "levels." Each level of a data flow diagram represents a more detailed view of a portion of the system at a previously described, higher level.

Different data flow diagrams are used to reflect different amounts of detail. For example, a "level 0 data flow diagram" provides insight only into the highest level of the system. Each item in a data flow diagram that reflects a transformation of data or a transaction between system components can be expanded into additional data flow diagrams. Each of these data flow diagrams can, in turn, be expanded into data flow diagrams at different levels until the system description is considered sufficiently clear.

Of course, there are other ways to model the behavior of software systems. Many compilers for object-oriented languages such as C++, Objective C, Eiffel, or Java are part of software development that support object modeling. These environments generally allow the user to develop an object model of a system, with the major structure of the objects specified as to the public and private data areas and the operations that can be performed on instances of these objects.

The more elaborate environments include an editor, a drawing tool, and a windowing system that allows the user to view his or her software from many different perspectives. Such advanced environments certainly qualify as CASE tools. An example of a typical advanced software development environment is given in Figure 1.13.

The more advanced software development environments go one step further. An object model created by a user has a description of the public and private data used by an object and the operations that can be performed on the object. The additional step is to allow the generation of frameworks for source code by automatically transferring the information for each object that is stored in the model in a diagram to source code files that describe the object's structure and interfaces. This generated source code can be extended to the creation of complete programs. This code generation is an excellent example of a common CASE tool capability.

Many compilation environments include language-sensitive editors that encourage the use of correct language syntax, debuggers to help fix errors in source code, and profilers, which help improve the efficiency of programs by allowing the user to determine the routines that take the largest portion of the program's execution time. These editors, debuggers, and profilers are all CASE tools.

Other tools allow the management of libraries, where several people are working on source code development at the same time. These tools ensure that only one person is working on any particular source code file at the same time. If a software engineer is editing a source code file, any request from others to edit the same document is denied. As discussed in Section 1.8.1, this technique is called configuration management. It is an essential part of the software engineering process and is almost always supported by CASE tools.

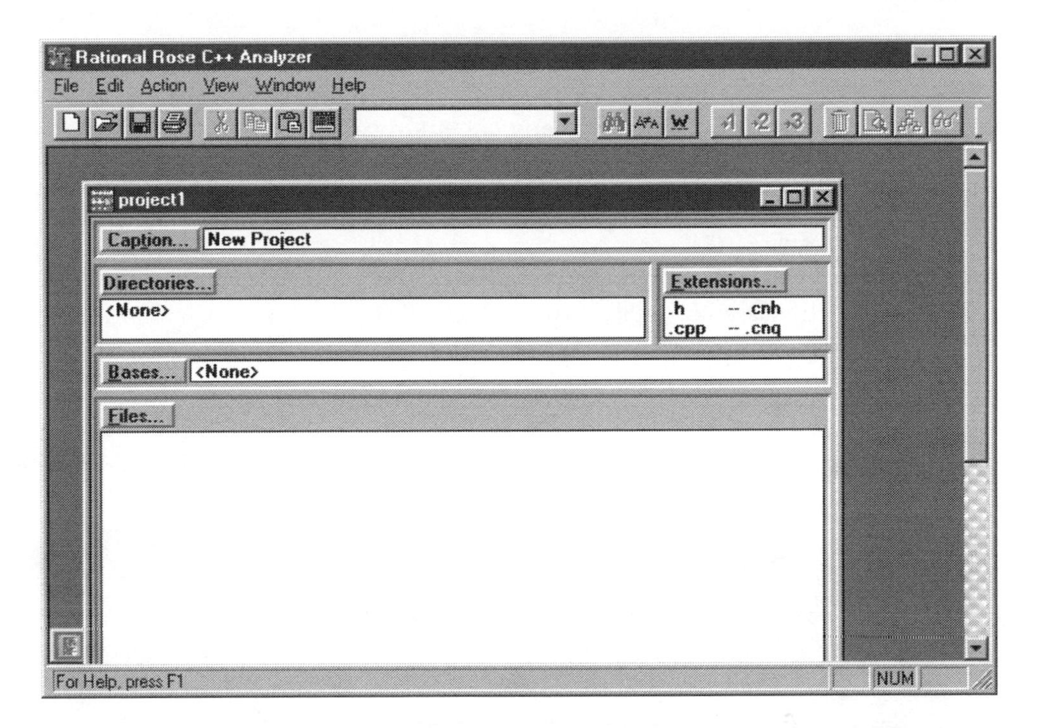

FIGURE 1.13 An example of a window-based software development environment for an object-oriented system. (From Rational Rose Software opening screen, Rational Software Corporation. Rational is now a subsidiary of IBM.)

Clearly, CASE tools are pervasive in software engineering. Software development tools have become so popular and inexpensive that many software engineers do not refer to a development tool as a CASE tool unless it is elaborate, expensive, and supports the entire life cycle. I choose to use the term more broadly, in the spirit of the examples described in this section.

1.8.4 Cost Estimation

Software cost estimation is both an art and a unique subfield of software engineering. It is far too complex to discuss in detail in any general-purpose book on software engineering. We will be content to provide simple rules of thumb and first approximations to determine the cost-benefit ratio of different system design and programming alternatives. The classic book by Boehm (Boehm, 1981) is still the best introductory reference to software cost estimation.

The basic idea is to estimate the cost of a software project by comparison with other, similar projects. This in turn requires both a set of "similar" software projects and a method of determining which projects are "similar" to the given one.

In order to determine the cost of projects, a baseline must be established. By the term *baseline*, we mean a set of previous projects for which both measurements of size and cost are available. The information obtained from these previously completed projects is stored in a database for future analysis. The baseline information should include the application domain; the size of the project, according to commonly agreed-upon measurements;

any special system constraints, such as the system having a real-time response to certain inputs; unusual features of the project, such as being the first time that a new technology, programming language, or software development environment is used; any interoperable software systems that may impact the cost of the software; and, of course, the cost of the software itself. In many organizations, the baseline information may be broken down into the cost of different subsystems.

The determination of "similarity" or "sameness" is made by examining the characteristics of projects in the baseline database and selecting the projects for which information is deemed to be relevant. The effective determination of "similarity" is largely a matter of experience.

Ideally, the cost estimation will be developed in the form of a cost for the system (and each subsystem for which cost estimates are to be made) and an expected range in which the costs are likely to be for the system whose cost is being estimated.

1.8.5 Reviews and Inspections

Different organizations have developed different approaches to ensure that software projects produce useful, correct software. Some techniques appear to work better in certain development environments than others and, thus, there is difficulty in simply adapting one successful approach to a totally different environment.

However, there is one thing that has been shown to be highly successful in every organization in which it is carried out properly: the design review. Reviews should be held at each major milestone of a project, including requirements, design, source code, and test plans. For a software life cycle that uses an iterative process, a review should be held at each iteration of the creation of major items within the software life cycle.

The basic idea is simple: each artifact produced for a specific milestone during development must be subjected to a formal review in which all relevant documents are provided; a carefully designed set of questions must be answered; and the results of the review must be written down and disseminated to all concerned parties.

The following will describe several kinds of reviews for requirements, design, source code, and test plans. In the sense in which they are used here, the two terms *review* and *inspection* are synonymous. In the general literature, the term *inspection* may be used in this sense or may instead refer to a procedure for reading the relevant documents.

There is one other point that should be made about reviews and inspections. This book emphasizes systematic methods that attempt to determine problems in requirements and design well before the software is reduced to code. This emphasis is consistent with experiences across a wide variety of software projects in nearly every application area.

The open source approach of the development of the Linux operating system avoids reviews of requirements and designs, and instead relies on the efforts of a large number of programmers who review source code posted on line. Thus each source code module is intensively reviewed, often by hundreds of programmers. Modifications to the source code are controlled by a much smaller team of experts. Of course, there is also configuration management and testing, just in different order. Consult the article by McConnell (McConnell, 1999) for an overview of the open source approach.

1.9 STANDARDS COMMONLY USED FOR SOFTWARE DEVELOPMENT PROCESSES

Many organizations have very specific software development processes in place. These processes are motivated by both organizational experience and the pressures brought on by outside customers from both government and industry. The most common formalized software development processes in the United States are

- The Capability Maturity Model (CMM) from the Software Engineering Institute (SEI) at Carnegie Mellon University. (Technically, this is not a process but an evaluation of how systematic an organization's software development process is and a guideline for assessing the possibility of process improvement.) Information on the CMM is available from the website http://www.sei.cmu.edu/cmmi/.
- The CMM model has been superseded by the Capability Maturity Model Integration (CMMI). The primary website for this model has been moved from its original home at the Software Engineering Institute (http://www.sei.cmu.edu/cmmi/) to the 100 percent-owned subsidiary of Carnegie Mellon University, http://cmmiinstitute.com/. This newer website has several examples of how appropriate measurement of software development processes can be used to improve quality and reduce costs.
- The Process Improvement Paradigm (PIP) from the Software Engineering Laboratory (SEL) at NASA's Goddard Space Flight Center. (As with the CMM, technically, this is not a process but an evaluation of how systematic a software development process is and a guideline for assessing the possibility of process improvement.)
- The Department of Defense standard MIL-STD 2167A.
- The Department of Defense standard MIL-STD 1574A.
- The Department of Defense standard MIL-STD 882C.
- The Electronic Industries Association (EIA) SEB-6-A.

Some international standard software development processes include the following:

- The European ESPRIT project.
- The International Standards Organization (ISO) standard ISO 9001.
- United Kingdom MOD 0055.

Each of these software process models has a heavy emphasis on the collection and use of software metrics to guide the software development process. The models from the SEI and the SEL stress the development of baseline information to measure current software development practice. These two models can be used with classical waterfall, rapid prototyping, and spiral methodologies of software development. The more rigid standard DOD 2167A is geared primarily toward the waterfall approach, but it, too, is evolving to consider other software development practices.

These are probably the most common nonproprietary standardized software development processes being used in the United States. Note that all these process models emphasize the importance of software reuse because of its potential for cost savings and quality improvement.

Many private organizations use models based on one of these standards. For example, the rating of a software development organization's practice from 1 ("chaotic") to 5 ("optimizing") on the CMM scale is now a factor in the awarding of government contracts in many application domains. The rating is done independently. The reason for rating the development organization's practices is to ensure a high-quality product that is delivered on time and within budget. The details of the CMM scale are given in Table 1.3.

As another example of the importance of software development processes, Bell Communications Research (1989) had used a process assessment tool to evaluate the development procedure of its software subcontractors for many years. An assessment of the subcontractor's software development process was an essential ingredient in the decision to award a contract.

Both Allied Signal Technical Corporation and Computer Sciences Corporation, among others, have an elaborate written manual describing their software development processes. The reference (Computer Sciences Corporation, 1992) is typical. You should expect that

TABLE 1.3 Description of the Levels of the Capability Maturity Model (CMM) Developed by the Software Engineering Institute (SEI)

Level 5 (Optimizing Process Level)
The major characteristic of this level is continuous improvement. Specifically, the software development organization has quantitative feedback systems in place to identify process weaknesses and strengthen them proactively. Project teams analyze defects to determine their causes; software processes are evaluated and updated to prevent known types of defects from recurring.
Level 4 (Managed Process Level)
The major characteristic of this level is predictability. Specifically, detailed software processes and product quality metrics are used to establish the quantitative evaluation foundation. Meaningful variations in process performance can be distinguished from random noise, and trends in process and product qualities can be predicted.
Level 3 (Defined Process Level)
The major characteristic of this level is a standard and consistent process. Specifically, processes for management and engineering are documented, standardized, and integrated into a standard software process for the organization. All projects use an approved, tailored version of the organization's standard software process for developing software.
Level 2 (Repeatable Process Level)
The major characteristic of this level is that previous experience provides intuition that guides software development. Specifically, basic management processes are established to track cost, schedule, and functionality. Planning and managing new products is based on experience with similar projects.
Level 1 (Initial Process Level)
The major characteristics of this level are that the software development process is largely ad hoc. Specifically, few processes are defined, and success depends more on individual heroic efforts than on following a process and using a synergistic team effort. (The term *chaotic* is often used to describe software development processes with these characteristics.)

Source: Software Engineering Institute.

any large software company that you work for will have its own software procedures and will probably have a standards and practices manual.

There are many other software standards that are still being followed to some extent. A count of standards activities of the IEEE indicates 35 distinct standards in the area of software engineering at the end of 1995. One of the most important IEEE standards is number P610.12(R), which is titled "Standard Glossary for Software Engineering." See also the SWEBOK website.

Any introductory book on the topic of software engineering would be incomplete if it did not mention the cleanroom software development process (Mills et al., 1987). This process was initially developed at IBM by Harlan Mills and his colleagues. The goal of the cleanroom process is to produce software without errors by never introducing errors in the first place. The term *cleanroom* was chosen to evoke images of the clean rooms used for manufacture of silicon wafers or computer chips. The workers in such rooms wear white gowns to reduce the possibility of introducing contaminants into their product. In the cleanroom approach, any errors in software are considered to have arisen from a flawed development process.

The major technique of the cleanroom process is to use mathematical reasoning about correctness of program functions and procedures. The idea is to create the source code with such care that there is no need for testing. Rigorous mathematical reasoning and very formal code reviews are both integral parts of the cleanroom process.

Chapter 6 will indicate that the purpose of software testing is to uncover errors in the software, not to show that the software has no errors. The practitioners of the cleanroom approach believe that their approach may be more conducive to the goal of the software development process: producing error-free software.

We will not discuss the cleanroom approach any further in this book. Instead, our limited emphasis on the role of mathematics in software engineering will be restricted largely to reliability theory and a brief mention of some other formal development methods that we will discuss in Chapters 2 and 3.

There is an approval process used by the Apple App Store before an app is allowed to be sold either for Macintosh computers or other Apple products such as iPads and iPhones. The App Store's Review Guidelines provide rules and examples for such things as user interface design, functionality, content, and the use of specific technologies. There are also guidelines for the use of push technology and for the secure handling of data. More information on these particular guidelines can be found at the site https://developer.apple.com/app-store/review/. (Note: It is good to be an Apple developer if you can get to tell your grandson that you can get discounts on several different Apple products.)

Not surprisingly, other app stores are beginning to have similar review processes.

1.10 ORGANIZATION OF THE BOOK

Each of the major phases of the software development life cycle will be covered in detail in a separate chapter after we discuss project management in Chapter 2. Beginning with Chapter 3, each of the chapters will consider a relatively large project that will be continued throughout the book.

The purpose of considering the same project throughout the book is to eliminate the effects of different decisions for the requirements, design, and coding of software. Some of the source code and documentation will be given explicitly in the book itself. However, most of the rest will not be made available, allowing instructors to have more flexibility in tailoring the assignments especially to individual groups within their classes.

A continuing case study of an agile software development project will also be given throughout the book to provide a complete picture of how a successful agile project can work.

The order of Chapters 3 through 8 is essentially the same as in the classical waterfall software development model. This order is artificial (something had to be presented first!) and the chapters can largely be read in any order. Some instructors may choose to cover several development methodologies rather quickly in order to focus on the development methodology that fits best with the educational objectives of the course.

The book will work best when coordinated with the creation of team software projects that reuse existing code, but require something more elaborate than a simple cut-and-paste of source code.

As you will learn, there are many nonprogramming activities necessary in software engineering. Some of these activities require cost estimation, or data collection and analyses. Spreadsheet forms have been included to illustrate typical industry practices and to provide you with experience.

Each of Chapters 3 through 8 will include one or more sections that provide typical views of software managers on the relevant technical activities. The purpose of the managerial view is to provide you with additional perspective on software engineering, not to prepare you to be a manager. In most organizations, a considerable amount of varied project experience is essential before a person is given a managerial position at any level.

Chapter 9 discusses how to read the software engineering literature and provides a large set of important references.

The three appendices on software patents, command-line arguments, and flowcharts can be read if necessary.

SUMMARY

Software engineering is the application of good engineering practice to produce high-quality software in an efficient manner. Good software engineering practice can reduce the number of problems that occur in software development. The goal of software engineering is to develop software that is efficient, reliable, usable, modifiable, portable, testable, reusable, easy to maintain, and can interact properly with other systems. Most important, the software should be correct in the sense of meeting both its specifications and the true wishes of the user.

The goals of software engineering include efficiency, reliability, usability, modifiability, portability, testability, reusability, maintainability, interoperability, and correctness. In support of these goals, software engineering involves many activities:

- Analysis
- Requirements

- Design
- Coding
- Testing and integration
- Installation and delivery
- Documentation
- Maintenance
- Quality assurance
- Training

Project management is essential to coordinate all these activities.

There are several common models of the software development process. In the classical waterfall process model, the requirements, design, coding, testing and integration, delivery and installation, and maintenance steps follow in sequence, with feedback only to the previous step. In the rapid prototyping and spiral development models, the software is developed iteratively, with customer or user feedback given on intermediate systems before a final system is produced.

A disciplined software engineering approach is necessary, regardless of the application domain: development for the Internet, personal computer applications, and health- or safety critical applications. The risk analysis that is typically part of any engineering activity might have prevented the Year 2000 problem from occurring.

The open source model involves a small group of experts who act as gatekeepers evaluating the submissions from a vast number of individual software engineers to help solve problems set by these expert gatekeepers. All software submitted is done so under the understanding that others are free to use or modify it as long as they do not charge for it; that is, the software must remain free, although charging for installation, setup, and consulting are allowed.

A major goal of the agile programming model is to provide fast solutions to problems where the team has significant knowledge of the particular application domain and there are a lot of high-quality software components that are available for reuse to help solve these problems.

The use of high-level programming languages can increase programmer productivity. However, reusing high-quality software components can have a greater effect.

KEYWORDS AND PHRASES

Software engineering, software reuse, classical waterfall software development model, rapid prototyping model, spiral model, agile software development, agile methods, open source software development, market-driven software development, software evolution, reengineering, CASE tools, metrics

FURTHER READING

There are many excellent books on software engineering, including Pfleeger and Atlee (Pfleeger and Atlee, 2010), Schach (Schach, 2011) Sommerville (Sommerville, 2012), and Pressman and Maxim (Pressman and Maxim, 2015). Some interesting older general software engineering books are by Ghezzi, Mandrioli, and Jayazerri (Ghezzi, Jayazerri, and Mandrioli, 2002), Jalote (Jalote, 1991), and Shooman (Shooman, 1983). The book *Software Development: An Open Source Approach* (Tucker and de Silva, 2011) is devoted to open source software development. The classic book by Boehm (Boehm, 1981) provides a good overview of software engineering economics. Fenton and Pfleeger (Fenton and Pfleeger, 1996) and Fenton and Bieman in a later edition (Fenton and Bieman, 2014) provide a detailed, rigorous description of software metrics. Beizer (Beizer, 1983, 1990) Howden (Howden, 1987), and Myers (Myers, 1979) provide excellent introductions to software testing that are still relevant today. An excellent introduction to the cleanroom process can be found in the article "Cleanroom Software Engineering" (Mills et al., 1987).

A large-scale effort by the software engineering community has created the "Software Engineering Body of Knowledge," commonly known by the acronym SWEBOK. This effort can be found in a collection of online PDF documents (SWEBOK, 2013).

The helpful 2014 Research Edition of the Software Almanac from Software Quality Management is available for download at qsm.com.

The classic 1978 paper by Ritchie and Thompson (Ritchie and Thompson, 1978) provides both an overview of the extremely successful UNIX operating system and an insight into the design decisions that were used. The Linux operating system is a modern variant of UNIX.

Information on computer languages can be found in many places. The Ada language is described in Ada (1983, 1995) and Ichbiah (1986). The rationale for the original description of C++ is given by Stroustrup (Stroustrup, 1994), and a detailed language manual is provided in the book by Ellis and Stroustrup (1990). Kernighan and Ritchie provide an original description of C (Kernighan and Ritchie, 1982), with a second edition (Kernighan and Ritchie, 1988) describing the ANSI C features. There are many other excellent books on various computer languages.

Nielsen (1994) and Shneiderman (1980) provide excellent overviews of human–computer interaction. A brief introduction to the role of color in user interfaces can be found in the article by Wright, Mosser-Wooley, and Wooley (1993). The older paper by Miller (1968) provides good empirical justification for limiting the number of choices available to system users. It is especially relevant to a designer of a computer system's menu structure.

Laplante (2014) provides an interesting perspective on an evolving position on requiring licensing for software engineers.

The article by Cusumano and Selby (1995) provides an excellent overview of some software development practices at Microsoft.

Information on the Capability Maturity Model (CMM) is available from the website http://www.sei.cmu.edu/cmmi/.

More information on the Capability Maturity Model Integration (CMMI) is available from the website http://cmmiinstitute.com/.

The paper by Basili and Rombach (1988) is a good introduction to the GQM paradigm. The book by Leach (1997) is a good introduction to software reuse.

EXERCISES

1. Name the major types of software development life cycles. How do they differ?

2. Name the activities that are common to all software development life cycles.

3. What are the goals of software engineering?

4. Explain the GQM paradigm.

5. Consider the largest software project you ever completed as part of a class assignment. How large was the project? How long did it take you to develop the system? Did it work perfectly, did it work most of the time, or did it fail the simplest test cases?

6. Repeat Exercise 5 for software that you wrote as part of a job working in industry or government.

7. Examine the classified advertisements of your college or university placement office for listings of jobs in the computer field. Classify each of the jobs as being for analysis, requirements, design, coding, testing and integration, installation and delivery, documentation, maintenance, quality assurance, or training. What, if anything, is said about open source software? What about agile programming?

8. If you have a part-time (or full-time) job in the computer field, ask to see your employer's software development manual. Determine which software development process standard is followed by your employer.

9. Several software development process standards were mentioned in this chapter. Find one or more of these in the library or on the Internet. Examine the details of the standard. Can you explain why they are necessary?

10. Examine a relatively large software system that you did not develop yourself and for which source code is available. (The GNU software tools from the Free Software Foundation are an excellent source for this question.) Can you tell anything about the several software development process standards used in this software from an examination of the source code? Why or why not?

11. We discussed some issues about the role of networking in the future development of word processors. There are other issues, such as the responsibility of system administrators responsible for software updates. Discuss the effect of the four options given in Section 1.1 from the perspective of system administrators. You may find it helpful to read the discussion on how an index might be implemented on a multicore

processor in Andrew Tanenbaum's recent book *Modern Operating Systems*, 4th edition (Tanenbaum, 2014).

12. Research Lockheed Martin's Skunk Works project, also known as "Advanced Development Programs," to obtain a description of its very successful quick-and-dirty development process. Compare this to the goals of the "Agile Manifesto."

CHAPTER 2

Project Management

In this chapter, we will discuss some of the issues involved with project management. As was pointed out in Chapter 1, most of the discussion of this topic will appear quite remote from the experience of a beginning software engineer. It is difficult to communicate the complexity of software project management to a person who has not been part of a multiperson software project in government or industry. Nevertheless, it is useful for every software engineering student to have at least an introduction to those software engineering activities that are typically associated with group projects. Therefore, we will present a high-level view of project management in this chapter. The intention is to introduce the subject, not to create instant experts. Keep in mind the point that most modern software development is done by teams.

Many of the most successful software companies, especially the ones that have achieved almost overnight successes, appear to take considerable risks in the sense that they produce software that performs a task or solves a problem that might not have been considered for a computer solution before the software was created. Software intended for word processing, spreadsheet analysis, or the Internet itself, with its collection of browsers and search engines, were clearly revolutionary breakthroughs. As such, they were considered risky investments until they achieved some success. All so-called killer apps have this feature. Risk is inherent in most software engineering endeavors.

Keep in mind, however, that an organization whose existence depends upon the continued success of its software products must take a systematic approach to management of its software projects. This approach must be more systematic than would be necessary for a student who is working on a simple class project that may require only one person or, at most, a small group of classmates. It is generally more systematic than what might be appropriate for a single programmer creating a small, stand-alone app. Software project management should attempt to reduce the risk in the way that the software project follows a schedule and is developed efficiently according to reasonable software engineering standards for quality and productivity.

The primary goal of this chapter is to introduce project management issues in sufficient detail so that you can understand the types of organizational cultures you might meet when you begin your career as a software engineer. Different organizations will have different software development processes and will structure their software development teams accordingly. However, each project will have a manager who will have to deal with many issues.

One of the most important issues for any project manager is the selection of the project team. An ideal team member is well versed in the application domain and is an outstanding software developer. It is very important that he or she be an excellent team player. Since perfect team members are hard to come by, they are often selected to have a mix of skills and experience. Think of the compromises that you had to make while working on team projects during your education!

There is one final point to be made about project management. Most formal software development processes that are commonly described in the software engineering literature are directly applicable to very large software systems written for a single customer, such as the government or a major company requiring a highly complex system of connections to databases.

There are many modifications necessary to apply some of the more formal of these processes in an environment that is highly market driven. Such development environments have a high degree of concurrency, with many activities, such as coding, design, requirements analysis, and testing, taking place at the same time in order to meet very tight delivery schedules. In this chapter, we will also discuss some of the issues that affect software development in such market-driven environments.

I have often found that at the start of their first course in software engineering, most students, especially those with little experience working on software engineering projects in industry or government, dislike the formalities of project management. I believe that this is because their primary experience is with relatively small programs that they develop themselves, and that they see the overhead of project management to be pointless, preventing them from getting to the point of coding. They begin to understand the importance as they gain more experience with team projects. During the assessment of program objectives at my university, recent computer science graduates often give high ratings to their experience with project management during their times as students. After their initial reluctance, they eventually learn to greatly value project management and team projects.

Students beginning a software engineering course often wish they could use what they have heard described as an agile software development process. They usually take this term to mean low overhead, with few, if any, meetings and a lot of coding. Agile programming becomes less appealing when students learn that they must have mastered the application domain, at least to the level of being able to determine large-scale reusable software components, in order to adequately move projects forward. It takes a while for them to appreciate the constant coding and testing, and reporting of status at regular formal meetings necessary for agile processes.

Management of agile processes is quite distinct from most of the other commonly used software development processes and will be discussed in more detail in a separate section at the end of this chapter.

As indicated several times in Chapter 1, open source software development is essentially only done well if there is a rather small set of experts who determine, using their own evaluation and that of the open source community on contributed software components, whether open source components should be included in the desired software application or utility being developed. Clearly, the management of open source projects is essentially limited to experts and I believe it is hard to expect relatively new software engineers such as those reading this book to perform the tasks that this highest level of authority will require. Therefore, we will not discuss management of open source projects any further in this chapter.

2.1 SUBTEAMS NEEDED IN SOFTWARE ENGINEERING PROJECTS

It is clear that the systematic development of large, modern, software projects requires teams. It is often helpful to think of a software project team as being made up of several "subteams." These subteams need not be as formally constituted as they are described here and, in fact, some of them will consist of a single person in many cases. In fact, the same person may perform all the actions of multiple subteams. However, the team's activities still need to be performed, regardless of the overall team organization and the size of the project. The activities performed by these teams occur regardless of the software development life cycle used, although the timing of these activities will almost certainly differ.

Several of these subteams are not likely to be in existence during the entire lifetime of the software project. For some software projects, one or more of the teams may consist of a single person whose duties may be split among several projects or activities. In other projects, the teams may be very large and geographically separated. Team members may even report to different companies in some of the larger cooperative software development projects.

Some typical software engineering subteams and their duties are listed next. Although all activities are important, this list highlights some cases where a particular software development life cycle model requires special knowledge from a subteam.

Systems analysis team—This team is responsible for determining if the project is feasible. The feasibility study includes cost analysis, estimated revenues, and an estimate of the difficulty of engineering the project. After it produces the feasibility study, this team should interact with the requirements team, receiving its feedback. If the software development process is iterative, as in the rapid prototyping and spiral models, then the interaction and feedback should be more frequent and may occur with additional subteams.

Planning team—This team is responsible for developing the overall management plan for the project and making sure that the project is proceeding within the expected time frame for various activities. This often involves staffing, which becomes especially critical for agile software development processes, to make sure that the team has adequate knowledge of the application domain to know the capabilities of existing software components and systems. The same is true for open source projects.

Requirements team—The duties of this team are to meet with the customer and determine a complete, precise set of requirements for this project. This will require a set of formal and informal meetings with the customer to finalize the requirements from relatively imprecise and incomplete initial requirements. If no customer is available, then the requirements team is to obtain the same information from one or more potential users. If no potential users are available, then surrogate customers may be used in their place. After it produces the system's requirements, this team should interact with the design team, receiving its feedback. If the software development process is iterative, as in the rapid prototyping and spiral models, then the interaction and feedback should be more frequent and may occur with additional subteams. This type of feedback is crucial to agile software development processes.

System design team—The duties of this team will be to produce a detailed design of the system after the requirements have been set by the requirements team. If the software development process uses the classical waterfall model, then the system design team should provide feedback to the requirements team about any difficulties encountered. After it produces the design, the system design team should interact with the implementation team, receiving its feedback. If the software development process is iterative, as in the rapid prototyping and spiral models, then the interaction and feedback should be more frequent and may occur with additional subteams.

Implementation team—The duties of this team will be to implement the software designed by the system design team. After they produce the implementation, this team should interact with the testing and integration team, receiving its feedback. If the software development process is iterative, as in the rapid prototyping and spiral models, then the interaction and feedback should be more frequent and may occur with additional subteams.

Testing and integration team—The duty of this team is the formulation of test cases for the modules and systems that are created by the implementation team. This team may take some modules from an incomplete system for testing by mutual agreement with the implementation team. After it produces the test plan and tests the software modules produced, this team will integrate the software modules into a working system. This team should interact with the implementation team, receiving its feedback. If the software development process is iterative, as in the rapid prototyping and spiral models, then the interaction and feedback should be more frequent and may occur with additional subteams. The integration team is responsible for an interface control document (ICD) that precisely describes the interfaces between major system components. This can be in the form of specifying APIs, or the interfaces between software subsystems.

Training team—This team is responsible for the development and production of training materials.

Delivery and installation team—This team is responsible for the delivery and installation of the software.

Maintenance team—This team is responsible for the maintenance of the software after it is delivered and installed. After the system is delivered and installed, this team should interact with the implementation team, receiving its feedback. If the software development process is iterative, as in the rapid prototyping and spiral models, then the interaction and feedback should be more frequent and may occur with additional subteams.

Quality assurance (QA) team—This team has two duties. The first is to set standards for the adherence of the project team to a set of agreed-upon processes for the system's creation and set standards for performance of the software produced. The second is to provide an evaluation of how well the project teams meet those standards. Standard industry practice is for the information obtained by this team to be kept internal and not shared with the customer. The information can be released in the event of a legal action and, thus, cannot be destroyed. This information is to be presented to the project manager who will use it to evaluate performance of the QA team.

Metrics team—This team is responsible for keeping statistics on the performance of the teams on the project. Depending on the organization's data collection procedures, some typical statistics kept might be the number of maintenance requests generated, number of maintenance requests serviced, number of lines of code written, number of hours performed on each task, and values produced by the tool on each new version of the system. This team will interact with the requirements, design, implementation, testing and integration, and maintenance teams, providing assessments of quality and efficiency, as well as feedback to these subteams.

Documentation team—This team is responsible for the project documentation. This includes external documentation of the requirements, design, source code, and other supporting documents.

System administration team—This team is responsible for ensuring that the underlying computer hardware, software, and network support are working as needed by the project team. This team often includes the network administration team.

Reuse and reengineering team—This team is responsible for selection and use of appropriate existing reusable software components. Reengineering may be necessary if the software project depends upon some old code that must be changed because of new advances in technology.

It is not surprising that there are several managerial tasks involved here, one for each subteam. Of course, if the teams are small enough, a manager may be one of the technical members of the team (or even the only team member).

The aforementioned tasks listed are required as part of most software projects. Therefore, it is reasonable to ask how they are scheduled. Of course, the scheduling of these tasks is highly dependent on the software life cycle model used by the project. For example, a software development process based on the classical waterfall software development model might have the tasks incorporated into the process as shown in Figure 2.1. You will be asked to produce similar diagrams for the rapid prototyping and spiral software development models in the exercises.

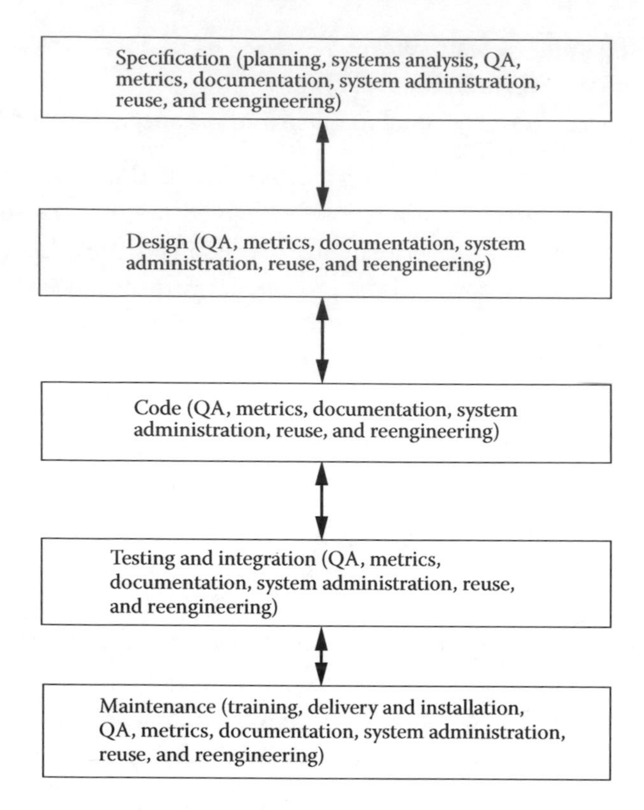

FIGURE 2.1 Incorporation of basic software engineering tasks into a simplified illustration of the classical waterfall life cycle model.

The plan for market-driven software development is interesting. The entire process is compressed, at least in comparison to the classical waterfall process. The steps usually include the following:

1. Determination of the market for the product
2. Requirements determination
3. Combined design and coding
4. Testing and integration
5. Delivery

At first glance, the process seems similar to others, except for the grouping of the design and implementation phases. The easiest way to see the difference is to consider the milestones that are typical in market-driven software development.

In this model, planning the product's requirements and its marketing are often done in concert. The planning interval rarely exceeds five years and is often much shorter. The goal is to have a project plan that results in a product that the marketing and sales teams can sell. Since technology is changing so fast, longer development periods are not acceptable.

Usability testing is often included in the design and early implementation phase. This is before the detailed testing phase begins.

The delivery process often includes determination of minimal, typical, and maximal installations, together with an installation suite that can select components depending on the user's desires. The software may be packaged in academic, regular, or professional versions, each of which has a different set of incorporated components or system configuration parameters and, therefore, different installation procedures.

In the highly competitive global economy, the product may be shipped to countries where the native language is not English. This means new versions of online help and manuals. It may also mean a different configuration of such things as the built-in security, since the United States government does not allow 128-bit encryption to be exported for nonmilitary applications without restriction. (There are different restrictions for military and nonmilitary products, and for products being sold to so-called rogue states.) See the Bureau of Industry and Security website at the URL www.bis.doc.gov/index.php/policy-guidance/encryption for the latest information.

You should note that there are several technical tasks that must be performed in addition to the tasks listed earlier. The persons performing these tasks are often given official or unofficial titles such as the ones listed next. Even without particular titles, software engineers often have to account for the percentage of their time that is spent on particular activities. This time accounting is often necessary to determine the cost of individual projects.

Human–computer interface evaluator—This person is responsible for evaluating the type of interaction between the software system and users. At the very least, he or she must have a mental model of the expected skills of both novice and experienced users of the final software product.

Tools support person—This person is responsible for making sure that supporting software environments are working properly. This person will work with system or network administrators.

Software economist—This person is responsible for development and use of appropriate models in order to estimate software costs, hardware and software resources, and the time needed for project completion.

Project librarian—This person is responsible for keeping track of all project documents.

There may be other personnel needed in specialized software development environments, particularly ones with extensive software reuse. For simplicity, we will ignore these specialists and concentrate on those responsibilities most likely to be encountered by the beginning software engineer.

2.2 NATURE OF PROJECT TEAMS

In typical classroom environments in an academic setting, a software project is done by either one person working alone or by a small team of students. The student teams work together on the same project for a semester or, occasionally, for an entire year. There is little turnover in upper-level computer science courses and, hence, it is relatively easy to have the same teams in class projects. Projects intended for graduate students, especially PhD students, will generally be larger and require more time, but the same team members are available.

In real-world software development projects, the teams are rarely kept together for the entire life cycle of a software system, usually with a major change after each release of the software, if not sooner.

There are several reasons for the relatively high rate of personnel changes in actual software projects:

- Project staffing needs change. More people may be needed to work during the coding or testing and integration phases than are needed for the initial requirements gathering activity. Often, fewer people are needed for maintenance, especially if the system is expected to have a short life.
- The project may continue for a long period. For example, NASA's International Ultraviolet Explorer satellite was expected to produce scientific data for one year and, thus, the original software maintenance team was expected to work for at most one year after the satellite was placed into orbit. Instead, the satellite's scientific life lasted for 19 years, producing as much scientific data and subsequent research in its last year of operation as it did in its first. Clearly, there were many changes in project personnel during this time. Several well-known database systems have been in operation for more than 30 years. Keeping the same level of staffing as in the initial development is extremely wasteful.
- People change jobs. Many software engineers prefer to be associated with new technology and leave projects after the technology used in a particular project ceases to be cutting edge.
- Companies lose contracts. Many organizations merge and business units are refocused. In an extreme case, companies go out of business. If a company was working on a project as a subcontractor that is now defunct, the project team will change.
- People retire or have major illnesses.
- Product lines, and companies themselves, may be sold to new companies.
- New people are hired to bring fresh ideas and new technology skills into organizations.

It is clear that most real-world projects have substantial changes in their personnel during their lifetimes.

Now that you understand that the composition of project teams often will change considerably over the life of a project, you might consider the ramifications of this relative instability for your own career as a software engineer. Clearly, you will need to be flexible, because different individuals have different learning styles and you will have to interact with many of these different styles over time. For instance, you might be paired with an individual who always follows the written company policies to the letter, refusing to submit any organization-supplied form unless the sentences are grammatically correct. Another person on the project will use shortcuts, writing phrases and sentence fragments, whereas the other might write formal, proper sentences. Both approaches might be very effective for the person who uses them and you might have to interact with each type of individual. Of course, the final deliverable product will have to be grammatically correct and adhere to the organization's standards. This is becoming even more important for safety-critical software, for security software, or for software for financial purposes, due to the likelihood that such software may be audited in the event of a software failure.

It is important to accommodate different personality and learning styles within a group and to avoid what James McCarthy of Microsoft called "flipping the bozo bit" (McCarthy, 1995). This colorful term refers to assessing a person's opinions and suggestions as being irrelevant, because the person is a "bozo," and hence their "bozo bit" is flipped (set to one). McCarthy (and I) believes strongly that this is a very bad idea. Be receptive to all ideas and consider them as being potentially reasonable, regardless of the source of the idea. Decide on the validity of an idea only on its merits. This will help make you a valuable team member and will make your team work better.

It is important to understand the differences between organizational cultures. Some organizations follow a rigid process, demanding that all procedures be carried out and all decisions be carefully documented. An individual who does not precisely follow the organization's procedures often will be ignored or given less demanding work.

Other organizations have much less formal structure in their process. They often give software development teams only the most general guidelines, with almost complete freedom to perform the essential activities. This freedom occurs most often when the team has a history of completing its projects successfully and within budget. A NASA technical report by Mandl (Mandl, 1998) describes an effective software development team that used relatively few formal procedures but was able to produce actual systems. It should be noted that this team had many years of experience in its particular application domain. We will discuss this example of agile programming that preceded the "Agile Manifesto" in more detail when we describe agile project management in Section 2.10.

You should make the distinction yourself in the kind of organization you choose to work for in order to match your temperament with the organizational development style that is right for you.

2.3 PROJECT MANAGEMENT

The term *project management* refers to the coordination of people, resources, and schedules that are necessary to successfully produce a software product that is of high quality, is delivered within the allotted time period, and is within budget.

Suppose that you were assigned to direct a project whose goal was to develop a particular software system. You would want the project to be a success and so you would try to assemble the best team available. You would want the team to have proper equipment and any software development tools that were appropriate. You would need to know about any other software systems that your software must be interoperable with. You would need to be able to predict when the project would be completed and what the total cost might be.

It is clear that the first task of a project manager is to obtain a good estimate of the size and complexity of the project. These estimates are necessary in order for the manager to obtain the proper amount of resources for the project. We will discuss software project estimation in Section 2.4.

Once you have determined the resources that are needed for the project, you would develop a schedule. The schedule would contain dates for starting and completing activities that are considered to be essential. These accomplishments are often called "milestones." Since the milestone events might require different levels of effort, different phases of the project might have different demand for resources. Certain project activities might require almost no effort. We will discuss software project scheduling in Section 2.5.

At this point, the project manager has planned the project's resources and schedule. He or she may have had pressure to create the schedule or resource list, but in any event some planning and thoughtful work went into this effort. Unfortunately, many other project activities do not involve the luxury of planning and, instead, can only react to situations that arise.

We illustrate this point next. The rest of the project management activities generally consist of many things, including but not limited to:

- Managing people
- Allocating resources
- Defining roles
- Motivating project personnel
- Dealing with inevitable slippage in the schedule
- Handling changes in the project's requirements
- Measuring the progress and quality of the system
- Reacting to unexpected events such as computer crashes
- Informing upper level management about problems and successes
- Ensuring that major milestone events have proper reviews
- Interacting with the prospective customer or customer representatives

Some software development methodologies, such as Scrum, have additional project management activities. A Scrum team will generally have a "product owner," who represents the primary stakeholders in the project. There may even be a "Scrum master," who acts as a facilitator for the project. (In organizations not using such agile methods, this role may be performed by the QA, or quality control, team.)

If a project manager only reacts to crises, as appears to be the case, then there is no time to be proactive, that is, to take actions that may head off problems rather than simply react to the problem du jour.

One of a project manager's major tasks is motivation of the project team. In industry or government, the motivation can be promotions, salary increases, bonuses, or simply praise and recognition. It can also mean demotion and termination. (The absence of such rewards and punishments in academic projects is a major reason that such projects are often considered toys. The size of projects and the academic calendar limitations are other contributors to this presumed unreality.)

One thing missing from the preceding list is any form of continuing education for the managers. Such education involves keeping up with current trends in project management both within the manager's organization and in the outside world. Assuming that the project manager wishes to advance his or her career, he or she may take continuing education or other courses and seminars.

In short, project managers are very busy. You should not expect a great amount of mentoring from a senior manager with responsibility for multiple projects or a large number of software engineers. There simply is not enough time.

2.4 SOFTWARE PROJECT ESTIMATION

Software project estimation is part of the general systems engineering process that takes place when a project is planned. It is always a part of major software projects for one basic reason: no reasonable person or organization would begin a large project without some belief that the project can be done with the resources that the person or organization is willing to commit to the project. Clearly, some form of process estimation is necessary. This section provides additional details that will extend the discussion begun in Chapter 1, Section 1.8.4.

Any project will need most of the following resources, with many people, computers, or software required to perform multiple duties in very small projects or organizations:

- Computers for software development
- Basic software for development such as compilers or linkers, ideally within an integrated development environment that includes configuration management and similar tools
- Methods of communicating between computers such as networks
- Computer-aided software engineering (CASE) tools
- Software packages with which the system must be interoperable
- Software and hardware emulators of both testing and development environments for devices such as smartphones and tablets
- Actual deployment devices such as smartphones and tablets

- Computers for testing
- Computers for training
- Clouds and cloud computing services for large-scale development
- Commercial off-the-shelf (COTS) products deemed necessary for system operation
- Documentation tools
- Copying devices
- Programmers
- Testers
- Managers
- Designers
- Requirements engineers

Note that projects of moderate size or larger may require multiple instances of each resource, including software tools, such as CASE tools or configuration management systems. In fact, every software team duty mentioned earlier in this chapter would have to be counted as a resource requirement of the project.

The term *size* has been used informally in this section. Determination of the actual size of a project is a nontrivial matter and we will return to it several times. For now, the size of a project is the number of lines of source code created by the organization in order to develop a project. A line of code is any line in a program that is neither blank nor consists only of comments. (Better definitions of the term *line of code* will be given later in this book.)

Understanding what is meant by the size of an existing system and being able to quantify this size in a consistent manner are absolutely essential if we expect to estimate the size of new systems. Thus, we will temporarily turn our attention to the subject of measuring the size of a software system.

Of course, the measurement of software system size can be very difficult. For example, suppose that you write a program 100 lines long that writes its output to a data file. Suppose that the data file is then imported into a spreadsheet and that the spreadsheet you created uses the spreadsheet's built-in statistical routines, which are then exported to another file, which contains the final result. How big is the system? Is it the 100 lines of code that you wrote? Is it the millions of lines of code that make up the spreadsheet (plus the 100 lines you wrote)? Is it the number of lines you wrote (100) plus the size of the code you entered into your spreadsheet program? Is it the number of lines you wrote (100) plus the size of the data output that is written to the spreadsheet?

The difficulty in measuring software system size requires precise definitions. Unfortunately, there are few standards that are common throughout the software industry. We will return to this point several times in this book when we discuss

TABLE 2.1 An Unrealistically Optimistic View of the Relationship between the Size of a Software Project in Lines of New Code versus the Number of Programmers on Development Teams

Lines of New Code	Approximate Number of Software Engineers
5,000	1
10,000	1
20,000	2
50,000	5
100,000	10
200,000	20
500,000	50
1,000,000	100
2,000,000	200
5,000,000	500
10,000,000	1,000
100,000,000	10,000

software metrics. For now, we will just consider the number of lines of code as the measure of the size of a software system and ignore exactly how the number of lines of code was computed.

There is a rule of thumb that says 10,000 is approximately the largest amount of lines of source code that a good software engineer who is experienced in both the application area and the programming language used can understand completely. (This rule of thumb is part of the folklore of software engineering.) Let us use this number to estimate the size of a team that would be needed for software development of various sizes. (An assessment of a smaller number as the maximum that anyone can understand would increase the number of software engineers needed, whereas a larger number would decrease this number.) For simplicity, we will assume that all source code for the project is created by the team. The results are summarized in Table 2.1.

Note that there are many software projects in each of the larger size ranges. For example, the typical word processor software for a modern personal computer consists of millions of lines of code. As stated before, one particular version of Microsoft Excel consisted of more than 1.2 million lines of code. The project to revise the United States air traffic control system, which was terminated in the late 1990s, was expected to consist of well over 10 million lines of code. These huge numbers make it clear that most products reuse a huge amount of code, simply because there are not enough good software engineers available.

Unfortunately, the enormous numbers shown in Table 2.1 greatly underestimate the number of people needed for extremely large software projects that consist of large amounts of new code. There are several reasons for this:

- Not all programmers can understand systems at the complexity level of 10,000 lines of code.
- Larger systems mean that the programmers developing the software must be physically separated, that is, on different floors of the same building, in different buildings,

different locations, or even in different countries. There is no way for the informal, one-on-one discussions that can solve problems quickly to occur as spontaneously as in a smaller environment.

- Coordination of efforts is essential. This means many meetings; many managers to coordinate meetings; and many support personnel to install, maintain, and configure the computers and software needed to support this project. It is extremely rare for a software manager to coordinate more than twenty people, with eight to ten people a much more realistic number.
- The number of middle-level managers increases exponentially with the size of the project. For a small team, one manager might suffice. For a larger team, there may be a first-level software manager for every eight to ten people, a second-level manager for every eight to ten first-level managers, and so on. These managers are essential for coordination of efforts and to ensure that one group's changes to a system are localized to that group and do not affect other efforts of the project. Even a flatter organizational structure with more "programmers" and fewer middle-level managers must develop higher levels of administration as the project size gets larger.
- The project team rarely stays together for the duration of a project. Every organization has turnover. There is often an influx of new personnel just out of school ("freshouts") who need training in the organization's software standards and practices. Even experienced personnel who are new to a project often need a considerable amount of time and effort to get up to speed.
- There are many other activities that are necessary for successful software development. There must be agreement on the requirements in the software being developed. The code must be tested, both at the small module level and at the larger system level. The system must be documented. The system must be maintained.

There are other activities, of course. Higher-level management wants to spend its resources wisely. It wants to know if projects are on schedule and within budget. It does not want to be surprised by last-minute disasters. Ideally, higher-level management wants feedback on process improvement that can make the organization more competitive. Feedback and reporting often require the collection of many measurements, as well as one or more individuals to evaluate the data obtained by these measurements. Project demonstrations must be given and these require careful preparation.

How much of a software engineering project's total effort is devoted to writing new source code? The answer varies slightly from organization to organization and project to project, but most experienced software managers report that development of the source code takes only about 15 percent of the total effort. This appears to be constant across all sizes of organizations. People who write source code in very small organizations may have to spend much of their time marketing their company's products.

The need for these extra activities suggests a more realistic view of a software team's size; see Table 2.2. As before, we only consider the case of new code. We assume that 5

TABLE 2.2 A Somewhat More Realistic View of the Relationship between the Size of a Software Project with Only Lines of New Code and the Number of People Employed on the Project

Lines of New Code	Approximate Number of Software Engineers
5,000	7
10,000	14
20,000	27
50,000	77
100,000	144
200,000	288
500,000	790
1,000,000	1,480
2,000,000	3,220
5,000,000	8,000
10,000,000	15,960
100,000,000	160,027

percent of a programmer's time is spent on measurements, and another 20 percent is spent on meetings, reporting progress (or explaining the lack of it), and other activities (requirements, design, testing, documentation, delivery, maintenance, etc.). We also assume that programmers have difficulty understanding 5,000 lines of code, much less 10,000.

It is clear from Table 2.2 that some of the very largest projects require international efforts. The reality is that, unless you work for a software organization whose primary activities are training, hardware maintenance, system configuration, or technical support as a part of customer service, most of your work in the software industry will be as a member of a team.

You are certainly aware of the oft-voiced concern in the United States and elsewhere about the trend of outsourcing software, thereby taking away jobs. The problem is not considered to be as serious as it once was, because of the relative shortage of software engineers in the United States and the difficulties in coordination of teams in radically different time zones. This has affected some international efforts.

The greatest issue in outsourcing now appears to be the independent programmer taking jobs "on spec," which often leads to unsatisfied customers who did not properly define their proposed project's requirements to the programmer working on spec.

Of course, the numbers will be very different if the projects make use of a considerable amount of reused code. We ask you to consider this issue in one of the exercises at the end of the chapter.

There is one final point to make on this issue of size measurement. As measured by the number of new lines of code produced, the productivity of the typical programmer has not increased greatly in the last fifty years or so and is still in the neighborhood of a few documented, tested lines of code written per hour. Yet software systems have increased tremendously in complexity and size, without requiring all an organization's, or even a nation's, resources. How is this possible?

There are two primary reasons for the improvements that have been made in the ability to develop modern complex software systems. The first is the increase in abstraction and expressive power of modern high-level languages and the APIs available in a software development kit (SDK) over pure assembly language. The second is the leveraging of previous investment by reusing existing software when new software is being developed.

The code of Example 2.1 is a good illustration of the productivity gained by using high-level languages. It also illustrates how different people view the size of a software system.

Example 2.1: A Simple Example to Illustrate Line Counting

```
#include <stdio.h>
main()
{
int i;
for (i = 0; i < 10; i++)
  printf("%d\n", i);
}
```

The code source consists of 9 lines, 16 words, and 84 characters according to the UNIX wc utility. The assembly language code generated for one computer running a variant of the UNIX operating system (HP-UX) by the gcc compiler consisted of 49 lines, 120 words, and 1,659 characters. The true productivity, as measured by the functionality produced, was improved by a factor of 49/9, or 5.44, by using a higher-level language. This number illustrates the effect of the higher-level language. We note that the productivity is even greater than 5.44 when we consider the advantage of reusing the previously written printf() library function.

Reuse is even more effective when entire applications can be reused. It is obviously more efficient to use an existing database package than to redevelop the code to insert, delete, and manage the database. An entire application reused without change is called off-the-shelf; the most common occurrence is the use of COTS software to manage databases, spreadsheets, or similar.

We note that many methodologies such as agile software development can greatly reduce the numbers of software engineers suggested in Table 2.2. With the agile team having a proper understanding of the application environment, specifically which existing large-scale components can be combined into programs with minimal coding to create the desired software solution, the numbers shown in Table 2.2 can be reduced immensely.

There are several components to a systematic approach to software project estimation:

- Resources must be estimated, preferably with some verifiable measurement.
- An “experience database” describing the histories of the resources used for previous projects should be created.

TABLE 2.3 An Example of an Experience Database for Project Size Estimation

Project Name	Domain	Elapsed Months	Effort in Person-Months	Size in Lines of Code
Application 1	Graphics utility	12	30	5,000
Application 2	Graphics utility	10	40	8,000
Application 3	Graphics utility	24	30	5,000
Application 4	Graphics utility	36	100	20,000
Application 5	Graphics utility	12	30	5,000
Application 6	Graphics utility	24	30	10,000
Application 7	Graphics utility	48	90	25,000

Note: Only a portion of the database is shown.

We address each of these issues in turn.

The first step in resource estimation is predicting the size of the project. The beginning software engineer often wonders how project size is estimated. The most common approach is to reason by analogy. If the system to be developed appears to be similar to other systems with which the project manager is familiar, then the manager can use the previous experiences of the managers for those projects to estimate the resources for the current project. This approach requires that the other "similar" projects are similar enough that their actual resource needs are relevant to the current project. The less familiar a particular project is, the less likely it is that a manager will be able to estimate its size by reasoning by analogy.

The reasoning-by-analogy approach also involves determining the actual resource needs of the other "similar" projects. This can only be done if the information is readily available in an experience database. An experience database might look something like the one that is illustrated in Table 2.3.

Of course, such a table is meaningless unless we have units for the measurements used. In this table the effort is measured by the number of person-months needed for the project, where the term *person-month* is used to represent the effort of one person working for one month. The size evaluation can be any well-defined measurement. The most commonly used measurement is called "lines of code," or LOC for short. We will discuss the lines of code measurement in several sections later in the book. For now, just use your intuition about what this measurement means.

How can the information in the experience database be used? Let us look at a scatter diagram of the number of months needed for different projects and the size of the projects (which might be measured in lines of code). A typical scatter diagram is shown in Figure 2.2. This diagram may be uninformative. A model may be created, based on the fitting of a straight line or curve to the data according to some formula. A straight line fitted to the data in Figure 2.2 would slope from lower left to upper right.

The fitting of a straight line to data is often done using the "method of least squares." In this method, the two coefficients m and b of a straight line whose equation is written in the form

$$y = mx + b$$

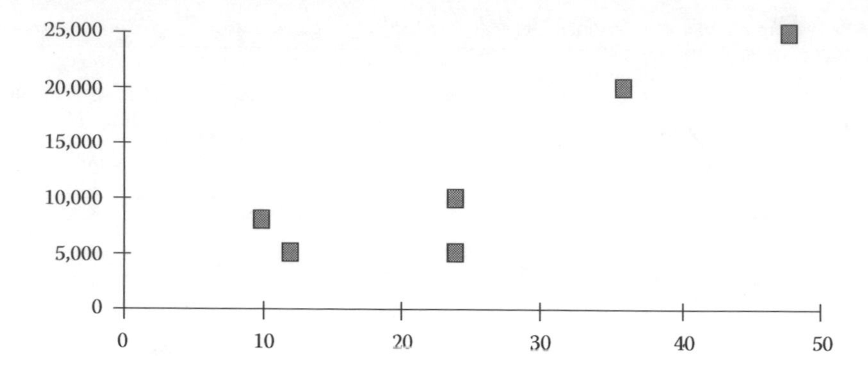

FIGURE 2.2 A scatter diagram showing the relationship between project size and duration in months for the experience database of Table 2.1.

are determined from the equations

$$m = \left[n\left(\sum x_i y_i\right) - \left(\sum x_i\right)\left(\sum y_i\right)\right] \Big/ \left[n\sum x_i^2 - \left(\sum x_i\right)^2\right]$$

and

$$b = \left[n\sum (y_i)\sum (x_i^2) - \sum (x_i)\sum (x_i y_i)\right] \Big/ \left[n\sum (x_i)^2 - \left(\sum x_i\right)^2\right]$$

These coefficients can be calculated using the built-in formula named *LINEST* in Microsoft Excel. The name of the function reflects that the line computed by this formula is called the linear regression line or the “least squares fit” to the data.

In the example illustrated in Figure 2.2, the values of m and b are approximately 0.002 and 1.84, respectively, and the equation of the line is

$$y = 0.002x + 1.84$$

This formula gives an approximation to the number of months needed for a typical project within the organization of any particular size. This implies that a project of size 15,000 LOC would take approximately 32 months. (This is clearly an inefficient software development process!)

One commonly used approach to software estimation is based on the COCOMO developed by Boehm (1981) in his important book *Software Engineering Economics*. There are several other commonly used methodologies used for cost estimation: SEER and SLIM, both of which are embedded within complete suites of tools used for project management. For simplicity, we will not discuss them in this book.

Boehm suggests the use of a set of two formulas to compute the amount of effort (measured in person-months) and the time needed for completion of the project (measured in months). Boehm’s formulas use data collected from an experience base that is a large collection of software projects in many different application domains.

Boehm developed a hierarchy of three cost models: basic, intermediate, and advanced. We describe the basic and intermediate models briefly in this section but will ignore the advanced model, referring the reader to Boehm's original book.

Boehm's models are based on an assessment of the size of the system to be produced. In the original COCOMO model, the first step is to estimate the size in lines of code. The assessment of the size of the project, often measured by the number of lines of code, is often made by using a "work breakdown structure." A work breakdown structure is created as follows:

1. Examine the list of detailed requirements.
2. For each requirement, estimate the number of lines of code needed to implement the requirement.
3. Ignore any requirements for which an existing component can be reused as is.
4. Compute the total of all new lines of code.

This total will be used as the variable K in the COCOMO formulas. It is measured in units of thousand lines of code. The approach is called a work breakdown structure because the project is broken into smaller portions.

You might object to this estimation process, because it replaces the estimate of the size of the entire system by a total of the estimates of the sizes of the individual components of the system. However, many practitioners of this approach believe that any errors in overestimating the size of individual components are likely to be balanced by other errors underestimating the size of other components. In any event, estimating the size of a project by a work breakdown structure is often used in practice.

Once the number of thousands of lines (K) has been estimated, the time and number of personnel can be estimated. We discuss the basic COCOMO model first. The formulas are

$$E = a_b \times K \times \exp(b_b)$$

$$D = c_b \times E \times \exp(d_b)$$

where the coefficients a_b, b_b, c_b, and d_b are based on relatively informal assessments of the relative complexity of the software. The computed quantities E and D are the amount of effort required for the project and D is the time needed for development of the project, but not maintenance.

The values of the constants a_b, b_b, c_b, and d_b should be taken from the appropriate entries in Table 2.4.

Note that the estimates for the quantities E and D are themselves based on estimates of the quantity K. Thus, it is not reasonable to expect an exact match between estimates and actual values for the size and resources needed for a project. At best, an approximation with an expected range of accuracy can be determined, and this range of allowable error is

TABLE 2.4 Coefficients for the Basic COCOMO Model

Software Project Type	a_b	b_b	c_b	d_b
Small project, with an experienced team and flexible requirements (commonly called a basic or "organic" system)	2.4	1.05	2.5	0.38
A system with hard real-time requirements and strict interoperability required (commonly called an "embedded" system)	3.6	1.2	2.5	0.32
A mixture of the other two types of projects (commonly called an "intermediate" level system)	3.0	1.12	2.5	0.35

Source: Boehm, B., *Software Engineering Economics*, Prentice Hall, Englewood Cliffs, NJ, 1981.

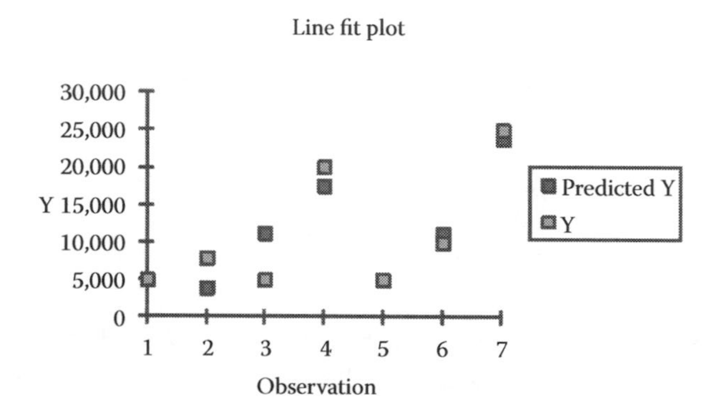

FIGURE 2.3 An attempt to fit a smooth curve to the data in the scatter diagram of Figure 2.2 using a COCOMO model approach.

heavily influenced by both the experience of the estimator and the quality of the information available in the organization for comparison with similar projects.

A typical relationship between the basic COCOMO model and some cost data is shown in Figure 2.3.

The basic COCOMO model can be extended to the "intermediate COCOMO model." The intermediate COCOMO model uses a set of "test driver attributes," which are given in Table 2.5.

The weights of these test driver attributes are to be determined by the person estimating the software's costs, based on his or her experience. The weights are then entered on a scale from 1 to 6 into a spreadsheet based on Table 2.5. The resulting sum is used to create a multiplication factor that is used to modify the results of the basic COCOMO model.

2.5 PROJECT SCHEDULING

Software project scheduling involves the allocation of resources to a project in a timely manner. As remarked earlier in this chapter, it is extremely inefficient to have a project that will take five years to complete and will need one thousand people at its peak and have all one thousand people on the payroll from the first day forward, when only twenty people might be needed for the first year.

The alternative is equally bad, however. If a project is severely understaffed at any critical time, it is likely to be completed much later than desired, if at all. Adding extra people late in the process usually does not help, because the more experienced project personnel

TABLE 2.5 "Test Driver Attributes" for the Intermediate COCOMO Model

Test Driver Attribute	Weight
Product Attributes	
Reliability requirements	
Size of application's database	
Software complexity	
Hardware Attributes	
Run-time performance constraints	
Memory limitations	
Other processes competing for virtual memory	
Personnel Attributes	
Analyst experience	
Software engineer experience	
Application domain experience	
Virtual machine experience	
Programming language experience	
Project Attributes	
Use of software tools	
Use of software engineering methods	
Required development schedule	
Total	

are forced to spend much of their time bringing up the levels of understanding of the new people. In addition, having more people always means more meetings to coordinate, as we saw in Chapter 1.

The efficiency of the system's software development is guided by its expected cost. A software economist will often have considerable experience with the pattern of staffing needed for projects of this size. He or she will often expect that the number of people employed on a software project will follow a relationship that might look something like that of Figure 2.4. The different line segments in the graph indicate different life cycle activities, each of which might require different numbers of people. Here the horizontal axis represents time and the vertical axis represents the number of personnel associated with the project between milestones.

The numbers at the beginning of the project represent the requirements team. The project size then increases to its maximum during the coding phase, which is intensive, but

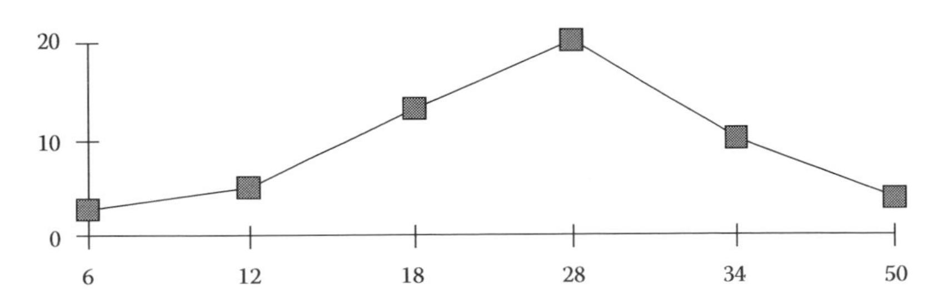

FIGURE 2.4 A typical pattern of project personnel over time.

limited in time. The number of personnel will decrease as the testing and integration phase comes to an end. The last number represents the people needed for maintenance of the software. If the system has a long operational life, then it is easy to see that the cost of the maintenance phase will be the largest cost in the system, as we stated before.

Several techniques are often used for project scheduling. Two of the most common are:

- Milestone charts
- Critical path methods

A simple example of a milestone chart was previously shown in Figure 1.4. (Other examples of milestone charts were given in Figure 1.6 and Figure 1.8 for the rapid prototyping and spiral models, respectively. See Chapter 1.) A slightly more elaborate one with milestones for two different releases of the same system displayed is illustrated in Figure 2.5.

Note the multiple releases of the software. In organizations, such charts will include both the projected and actual dates of the individual milestones. Note also that reviews must be held for each milestone that is a deliverable product.

A less common method of planning and representing schedules is to use so-called "critical path methods." Critical path methods are based on the assessment that certain events are critical in the sense that they fundamentally affect several other decisions. Once these critical decisions have been determined, any events not affected by the critical ones can be scheduled at any time. For example, an ICD that describes precisely the interfaces between component subsystems must be produced and reviewed before a major effort in subsystem design and implementation begins. (We will describe ICDs in Chapter 4 when we study detailed design.)

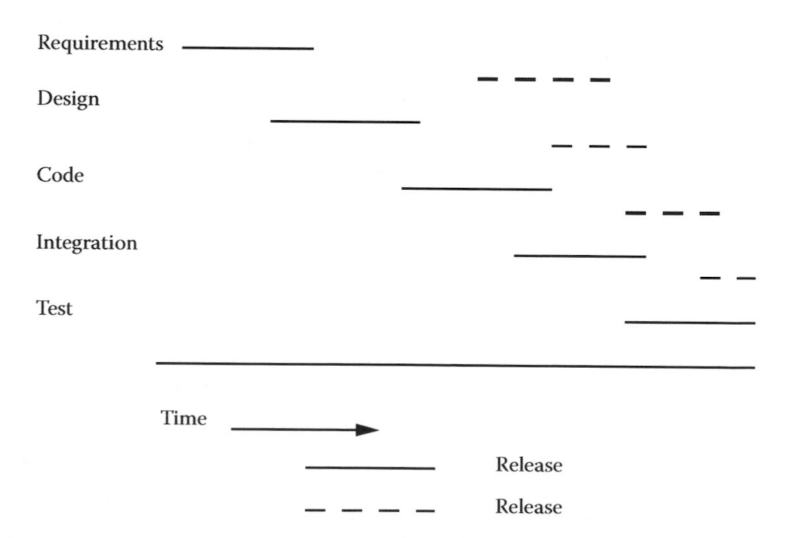

FIGURE 2.5 A typical milestone chart for a software project with multiple releases, following the classical waterfall software development process.

You should note how object-oriented design is consistent with the determination of critical paths in software development. If we can describe the methods of an object in terms of an abstract class, then we know the class interface, and hence anyone developing a class that will interact with this class has enough information to develop code. That is, if we develop a class with complete descriptions of the number and type of arguments to methods of the class, as well as the type of return values of methods in the class, then the design of the implementation of the abstract class's methods can proceed concurrently with the implementation of the details of the methods of the class during coding.

Clearly the determination of the abstract interface of a class that is to be used by another class is on the critical path, while coding the details of a member function that uses this interface.

A similar situation occurs at many places in every software development life cycle. You will be asked to study this point in the exercises.

2.6 PROJECT MEASUREMENT

Systematic project management usually requires measurements of the software development process. Recall that the Capability Maturity Model (CMM) of the Software Engineering Institute specifies that the software development process must be measurable. The higher levels (levels 4 and 5) of the CMM specify that "the software development organization has quantitative feedback systems in place to identify process weakness and strengthen them proactively" (level 5) and "detailed software process and product quality metrics are used to establish the quantitative evaluation foundation. Meaningful variations in process performance can be distinguished from random noise, and trends in process and product qualities can be predicted" (level 4). You might compare this to what is stated in the CMMI model.

How can a process, as opposed to a tangible product, be measured? As before, using the GQM paradigm of Basili and Rombach is helpful. Some obvious goals are to produce the software to meet the system requirements, on schedule, and within budget.

Related questions for these goals include:

- Have milestones been met?
- Which items are behind schedule?
- When is the project likely to be complete?

A related question is the horrifying one:

- Is the project likely to fail?

Of course, many of these questions require detailed measurements before they can be answered.

It is easy to tell at a glance if a project is behind schedule. Unfortunately, this simple piece of information can tell us little without additional information. The additional information might include a graph of how the actual and predicted amounts of time spent on various life cycle phases compares to previous data in the organization's baseline. It may be that the project is actually on schedule for completion, with the only difficulty being the relative amounts of time assigned for the completion of each activity. Alternatively, the project may be in serious trouble. The only way to tell is to develop the proper measurements.

What are the proper measurements? There is no absolute answer, given the wide range of software development environments. However, a manager is expected to have measurements of where the project is according to the schedule, and what costs have been expended so far.

2.7 PROJECT MANAGEMENT TOOLS

Project management is clearly an essential part of software engineering. As such, it can benefit from the application of high-quality software tools that automate some of the steps needed to coordinate projects. This will be our first discussion of an application of CASE tools. Later in this section we will discuss some less elaborate tools that are widely available and relatively inexpensive or even free.

Modern CASE tools for project management should help with the following tasks:

- Create and modify the project schedule
- Allow the schedule to be posted so that different project administrative personnel can view it and even make changes as necessary
- Support efficient resource allocation
- Identify critical items on the schedule and those items that depend on the critical ones
- Allow documents to be viewed by all authorized personnel
- Support different approaches to project management, so that the tool can be tailored to meet the organization's way of doing business, rather than the other way around

Some typical commercial CASE software packages include tools for project management. Perhaps the most common stand-alone tools used for project management and planning are Microsoft Project and SPR KnowledgePLAN from Software Productivity Research. GanttProject is another project management software package that is free in some versions that may be adequate for small student projects. Information on these products can be found at the websites:

- http://www.microsoft.com/project/
- http://www.spr.com/home
- http://www.ganttproject.biz

The tools are flexible and interface with several existing applications for data collection, reporting, and statistical analysis.

What should you expect from software that supports project management? The answer is as varied as there are software development practices. However, you should expect to have software support for cost modeling, project planning, project scheduling and reporting, resource allocation, configuration management, management of change, and anything else that is appropriate to the project. In any event, the software should be configurable to be able to work with the reporting and control processes that are commonly used in your organization.

Configuration management support will be available for text-based documents using something such as Google Docs but not necessarily for other types of documents. In such cases, it would be nice to have a configuration management tool that works with the different types of software artifacts associated with a project.

2.8 ROLE OF NETWORKS IN PROJECT MANAGEMENT

In addition to the project management tools discussed in the previous sections and the CASE tools that will be briefly discussed in future chapters, there are software tools and technologies that help the software development process that are beginning to be put into common use. Since the rate of technology advancement and movement into the marketplace is so rapid, we will be content with a brief overview of some important examples.

The most prominent force in changing the requirements process is, of course, the Internet and its resulting standards, such as Hypertext Markup Language (HTML), various word processors, PostScript and Portable Display Format (PDF) files, jpeg (Joint Photographic Expert Group), and mpeg (Motion Picture Expert Group). Standard file formats, graphical browsers, and powerful search tools allow relatively easy access to information.

Think of the advantages of project coordination over the previous methods. Project documents filled multiple loose-leaf binders, which made some offices impossible to work in because of the sheer size and quantity of the paper documents. There were sets of documents for each release of a software system, often making it impossible to store all of them. If a key member of a team was moved elsewhere within the software development organization, many essential documents were lost. The situation was even worse if a key person left the organization.

One of the most explosive growth areas is in what are commonly called "intranets," which are systems of computers that are networked together using Internet standards but are effectively cut off from the entire Internet. In this approach, the restricted portions of the Internet can communicate with one another using Internet standards but with relative confidence that their work stored on this intranet is secure from unwanted outside access to sensitive materials. Thus, organizations are able to conduct their business with little worry about the security of commercial application tools. However, ensuring complete isolation of company intranets from unprotected external networks is hard to enforce in reality.

One compromise I have run across is a two-level security process. A company allowed me access to a portal for initial access to specialized, confidential information, and then another process was required to obtain details of a specific set of projects using a secure depository called a workforce collaboration group. You may be familiar with a similar, but not identical, process of authorization in the way that a bank might require a user of its online banking system to use a site key (which typically contains a graphical image for the user to examine to make sure that he or she has not gone erroneously to a spoofed website) before entering his or her password to obtain access to a bank account.

I should note that many intranets prohibit the use of wireless connections to their secure areas. This is because of the relatively greater security inherent in well-managed hard-wired connections and the tendency of many people, even computer professionals, to use insecure wireless networks, such as those in hotels and coffee shops. The tendency of many companies and organizations to allow, or even encourage, a "bring your own device" has additional security risks. Setting up a virtual private network (VPN) provides a strong level of security but requires a degree of effort and coordination on both ends of the communication, together with cooperation of network managers and, therefore, is not always done. Using encrypted email can also help.

When I lecture on identity theft, I mention the movie *Ocean's Eleven*, because I personally know eleven people whose email accounts have been hacked when they were using "free" Wi-Fi accounts in hotels or coffee shops. Free Wi-Fi is not free if you have to pay for it later in terms of reinstalling all the applications and data on your computer. The Wi-Fi really is not free if using it means that critical confidential data belonging to either you or your organization has been compromised.

As indicated earlier, it has now become commonplace to have requirements documents and project schedules placed on an organization's internal network, in cloud storage, and occasionally on the open Internet itself. Several persons can work on portions of the same document at the same time, sharing it with all others who need access. The inspections of the requirements and design should be based on written documents and the inspections should be conducted as usual, without regard to the documents being available on the Internet.

Of course, there are some problems associated with the use of electronic files. These problems can be reduced, if not eliminated entirely, by using the following guidelines:

- Determine the standard format to be used for storage of graphical files.
- Make sure that the format is compatible with the browser to be used.
- Determine a mechanism for providing feedback about deficiencies in online documents.
- In particular, determine if both paper and electronic versions of documents will be required.
- Make sure that the online documents are subject to configuration management and revision control. This can be done by having different directories for each iteration of requirements and design documents.
- Standard text-based configuration management tools should be used for both source code and textual requirements documents.

- Provide easy feedback by means of the mailto facility of HTML. This makes communication much easier.
- Use reviews and inspections as they would be used in a software project that was developed without the use of the Internet.
- Use email to coordinate meetings and to notify members of the project team that documents have been changed.
- Determine if chat rooms will be employed for project management.
- Use wikis for concurrent access to documents by multiple users if appropriate. (We briefly discuss wikis in Section 2.9.)

As with many things in software engineering, it is difficult to quantify the effect of certain technologies or techniques. However, preliminary indications are that the improvements in efficiency are enormous. In any event, an organization that is not using networks to assist in project management is generally perceived as being technologically obsolete.

As an indication of the perceived efficiency of Internet and intranet publishing, many new software development projects at NASA's Goddard Space Flight Center post all requirements documents, designs, minutes of project meetings, project schedules, and major action items on the Internet for easy access by project personnel. This reduces the clutter in many offices, where entire shelves of bookcases were used previously for storage of large loose-leaf notebooks that contained project documentation. In addition, few documents are lost. Mandl (1998) describe some of the positive effects of using the Internet to reduce the number of project meetings.

2.9 GROUPWARE

There is one particular tool that is used in many organizations to help coordinate meetings and various project reviews, namely, groupware. The term *groupware* refers to the ability to have a group of people viewing and working on the same document at the same time. There is usually more access control than is available with many network browsers. This can be very useful in certain situations.

For example, the requirements engineers for a project can meet together in a room with a large monitor, with each of the requirements engineers having access to his or her own computer or workstation. Alternatively, the locations can be remote, with network access. All the individual computers are networked and the software that controls the document access is known as groupware.

You may be familiar with the Google product Google Docs, which is an excellent, free repository that encourages collaboration on documents stored in a cloud-based repository. Although extremely helpful, especially for the sharing of text-based documents and standard applications such as Microsoft Word and PowerPoint, this free product is not perfectly suited to all software projects. Hence we will discuss other groupware projects. I use a different Gmail address for collaborative projects involving groupware such as Google Docs that use Gmail accounts in order to keep information from different projects separate. I have done the same for projects using Yahoo Groups freeware and for the same reason. (I was recently informed that my grandson Matt, who is in the sixth grade in a public school in Howard County, Maryland, must follow the countywide requirement to use Google Docs in all language arts projects. This surprised me, until I learned that this sort of connectivity is used in even earlier grades in many schools.)

At one time, the leader in the field of commercial groupware was Lotus Notes. This product is now part of IBM's offerings, under the name IBM Notes and IBM Domino. IBM Domino is the name primarily used for IBM's groupware coordination server, and IBM Notes is the version that runs on client computers.

Groupware goes somewhat further than Internet access in the sense that it is easy to allow everyone on the project to change a common document or to restrict the ability to change the document to a single person. This is more transparent than the typical use of operating systems level access permissions for Linux or various releases of Windows. This use of groupware is also smoother than the operating system level approach in that there is no need to create new groups of users to allow certain groups of software engineers to change the common document.

One form of groupware that has become popular is known as a wiki. Think of Wikipedia as a primary example. Anyone can add or delete any content on a wiki. Of course, there are configuration management features that allow changes to be undone. The concept of a wiki often means that the wiki creation software itself is free. Most wiki systems interface well with components from free or low-cost software sources, often on Linux systems.

Of course, widely separated software teams frequently will use some tools to support technical meetings. One of the most popular commercial tools is GoToMeeting, produced by Citrix Systems. A thirty-day free trial of this commercial software is readily available as a download. Freeware solutions to groupware creation are also available. Note that Citrix Systems is a well-known company, whose first wildly successful product was software to turn older PCs that had insufficient power to run the newer versions of Microsoft Windows and Office into machines that would communicate with UNIX and Linux servers. There are other such products, and I have participated in meetings using a suite of open source tools.

The use of the Internet or an internal intranet for video conferencing has some of the same features as groupware, particularly if there is a mechanism for playback.

2.10 CASE STUDY IN PROJECT MANAGEMENT FOR AGILE PROCESSES

Proponents of the use of agile methods for software development nearly always claim that it reduces costs and speeds the development schedule by simplifying the management of agile projects. We will examine these claims by considering one particular case study with which I am highly familiar. Of course the results of a single case study, or even a group of them, cannot prove the claims of reduced costs and development times for all agile projects, but these results can provide some guidance. The lessons learned from this case study are believed to be typical of projects that are well suited to the use of agile methods because of the experience of the agile development team and the support of senior management for using agile development.

The particular case study of agile development discussed in this book took place in the application domain area that can generally be described as space satellite operations.

As is typical in case studies involving a series of actual software projects in industry and government, names of participants, their companies, and specific project details will not be provided. This limitation is almost always due to issues of confidentiality. A search for the name of the project leader provided over 565,000 hits on Google in February 2014, which is a large number for anyone not named Kardashian.

The project team had seven members: two government employees and five contractors representing several different companies. The most important attributes of the team were that it had a wealth of experience in this application domain and understood that the budget pressures required radical process improvement.

As is true in almost any organization that has existed for many years, team members had somewhat of a blind spot for not using existing software components that they could they could have used readily in their projects. This blind spot is often referred to as the "Not Invented Here Syndrome" (Webb et al., 2010).

In the first use of this agile process, many people complained about the inability to identify "who was in charge." This was troubling to senior management at the time. Fortunately, due to senior management's confidence in the ability and experience of the team, the experiment in this agile process was allowed to continue, even though initial results were not promising.

With the advantage of hindsight, this difficulty was a good thing, because it indicated that a transition was occurring from a rigid, closed, hierarchical control mechanism, which tends to be less efficient at learning new things, to a more open, distributed team-based control mechanism, which allows for much more learning.

Learning was important. The seven-member team was part of a larger effort in which multiple teams and internal organizations competed to get their ideas implemented. The working premise was that maximum learning capacity would have no management or coordination, just progress toward the product.

We can measure the degree of self-organization by a ratio of the activities devoted to management, coordination, and control over the remaining activity. The closer the ratio is to zero, the more self-organized and less hierarchical is the team. This ratio, described as the "self-organizing factor," or SOF, is

SOF = (Management + External test)/(Management + Systems engineering + Requirements gathering + Implementation + Internal test + External test + Software maintenance + System administration + Documentation)

A factor based on size was derived for each project to further normalize the numbers based on percentage of reuse and size of the code produced.

In this study, the projects with an SOF closer to zero tended to make larger improvement in costs and reduction of testing effort, as is illustrated in Figure 2.6. It is believed that innovation was enabled because obstacles had been removed and the teams could manage their "team model" more synergistically and with less effort expended on administrative control.

The process used two tools to aid in the development process: a Concurrent Version System for configuration management and a Comprehensive Discrepancy System for keeping track of errors and how they were removed during development. Although software tools for configuration management and error tracking have been available either as inexpensive applications as stand-alone systems for more than thirty-five years or as part of integrated development environments (IDEs) for nearly as long, there were some aspects of their use that merits discussion.

The Concurrent Version System for configuration management allowed team members to configure both software and documents online. This is typical in any type of modern collaborative distributed software development, such as Eclipse, but the extension to documents, which were often created in a word processor, was unusual at the time. Each word processing document had a tag in the form of a few lines of text indicating if the document had been modified.

The Comprehensive Discrepancy System was an online database that helps to manage discrepancies identified in testing and helps to track resolutions. As a special feature, the Comprehensive Discrepancy System could provide assorted metrics that would allow tracking of progress and assessing lessons learned (e.g., where and what kind of errors occurs in development). This also is typical in any type of modern collaborative distributed software development but was not typical of projects at that time.

There may be another confounding factor that supported the agile team's performance. NASA had used the Myers-Briggs Type Indicator Test to help determine personnel allocations for many years. This test rated people as introvert or extrovert, sensing or intuiting, thinking or feeling, and judging or perceptive, after which they were placed in one of four quadrants. A highly creative team member I knew extremely well scored high on being an intuitive extrovert. The team was highly balanced across the Myers-Briggs spectrum.

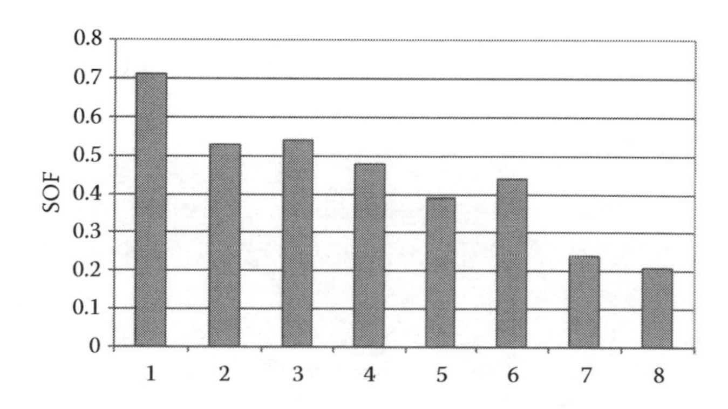

FIGURE 2.6 Self-organizing factor (SOF) for development of eight projects.

The well-known computer science professor Lisa Kaczmarczyk wrote a fascinating article on the experiences of introverts in computing, especially as pertains to agile processes (Kaczmarczyk, 2013). Here are some quotes from the article.

> A relatively senior developer said, describing the use of agile processes in a large multinational company:
> "… where there are language barriers between team members, agile still means daily communications between many members of the team, but typically across a chat interface. This suits introverts in many cases, although (in my opinion) doesn't really work well in this scenario …"
>
> A different, more junior, software developer at a different company said:
> "I don't mind working with someone when I'm working through a difficult problem or out of ideas on something, but if I'm just blasting my way through a task, I don't see the point of having someone else there …"

SUMMARY

Software project management is the systematic application of general management tools and techniques to improve the efficiency of the software development process. Software team activities include

- Systems analysis team
- Planning team
- Requirements team
- System design team
- Implementation team
- Testing and integration team
- Training team
- Delivery and installation team
- Maintenance team
- Quality assurance team
- Metrics team
- Documentation team
- System administration team
- Reuse and reengineering team
- Human–computer interface evaluator

- Tools support person
- Software economist
- Project librarian

All these activities have to be coordinated by a project manager. He or she will also be responsible for ensuring that the resources available to the project will be used appropriately.

Agile software development processes require management even though the participants are likely to be highly experienced professionals already working in the project's application domain. It may take considerable time for an agile process to become effective and, therefore, agile processes need the strong support of the organization at sufficiently high management levels. Agile processes often make use of self-organizing teams.

KEYWORDS AND PHRASES

software project management, software teams, project plan, project staffing, deliverables, software cost estimation, CASE, CASE tools, COCOMO, groupware, wiki

FURTHER READING

There are many excellent books on software project management. One of the best is Watts Humphrey's classic book *Managing the Software Process* (Humphrey, 1989), which is still a big seller even though it is more than twenty-five years old. Tom DeMarco's book on this topic (DeMarco, 1982) is also very useful, and is surprisingly up to date for a book written in 1982.

Another book by Humphrey (1995) describes the "personal software process" that can help a programmer determine the software engineering activities (requirements, design, implementation, testing, etc.) that consume most of his or her time. A major feature of this book is a set of graded exercises that are intended to provide a database of quantitative measurements of project experiences for individuals to improve their capabilities as software engineers.

A somewhat different view of software project management is presented in Jim McCarthy's book (McCarthy, 1995) titled *Dynamics of Software Development*. It describes experiences with software projects at Microsoft.

The 2014 Research Edition of the *Software Almanac* from Software Quality Management contains a discussion of some metrics for managing projects, along with some case studies. It is available for download at qsm.com.

The 1981 book by Barry Boehm (1981) is still one of the best references on software project resource estimation. It presents the COCOMO model in detail. The COCOMO 2 model, also created by Boehm, extends the COCOMO model. Other commonly used cost models are SEER (Galorath and Evans, 2006) and Win-Win (Boehm, 1998). The SEER and SLIM software cost estimation tools are part of larger suites used for software project estimation. They can be found at galorath.com and qsm.com, respectively. See my encyclopedia article on cost estimation (Leach, 1999) for an overview of this subject. Also see the article by Terry Bollinger and Shari Lawrence Pfleeger (1990) for information on the use of return on investment (ROI), which is often used to determine the costs and benefits if a particular software project is to be started.

Read the recent articles by Lisa Kaczmarczyk (2013) and Phillip Armour (2013) in their entirety for some perspectives on the use of agile and other fast software development processes.

The short, inexpensive book by Chris Sims and Hillary Louise Johnson provides a fast but comprehensive introduction to agile project management (Sims and Johnson, 2014).

For more about disruptive business practices, of which cloud computing is just one example, see the book by Nicholas Webb and Chris Thoen (2010).

EXERCISES

1. Examine an IDE that you have used before and determine if it is possible for two or three people to work on the same project concurrently. If no capability exists within the IDE, describe how you would organize communication between the team members.

2. What are the "test driver attributes" for the COCOMO model?

3. What activities should be present in a project plan? Does your answer depend upon the software development life cycle used?

4. What milestone events should be indicated in a project management plan?

5. Examine the Internet web pages of some government organizations that develop software. Choose one project and determine which software requirements or designs are available to you. Then estimate the amount of effort and resources needed to complete the software projects you have found.

6. Examine the Internet web pages of some companies that develop software project management tools. Determine if these tools allow a user to estimate the amount of effort and resources needed to complete the software projects you typically develop.

7. Examine several of the projects you did in your previous computer science courses. Use these projects to determine your productivity by comparing the number of lines of code of each of the projects with the amount of time used for each project, which is the difference between the date the project was assigned and the date you turned it in, multiplied by a factor that represents the average number of hours you spent on the project each day. Did the average amount of time represent an accurate view of how you spent your time or was most of your work done very close to the project deadline?

8. Draw a diagram showing the major software engineering tasks for the rapid prototyping software development model. Use Figure 2.2 as a model.

9. Draw a diagram showing the major software engineering tasks for the spiral software development model. Use Figure 2.2 as a model.

10. This question concerns the COCOMO model of software cost estimation. Examine a software system for which there have been multiple releases and for which source code is available to you. (The GNU software tools from the Free Software Foundation are an excellent source for this question.) Describe the system as organic, embedded, or intermediate. Examine the amount of new and reused code in the latest two releases. Does the time estimate predicted by the COCOMO model agree with the time between these releases? Explain.

11. Obtain a copy of an organization's software project management manual. Find out how staffing and resource allocation is determined. Are any CASE tools mentioned in the manual?

12. Examine a project you recently completed. Determine which activities were on the critical path of development and which ones could have been done concurrently. Would having this knowledge before the project started have made the software development process more efficient?

13. Develop a project plan for treating the "Year 2038 Problem" where the date fields used in UNIX systems will overflow sometime in the year 2038. The overflow is due to the maximum size of an integer. Many functions that use time in UNIX systems use the time as measured in seconds from 1970 and this number will eventually be too large to be stored in a 32-bit integer.

14. Develop a plan for the Social Security number overflow problem that will occur when the increased population of the United States together with the number of Social Security numbers already issued exceeds the capacity of the nine digits currently used for storage of the Social Security number. In your project plan, consider if Social Security numbers should be reused if they were issued to a person who is now deceased. Be sure to consider all possible classes of Social Security benefits.

15. This is a thought question, since little actual data to validate your answer will be available to you. In Section 2.4, we considered the relationship of the size of a hypothetical software project and the number of software engineers that would be needed to complete the project. We made the assumption that the projects would consist solely of new lines of code. (The data was presented in Tables 2.1 and 2.2.) Now you are to consider the same project but under the assumption that one half of the code would be existing reusable code by this hypothetical project. What would you expect to be the effect of this reuse in terms of the number of software engineers needed? Does your answer depend on the nature of the software development team, on when in the development life cycle a decision is made to reuse the source code, or what can be expected in the quality of the code? Give reasons to justify your answers. (You might consult a book on software reuse.)

CHAPTER 3

Requirements

IN THIS CHAPTER, WE will discuss what many practitioners believe is the most important part of software engineering: development of a software system's requirements. We will begin by introducing some typical problems of requirements engineering. We will then illustrate how these problems occur regardless of which software development life cycle model is used, whether there is a set of known customers who will provide feedback during requirements gathering, or whether it is hoped that the product will determine its own set of customers after it is delivered. Basic techniques such as information hiding, formal representations, and requirements reviews will be discussed, as will a typical managerial viewpoint of the software requirements engineering process.

We will then evaluate the requirements we develop. We will discuss requirements for agile software development a bit in the next section and again in Section 3.15 in a continuing case study. (One of the major virtues of well-run agile processes is that development of critical requirements and programming the software to meet these requirements occur simultaneously.)

After the basic techniques of requirements engineering are presented, we will begin the discussion of the major software project that will be considered throughout the remainder of this book. A set of requirements will be developed, using the problem statement as a basis.

3.1 SOME PROBLEMS WITH REQUIREMENTS DETERMINATION

All computer science students get extensive experience in software coding as part of their undergraduate and graduate education. Students often have to supply both internal documentation (in the form of comments within their programs) and external documentation (often in the form of project reports). They may also have some experience with software testing within their academic setting. Some students even get experience in software maintenance during their academic careers, although this is relatively rare. Students may see testing and maintenance during work experiences outside of their formal education setting. Although such work experience is highly desirable, not all students have this experience at this stage of their academic careers, and hence such familiarity will not be assumed in the discussion in this book.

Two topics are often missing from the typical education of computer science students (prior to taking a course emphasizing software engineering principles): detailed, practical instruction in (1) requirements and (2) design.

Unfortunately, most of the other software engineering activities are irrelevant if the requirements are wrong, incomplete, or vague. One way in which the term "wrong" here is used is the sense of being inconsistent; one requirement contradicts another. Let us consider what can happen in each of these three cases.

If the requirements for a project are wrong, then the rest of the software engineering activities will have to be redone, assuming that someone recognizes the error or errors in the requirements. This is extremely expensive, and often means that much of the subsequent work is wrong and has to be discarded. In the worst case, all the project effort must be scrapped.

What else can "wrong" mean? It may mean that the requirements do not fit the desired system's initial use. This is called "building the system wrong." "Wrong" might also mean that the requirements are for a system that is obsolete or completely the opposite of what most customers want. This is called "building the wrong system."

There are two possibilities if the requirements are incomplete. The best-case scenario when the requirements are incomplete is that the requirements are so modular that the project's design and code also can be developed in a modular manner. In this case, it is possible that the missing requirement can be fulfilled by additions both to the design and the source code implementation with relatively few changes needed to incorporate the new requirements. This flexibility is one of the main advantages of iterative approaches such as the rapid prototyping or spiral development life cycle models. The flexibility is even greater in projects using agile methods.

If the requirements were complete but not modular, then it is unlikely that the design and the resulting source code will be modular. In this case, major portions of the design and source code will have to be scrapped because they are difficult to create given the lack of consistency with modern programming techniques and the software engineering goal of modularity. The experience of many failed software projects strongly indicates that the lack of modular requirements is very expensive.

The third category of poor requirements occurs when the requirements are vague. In this situation, it is difficult to know if the design is correct, because the software designers do not know precisely what requirements mean. The designers may make some unwarranted assumptions about the intentions of the requirements engineer. These assumptions, if not correct, can be the major source of disaster for projects. The most common problem is that the unwarranted assumptions become part of the culture of the project and therefore are never questioned.

There is one more thing to keep in mind about vague requirements. If the requirements are vague as stated, but the software engineers working on the project are highly knowledgeable in a particular application domain and have the unquestioned capability to fill in the gaps and complete them in a clear, unambiguous manner, an agile software development process may work, provided there is support from senior management.

In some cases, the term *nonfunctional requirement* is used. The term usually is taken to mean that some particular software attribute is desirable, but there is no description of how the attribute is to be obtained or if there are, say, any performance requirements. Often, such requirements are only employed by highly experienced software engineers knowledgeable in a particular application domain. I have often seen student project proposals where a software system is promised to be "user-friendly," a term that always indicates little knowledge of human–computer interaction. For which class of users is the system to be user-friendly? Is "user-friendly" the opposite of "user-nasty"? We will not discuss nonfunctional requirements any further in this book.

The end result of vague requirements are software systems that either do not quite work as the customer wanted or do not work at all, often because the interfaces with other software were not properly specified. Vague requirements may lead to problems that cannot be solved on technical grounds and, therefore, must be resolved in the legal system. Having software problems resolved in court means that a jury of relatively nontechnical people has to determine the precise meaning of requirements that were too vague to be understood by the technical software people involved. This is hardly an appealing scenario.

In fact, the problem is more serious than meets the eye. The cheapest time to correct errors in software is in the requirements development phase (see Table 3.2 later). This is true regardless of the model of development: classical waterfall, rapid prototyping, spiral, or other. On the other hand, the most expensive time to correct problems in software is during the maintenance phase, because all previous life cycle activities (requirements, design, coding, testing, integration, documentation, delivery, and installation) have been performed. Software maintenance also requires a high level of understanding of the code. We will fully discuss software maintenance in Chapter 8.

You might think that using an iterative software development approach might eliminate most of these problems. It is true that the iterative models allow incomplete requirements to be discovered earlier than in the classical waterfall model. Inconsistent, or wrong, requirements tend to be discovered earlier with the iterative methods.

However, vague requirements may not be discovered unless the various evaluation reviews are done carefully. Without healthy skepticism at reviews, the vague requirements are still likely to become part of the project's culture, with the vagueness never being resolved.

An important study by Dolores Wallace of the National Institute of Science and Technology (NIST) and Herbert Hecht and Myron Hecht of SoHar Corporation in California shows that there is a pyramid effect due to the nature of errors in software systems (Hecht et al., 1997). They examined a large number of projects and classified the errors according to the life cycle phase in which the errors occurred. An informal summary of their results is shown in Table 3.1. Clearly, many errors occur during the requirements phase.

How important is it to discover errors or inconsistencies at requirements time? There are many studies that consider the cost of fixing a software problem at different phases of the software life cycle. The results of one (proprietary) study are shown in Table 3.2. Note the huge advantage to finding errors early in the software life cycle, during the requirements phase.

It should be clear that accurate determination of a software system's requirements is essential. Even in an agile process, or indeed any iterative software development process,

TABLE 3.1 Breakdown of Sources of Errors in Software Projects

Life Cycle Phase	Source of Software Errors
Requirements	Many
Design	Somewhat fewer
Code	Somewhat fewer still
Testing	Somewhat fewer still

Source: Hecht, H., Hecht, M., and Wallace, D., Toward more effective testing for high assurance systems, *Proceedings of the High Assurance Systems Engineering Workshop*, HASE '97, Washington, D.C., 176–181, August 11–12, 1997.

TABLE 3.2 Relative Cost to Fix Software Errors at Different Phases of the Software Life Cycle

Life Cycle Phase	Relative Cost to Fix Software Errors
Requirements	1.0
Design	5.0
Code	10.0
Testing	30.0

requirements must be determined eventually, and the more accurate and complete the initial requirements are, the fewer and less costly the iterations will be. We will discuss some techniques for the determination of requirements in an efficient manner.

Of course, many projects will have their requirements change over time due to changes in technology and new directions for the organization. This is especially true of larger projects, which might have their development times extend over several years. An initial set of high-quality, modular requirements and an effective software development process can support incorporation of changes in a systematic, efficient manner.

What about agile processes? I believe that the most important advantage of an agile process is that well-chosen project teams whose members have considerable experience in developing software in a particular application domain have an intuitive feel for what detailed requirements arise from a higher-level requirement. As such, they can shorten the development process by shortening the requirements process, because they can precisely determine the large-scale components and subsystems that can be used in the desired system. The constant improvement of prototypes reduces the risk of the informality of agile processes. You will see this during the continuing description of a case study of an agile software development process throughout this book.

It has been my experience that without this essential application domain expertise, agile processes can lead to disaster in critical projects and are best left to back-burner efforts whose failure will not be catastrophic to the organization. The literature suggests that agile processes should not be used for applications in which security is essential. See the 2013 article by Carol Woody of the Software Engineering Institute (Woody, 2013) for a more detailed discussion of this issue.

3.2 REQUIREMENTS ELICITATION

There are several approaches to requirements engineering. Some of these are classical (Davis, 1990). Others are motivated by object-oriented approaches (Berry, 1995). Still others

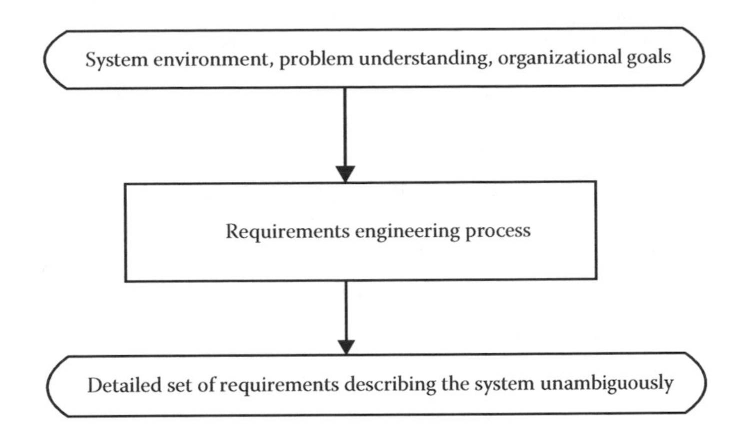

FIGURE 3.1 A stylized view of the requirements engineering process.

are motivated by cost issues, such as the reuse-driven requirements approaches (Kontio, 1995; Leach, 1998b; Waund, 1995), or the systematic process changes (Mandl et al., 1998). The agile approach known as "scrum," as suggested by scrum.org, focuses on determining the needs of so-called product owners that can be fulfilled within thirty days.

In this section, we will describe some basic principles that are common to most approaches. The individual differences will be expanded upon in the next few sections.

The requirements engineering process should be thought of as beginning with the things that are known about the project's purpose and ending with a set of requirements. These include such things as the environment in which the software system is to operate, the purpose of the system (as perceived by its potential users or customers), and the organization's basic goals.

The output of the requirements process is a complete, detailed, unambiguous description of what the system is supposed to do. The requirements should be so unambiguous that a design may be developed to match their requirements and it should be possible to determine if the design matches the requirements. The basic inputs to the requirements process are illustrated in Figure 3.1.

Let us illustrate the basic problem of requirements engineering by a simple example: determination of requirements for a software package that is to be run on a personal computer. We will not discuss the rest of the problem statement, because even this small portion of the problem description suffices to illustrate several points. The system requirements must answer the following:

- Which hardware platform will be used?
- Which operating system will be supported (Windows, UNIX, Linux, iOS, Yosemite, Android, etc.)?
- Which operating system versions will be required?
- Will the different versions of software run on different computers and operating systems? If so, which ones?

- Will the file formats be consistent across operating systems?
- Will the software have a graphical user interface?
- Will the software's graphical user interface be consistent with that of the operating system and with other applications running in the same environment?
- Are there required interface standards, such as those needed for software to be sold in the Apple App Store?
- Will there be multiple implementations, such as a minimal size for laptops with small amounts of memory and disk space and larger implementations with more support files and features, to be used in computers with fewer limitations on memory and size?
- What are the minimal system requirements in terms of memory size and available hard disk space?
- Will the program's existence in memory cause problems for other applications that are also running on the computer?
- Are there any other software packages, such as word processors, spreadsheet programs, database managers, data visualization tools, or drawing tools, with which the program must be able to share data?
- Will online help be provided within the software?
- How will the software be delivered? Will it use CDs, DVDs, or be available only from online downloads?
- Will the software be provided in compressed or uncompressed form?
- Will installation and setup software be provided?
- Will training be required?
- Are there any time constraints, such as completion of certain operations within specified time limits? (These constraints are called "hard" constraints, in contrast to the "soft" constraints described in the next bulleted item. A system with hard timing constraints is called a "hard real-time system.")
- Are there any time constraints, such as completion of a number of certain operations within a single specified time limit, so that the constraint is on the average number of operations per unit time? (These constraints are called soft constraints, in contrast to the hard constraints described in the previous bulleted item. A system with only soft timing constraints is called a "soft real-time system." A system that has either hard or soft real-time constraints is called simply a "real-time system.")

You can see how detailed the software's requirements must be in order to write a complete description of the design. All of this gets much more complicated if the software is supposed to *do* anything! The apparent complexity for any real system seems to be so great that the problem appears to be hopeless. The only way out is to follow an approach that is familiar to computer science students: use stepwise refinement and abstraction as much as possible. Many software projects in the real world also attempt to reuse existing software components and systems as much as possible.

In this case, the stepwise refinement process involves a systematic process of translation from the initial, rather incomplete problem statement, which describes the system at a high level, to a detailed set of requirements that specifies the functionality of the desired system. The result of the translation is simply the restatement of the list of questions into a set of unambiguous statements, with the decisions made as to platform, operating system, available APIs, and so forth.

Suppose that these steps have been carried out. At this point, we have a good understanding of the types of requirements that our hypothetical software project must satisfy, at least in the area of the operating systems support and environment. This is the sort of background information and knowledge that a requirements engineer is expected to have *before* the requirements process begins. The common terminology *domain knowledge* is used in the software engineering industry to describe this assumed level of expertise. This domain knowledge facilitates both abstraction and the reuse of existing large-scale components and systems.

What has not been done yet is the selection of the specific requirements from the set of all possible options. In our example, it is likely to assume that there will be many customers for our software product, but that none of these customers is known directly by the requirements team. In this case, the requirements team will probably meet with several people, including the marketing and sales teams, in order to make the initial decisions about the requirements for the system.

To fix our ideas for the rest of the section, let us assume that the following decisions have been made:

- The software will run on the Windows 8.1 operating system. Other versions will not be supported.
- The software will have a graphical user interface.
- The graphical user interface will be consistent with the interface of the Mystery system, version 3.1, in the menu organization.
- The commands used to create a new file, open an existing file, close a file, save a file in standard format, and save a file in another format are to be chosen from a list that includes pure ASCII text.
- There will be only one version of the system that will be implemented in all installations.
- The system will require less than 200 MB free memory and 500 MB free disk space.
- The software must be able to share data with the Catchall database, version 7.3.

- Online help will be provided.
- The software is to be delivered on CD in compressed form.
- The software is to include an installation utility that includes a software decompression utility.

One thing you should note is that requirements statements are part of an agreement between the software developers and the framers of the problem statement. This agreement may in effect be a contract, with the full force of the law behind it in some instances. Software requirements should be precise and complete, for the reasons given earlier in this chapter. Requirements documents are kept on hand long after the initial product has been delivered.

Here is an example of the long life of requirements documents. Several of the computer programs designed to operate spacecraft for NASA were produced by outside commercial organizations, rather than by government employees. Some of the spacecraft lasted far longer than originally intended, producing useful scientific data after more than twenty years in some cases! Multiple companies often worked on different portions of the same software. Every time an operator of the software found a problem, the occurrence of the problem was verified by comparing the actual operations of the software with the output that was specified by the requirements. When the actual required outputs differed from the output that was specified for this input, then the problem was certified as having happened and the problem became someone's responsibility to fix if it was deemed to be of sufficient importance.

As was indicated earlier, fixing errors at this late stage was expensive, and none of the companies wanted to incur this expense unless they had to. Therefore two things had to happen:

1. The problem had to be verified.
2. The company responsible for the software problem had to be determined.

Both required the careful reading of the software's requirements documents. This illustrates that requirements documents have long lives and that precise individual requirements must be met by the software that is to be developed. Of course, the same situation occurs many times in private industry.

By this point in our discussion, you should be aware of the importance of requirements and you should be concerned about ways of dealing with their complexity. You might also wonder just how requirements are developed for actual software systems.

One technique is often described as the process of elicitation. That is, one or more potential customers are interviewed with the intent being to discover precisely what the customer wants. If there is no actual customer known to the requirements team, as would be the case with most personal computer software, a stand-in can be used for the customer. In either case, the requirements team will conduct a sequence of meetings with the "customer."

The requirements team must do a considerable amount of work between each pair of these meetings with the customer. The team must refine the previous set of requirements that was presented to the customer, and add, delete, or modify any requirements that were objected to by the customer. Since the customer may not have indicated his or her requirements explicitly, there must be an analysis of what the customer really wants, not just what he or she has said.

Of course, these requirements-gathering meetings with the customer will be more efficient if the customer can see an actual system prototype and determine if the system meets his or her needs. If the existing prototype system does not meet the needs of the customer, the prototype's requirements can be changed and another prototype developed. This flexibility and the reduction of the likelihood of producing an unsatisfactory system are two of the most important reasons for choosing the rapid prototyping, spiral, or other iterative models of software development. Using an object model that automatically generates framework code for the various actions (member functions) that can be performed on objects can also facilitate the requirements process by giving the customer something to see at every iteration of the requirements process.

The requirements-gathering process continues until the customer is satisfied with the system's requirements. The end result must be a completely unambiguous description of the system to be designed. In many cases, the requirements process is like the process of writing source code to implement a design: stepwise refinement. After the requirements team is satisfied that it knows the customer's requirements, it then writes a requirements document, which may be given to the customer for review. Unfortunately, the work of the requirements team is not over. We will return to the discussion of the requirements process in Section 3.4.

3.3 REQUIREMENTS TRACEABILITY

One of the most essential portions of a requirements document is a requirements traceability matrix. Its purpose is to allow easy tracking of requirements throughout a software system's complete development.

The format of a requirements traceability matrix varies from organization to organization; most organizations have standard forms that they use. They all have several things in common: the individual requirements, places for entries for other life cycle phases, and places to sign off. The concept is illustrated in Table 3.3.

(The focus is on Microsoft Windows 8 and 8.1, because their user interfaces were radically different from earlier versions of Windows. Of course, there was a similar radical change in user interface of Windows 10. However, Windows 8 allowed two distinctly different user views.)

For simplicity, the requirements are numbered 1, 2, 3, and so on. Unfortunately, this simplistic type of organization of requirements is not typical of industry practice. More realistic software projects will have multiple levels of requirements, generally grouped by functionality. For example, a current software project at NASA has 1156 requirements, which are grouped into four levels of a hierarchy. Thus the index of the requirement numbered 4.1.3.2 reflects that this is the second requirement in unit 3 of sub-subsystem 1 of subsystem 4.

TABLE 3.3 A Requirements Traceability Matrix

Number	Requirement	Design	Code	Test
1	Intel-based			
2	Windows 8			
3	Consistent with User Interface of Windows 8			
4	Graphical User Interface			
5	Consistent with Mystery 3.1			
6	System One Size Only			
7	<200 MB System, <500 MB Disk			
8	Share Data with Catchall 7.3			
9	Online Help Provided			
10	Delivery on CD			
11	Include Installation and Decompression			

The next step in the requirements engineering process is to examine the state of our requirements. (Technically, this should have been done before the requirements traceability matrix was created.) We need to look for any inconsistencies or missing requirements. Of course, we also need to look for any that are vaguely stated, which should show up at this point because vague requirements cannot be tested. The presence of the last column in the requirements traceability matrix shown in Table 3.3 serves as a reminder that testable functional requirements are essential to software projects. It is hoped that most decisions made in the requirements-gathering process will be straightforward and can lead to software projects that can be completed.

A moment's reflection should indicate some potential concerns in our requirements. There are difficulties in the area of the user interface. The statement "be consistent with" the user interface of the Windows 8 operating system is entirely too vague. It is simply not testable. There are several potential problems with the meaning of this phrase:

- The software's user interface will not have any conflicts with the user interface of Windows 8. Which interface: the tiled one or the classic one? Are there any conflicts with Windows 8.1?
- The software's user interface will use the same conventions for keystrokes and menu selection that are used in the Windows 8 operating system for identical purposes. Which conventions (this operating system has two user interfaces)?
- The software's menu will have the same organization of options as does Windows 8.
- The software will use the same system calls and APIs as Windows 8.

It is not clear which of these was meant. The rule of thumb is "more detail is always appropriate." So is the use of clear verbs, such as "must," "shall," and "will," instead of "might" or "may." Unclear requirements such as these can cause disasters in an actual project. With such variation, it is difficult to trace any decisions about design or implementation back to requirements in order to ensure that the requirements have been met. As you organize requirements into a requirements traceability matrix, watch for statements that cannot be tested. In fact, some organizations require that test cases be written for critical requirements before the requirements are approved.

3.4 SOFTWARE ARCHITECTURES AND REQUIREMENTS

It is time to evaluate our understanding of the requirements process. Since we have gotten to this point in the requirements process, we have completed the use of a stepwise refinement process and our knowledge of the application domain to write a set of requirements that are appropriate for our software. In Table 3.3, we showed a set of eleven distinct requirements for our hypothetical system. However long the requirements-gathering process took, we did manage to get an initial set of requirements.

Unfortunately, the process is unsatisfactory for more complicated software. A software system with over 1150 requirements will take at least 100 times as much effort as the trivial example we have discussed so far. The larger system is likely to have much more severe requirements for systems that the software will have to interface with. Many of the requirements will affect other requirements, causing some inconsistency. There may even be some real-time requirements. The requirements effort for this system is likely to be closer to 200 times as complex as our example. The resulting effort may be much larger for realistic systems. Often, a software architecture, showing the major components and functionality in the high-level structure of a software system, can be useful. We discuss software architecture more fully in Section 4.5.

How can we handle this increased complexity? Stepwise refinement and abstraction are the obvious approaches, but there is a need for more guidance.

There are four general techniques in common use:

1. Use principles of data abstraction and information hiding to produce a complete description of the system's functionality. This approach leads naturally to an object-oriented set of requirements but can be used with systems that are completely procedurally oriented in nature.

2. Regroup requirements to be consistent with the requirements and design of both existing and planned systems.

3. Reuse requirements in order to be able to reuse existing designs and source code that previously have been developed to meet these requirements.

4. Automate the requirements engineering process.

For the purpose of consistency of exposition, we will assume that a hypothetical graphical database system has had a set of requirements determined. (A graphical database allows the organization, search, and retrieval of two- and three-dimensional objects together with

TABLE 3.4 Initial Requirements for a Hypothetical Graphical Database System

1. The system must allow up to 100 polyhedra to be represented.
2. The system must allow objects to be either a single polyhedron or combinations of up to three polyhedra.
3. The system will allow combinations to be formed by using the standard set operations of union or intersection on one, two, or three polyhedra.
4. The system will keep track of the volume of all objects.
5. The system will keep track of the surface area of all objects.
6. The system will allow the most recent combination of two or three objects into a single larger object to be undone.
7. The system must display objects using a world-to-screen transformation.
8. The system must allow pointers to each face or edge of each polyhedron to be kept.
9. The system must be able to determine the length of each edge of each object.
10. Other requirements related to display of the objects (these will not be mentioned here for simplicity).
11. The system must run on the XYZ computer with 200 MB RAM and 500 MB disk space available, using the operating system version 3.72, DISPLAY Master 3.32, and the WWM database system, version 3.4.

a set of allowable operations. Such databases are often included with solid modeling systems, such as those used in computer-aided design software.)

The set of requirements is given in Table 3.4. Different approaches to the refinement of this set of requirements to a complete and unambiguous set will be discussed in a separate section.

3.4.1 Use of Data Abstraction and Information Hiding in Requirements Engineering

Data abstraction and information hiding are two of the most important tools of a software engineer. They encourage the reduction of complexity that is necessary for good software engineering. As such, they are present in every successful software project of any size.

Unfortunately, they are also perceived as being impossible to achieve without an object-oriented approach. Although object orientation encourages data abstraction and information hiding, it is not necessary in order to achieve them. To emphasize this point, we will develop a completely procedurally oriented set of requirements for the system that has its initial requirements given in Table 3.4. The discussion in this section is influenced heavily by a paper by Daniel Berry (1995) in the *Journal of Systems and Software* and its explication in a seminar by Berry.

The paper by Berry (1995) is essential reading for requirements engineers. There are two essential participants in the process: a requirements engineer, who is extremely knowledgeable about the requirements process but who is ignorant about the application domain; and a domain expert, who understands the application area but is not especially knowledgeable about requirements engineering. The essential message of the paper for the requirements engineering process is clear and may be described by two complementary principles:

1. The requirements for the system must be modified and refined until the system can be developed by software designers and coders who are familiar with the fundamental algorithms of the application domain but who are ignorant about the requirements

engineering process. The eventual requirements will be so complete and unambiguous that the design will be clear and the coding can be easily done by implementing standard, well-known algorithms.

2. The requirements for the system must be modified and refined until the system requirements can be understood by requirements engineers who are knowledgeable about requirements engineering but who know nothing about the application domain. The eventual requirements will be so complete and unambiguous that the requirements engineer will understand them completely, even without knowledge of the application domain.

The requirements for the system must be iterated until each of these two principles is true for each of the requirements. The approach is called "ignorance hiding" by Berry (1995) because the lack of understanding (ignorance of requirements engineering by the designers and coders, and ignorance of the application domain by the requirements engineer) forces both the requirements engineer and the domain expert to communicate precisely, with no preconceived notions. This works only if both sides are persistent in getting unambiguous requirements.

Let us look at the first two requirements listed in Table 3.4. The first requirement is "The system must allow up to 100 polyhedra to be represented." There are two nouns in this requirement: "system" and "polyhedra." The verbs are "allow" and "represent." Other important words are "up to" and "100."

Imagine that you were a requirements engineer with no knowledge of the application domain. The "system" clearly refers to the software to be produced by this project. No confusion there. But what about the term "polyhedra"? Assuming that we recognize this term as a plural, we have to ask the application domain expert what a polyhedron is. He or she will reply something like: "A solid figure whose boundary consists of polygons." The discussion would continue, until the requirements engineer is sure about exactly what a polygon is. Clearly, the objects indicated in Figure 3.2a and b are polyhedra, if we assume that we are seeing the visible boundary of some three-dimensional object. However, the degenerate case in Figure 3.2c is more difficult to characterize.

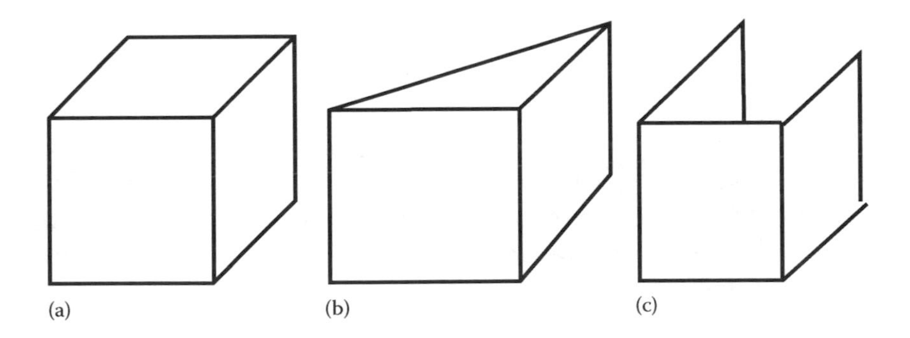

FIGURE 3.2 (a) Cube. (b) Prism. (c) Polyhedron?

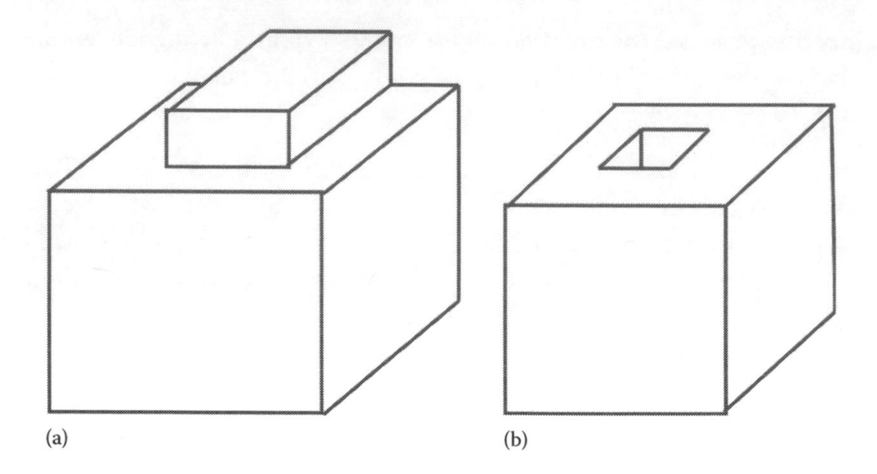

FIGURE 3.3 (a) Combination. (b) Removal.

Clearly a precise definition of polyhedron must be obtained before any more of the system can be designed. In order to conserve space, we will assume that this term has been defined properly and will turn our attention to the second requirement. The primary difficulty there is with the term "combination." A three-dimensional geometric figure can be considered as a solid object with a boundary, as a solid object without a boundary, or as a region in space described by its boundary surfaces. The set of operations that eventually will be performed on these polyhedra include construction, destruction, determination of orientation in space, and determination of which faces of the polyhedra are visible to an observer at a certain position and looking in a particular direction. Thus, the term "combination" can refer to either of the cases illustrated in Figure 3.3. The requirements need to indicate which, or both, of these combinations will be allowed. Unusual cases, such as removing a rectangular solid from itself, must also be considered. We note in passing that proper treatment of these "unusual cases" can take up most of the programming effort needed by a computer-graphics-based solid modeling system.

Several other of the initial requirements of Table 3.4 are lacking in clarity according to the ignorance hiding principle. We will leave the discussion of the improvements to the other requirements to the exercises.

3.4.2 Regrouping Requirements in Requirements Engineering

By regrouping requirements we mean that the individual requirements are grouped into sets that can be considered as an entirety. This is a higher-level view than that of individual requirements. Now we are focusing our attention on higher-level components and on a view of the system that is called the system's software architecture. This regrouping process is often considered to be software reengineering when it is applied to existing, complete software systems. (Software reengineering is much broader.)

What would the requirements for such a reengineered system include? Consider the example of an initial set of requirements that was shown in Table 3.4. The requirements listed in Table 3.4 seem to have been determined in a rather haphazard fashion. This was done deliberately to illustrate the reorganization of the requirements. We now consider a more realistic, efficient organization of the requirements.

There are two other issues to be resolved: How do we use the hypothetical Display Master software for graphics display and how do we use the WWM database system for storage? The answer is that we regroup the requirements. For example, the world-to-screen transformation requirement appears to be redundant, because the Display Master software probably has its own routines and programming interfaces for this. (If not, it might be better to use a software utility that is more standardized.)

Use of a hypothetical database package implicitly indicates that several of our requirements can be met by the database package, using standard query facilities. Thus, the requirements for the hypothetical system described in Table 3.4 can be regrouped into those of Table 3.5.

Here we have placed all the database-related requirements at the top of the list in requirements 1 through 5. The display requirements information is kept in items 6 through 8. Notice how much easier it is to determine if a database package has the proper facilities. We could go one step further and design the database itself. For simplicity, we will omit this discussion, since we will return to the point later in a more realistic example.

Note that there is another way to organize the requirements. Most software engineers would agree that in many application domains the requirements for nearly all current software systems are described using hierarchical listings such as those used so far. These hierarchical listings are implicitly geared toward eventual implementation of source code in a procedural programming language such as C, Ada, or even some scripting languages.

Therefore, a mapping from the essentially procedural perspective to an objective oriented one is necessary if the software is to be written in an object-oriented programming language such as Java, Swift, Smalltalk, or Eiffel, or a hybrid language such as C++. We will not do the mapping at this point, postponing a more complete discussion of object orientation until the design phase. At that time, we will note that the "has-a," "is-a," and "uses-a" relationships will be important in their mapping, regardless of the life cycle phase where it takes place.

TABLE 3.5 Reorganized Requirements for a Hypothetical Graphical Database System

1. The system must allow up to 100 polyhedra to be represented in a database using the WWM database software.
2. The system must allow objects to be either a single polyhedron or combinations of up to three polyhedra.
3. The system will allow combinations to be formed by using the standard set operations of union or intersection on one, two, or three polyhedra.
4. The system will allow the most recent combination of two or three objects into a single larger object to be undone.
5. The system must allow pointers to each face or edge of each polyhedron to be kept.
6. The system must display objects using a world-to-screen transformation.
7. Other requirements related to display of the objects (these will not be mentioned here for simplicity).
8. The system must run on the XYZ computer with 16 MB RAM, 42 MB disk space available, using the operating system version 3.72, DISPLAY Master 3.32, and the WWM database system, version 3.4.

3.4.3 Reuse of Requirements in Requirements Engineering

Reuse of requirements presents one of the greatest opportunities for improving the efficiency of the requirements process. The idea is to carry the process one step further than the regrouping of the requirements. Reusing a set of requirements for a subsystem can ensure that no requirements are missed for this subsystem. In addition, software reuse can reduce further software life cycle costs by allowing an entire subsystem (design, source code, and so on) to be put into place in the final software system.

Reuse of requirements is only possible if there is some sort of repository for requirements that have been used successfully in other systems. One or more persons who are familiar with either the application domain (a so-called domain expert) or the available reusable requirements and other saved software artifacts (a domain engineer) will be necessary.

In this case we can have the best of both worlds. The system we are trying to set requirements for has already been created. That is, there is another system whose requirements are identical either to the entire system we are trying to create here or at least to a substantial portion. The existing system has been developed for another project and can, in fact, be found in the source code available with the book *Object-Oriented Design and Programming in C++* (Leach, 1995). Since I am the author of that book, there is no obstacle (other than obtaining permission from the publisher) to my reusing that code here in its entirety. This code was designed, implemented, and tested, thereby saving considerable development effort.

In more realistic situations, a search of sources of reusable software would be necessary to locate the portion of the system that will be reused. In addition, there may be some questions about the ownership of the software that is to be reused.

3.4.4 Automation of the Requirements Engineering Process

Automation of at least a portion of the requirements process is also appealing. The idea is that a single high-level requirement is entered into a "requirements generation system." The requirements generation system then produces a set of additional requirements, which generally will have more detail. The requirements generation system can be applied to all or a portion of the requirements.

The automation can be either complete or partial. In a complete automation process, entering one requirement, such as the target computer environment, will generate all the other relevant requirements that can be determined for this software.

In a partially automated requirements generation process, entering a single requirement will result in a group of questions that must be answered before the complete set of relevant software requirements can be met.

As an example of the operation of a partially automated requirements system, the requirements engineer might enter the phrase "personal computer" and the system would respond with something like:

If the user enters "Intel," then the system might respond with

Operating System: Windows 8, Linux, OS X, Yosemite (select one)

After the user has indicated an operating system, he or she might be prompted for a version. The selection of one of these two versions might have an impact on the version of the associated dynamically linked library (DLL). The dialogue might continue until the user selects a particular hardware device or network connection, in which case a requirement such as

Must interface with DLL named ABC123

would be generated by the requirements generation system and so on.

Ideally, a requirements generation system should be organized to make optimal use of both requirements regrouping and reuse.

3.5 USE CASES IN REQUIREMENTS ENGINEERING

It is sometimes convenient to develop a set of requirements with the aid of what are called *use cases*. A use case is a description of how a system is to perform in a certain scenario that describes common action. Figure 3.4 illustrates a simple example of how an administrator would interact with a web crawler to gather information from a collection of websites in order to provide for later analysis. We will meet these use cases later when we consider the major software project that will be discussed throughout the remainder of this book. (The major project will involve collecting and analyzing metrics that describe the contents of several large open-source software repositories.)

Greater detail in the typical user's interaction with this software system is shown in Figure 3.5. The "Analysis: Equivalencies" will be the result of an analysis of the metrics

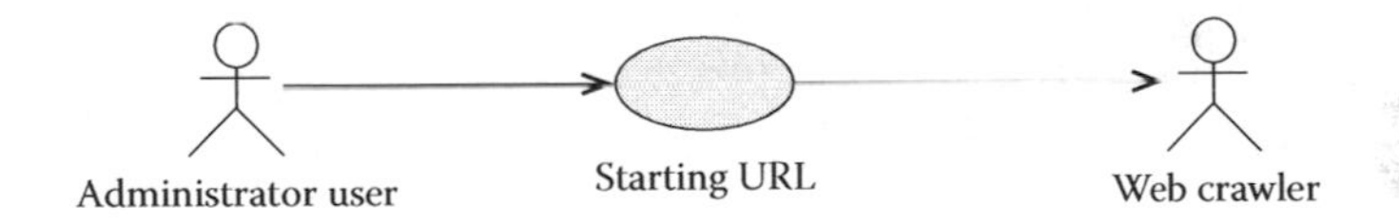

FIGURE 3.4 A simple diagram of a use case.

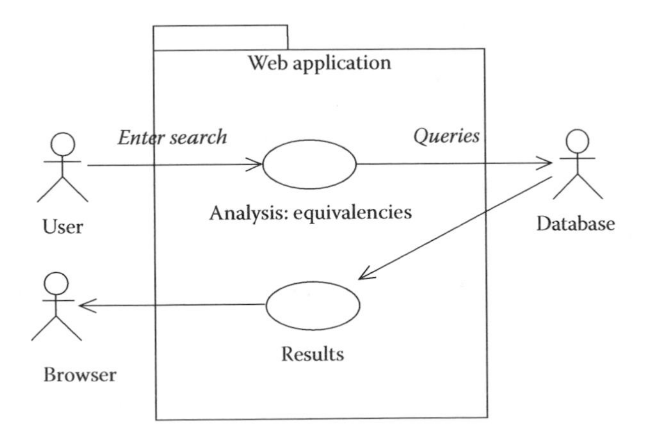

FIGURE 3.5 A slightly more complicated diagram illustrating a use case.

obtained by many source code software evaluations. Note the compartmentalization of different system aspects relevant to different stakeholders. Note also that this use case representation is different from that of a flowchart or dataflow diagram.

The purpose of use cases is to describe how the proposed software system is to be used. The intention, of course, is to aid in the development of requirements. Clearly, many different use cases will be necessary for the development of any nontrivial software system.

There is another concept that is related to use cases and is also especially relevant to safety-critical or fault-tolerant software systems. This concept is often referred to as a *misuse case*. A misuse case is intended to emphasize ways that a proposed system can be incorrectly used by either human users making errors in such things as, say, the human–computer interface, or if a subsystem does not respond to an expected query or does not respond at all. As such, a misuse case can guide the system's developers to provide some sort of fault tolerance by perhaps providing exception-handling capabilities at various places in the software system.

3.6 REENGINEERING SYSTEM REQUIREMENTS

As shown earlier in this chapter, determination of a system's requirements is an iterative process. When the requirements are set as to meet the (perceived) needs of clients and potential users, then the requirements-gathering process usually terminates. We will call the result of this standard activity the "ideal requirements." In many software development organizations at present, cost pressures require as much use as possible of commercial off-the-shelf (COTS) products and available building blocks. In this section we will show how a systematic reuse program encourages the potential reuser to change ideal requirements to meet cost pressures.

Suppose that the ideal requirements specify that a complex database entry be updated within some specific time requirement. Suppose also that the organization already licenses a commercial database that misses meeting this requirement by 10 percent.

The organization now has several options:

- It can reconfigure the database software to obtain better performance.
- It can test other commercial database products to see if any meet the ideal requirements.
- It can purchase faster hardware.
- It can reduce the computing load on the computer system.
- It can provide a performance analysis of the entire system to locate places where system performance can be improved.
- It can change the requirements from the ideal requirements to determine if lesser performance would be acceptable, especially since the existing software is essentially free.

Clearly some organizations will select the last alternative in many situations. This is an example of how the drive for reuse cost savings, which in this case are due to COTS products, can cause changes in requirements.

The high-level description of this "reuse-driven" requirements process is simple:

1. Develop an initial set of requirements. This should be done in concert with the customer. If no customer is known, then the requirements should be chosen according to the perceived needs of the system's end users. For simplicity, we will only describe the interaction of the development team with a known customer. The modification for new systems with no fixed customer but likely end users is similar and will not be discussed.

2. Determine if there is an existing reusable system that meets the set of requirements. If there is such a system, stop the requirements process and return the existing reusable system.

3. Determine if there is an existing reusable system that meets "nearly all" the requirements. If such a system exists, provide the customer with a description of the existing system's requirements, how they differ from the original requirements, and the expected costs of using the existing system to "nearly meet" the customer's requirements. If the customer accepts the modified requirements and is willing to accept the reused existing system at the estimated cost, stop the requirements process and return the existing reusable system.

4. If no existing system meets or nearly meets the customer's requirements, then the set of requirements should be separated into sets of requirements for subsystems. The decomposition into subsystems should be guided by the process of domain analysis, since the goal is to determine those subsystems that have the greatest probability for being available as COTS products.

5. Steps 2 through 4 should be carried out for each subsystem. The process will terminate for each subsystem as specified in these steps. The only additional activity is to determine if the reused subsystems meet appropriate interface standards. This should be done during a check of the certification of the reused subsystem. (It is assumed that each reused subsystem was previously certified as to its interface standards.)

6. New software development is limited to subsystems in which no agreement can be made between the customer's fixed requirements and the existing reusable subsystem's requirements.

7. After agreement between the customer and the software team on the final set of requirements for the subsystems, the existing subsystem building blocks are integrated with any new code into the new system, which is then configured, tested, documented, and delivered to the customer.

Incidentally, this is not an unrealistic academic scenario. Several of NASA's software systems for spacecraft control and handling data that is transmitted from satellites were designed with reuse cost savings factors influencing the requirements process. A paper by Bracken reports on experiences in the innovative IMACCS (Integrated Monitoring,

Analysis, and Control COTS System) project with a software development process based heavily on COTS and reuse to drive the requirements (Bracken, 1995; see also Leach, 1997 and Waund, 1995). The team won a centerwide award for cost saving and quality.

3.7 ASSESSMENT OF FEASIBILITY OF SYSTEM REQUIREMENTS

Engineering is often described as the systematic employment of scientific principles to develop systems in an efficient, cost-effective manner, with proper attention to safety. So far, we have concentrated on describing a systematic employment of scientific or engineering principles. We have not considered efficiency, cost-effectiveness, or safety.

It is time for an assessment of the system. Often this assessment is used as part of a proposal to management. Of course, any proposal or feasibility study must address cost. At this point, many requirements documents are given to a "software economist" who will estimate the total cost of producing the system. The software economist is generally experienced in the application domain and can predict the size of the software based on the detailed requirements. He or she is aware of any special features of the application domain such as the need for real-time processing, the complexity of the interface to existing or project applications software or operating systems, and the rate of change of related software and hardware technology and standards. In short, the software economist performs many of the activities that were described in Chapter 2 when we discussed project estimation.

The safety of software that can affect human lives or have major impact on financial records or people's privacy should also be considered. The requirements may spell out the need for special techniques, including always testing divisors for being nonzero before a division occurs, testing permissions before allowing database access, encrypting certain portions of the software's data, or the use of redundancy techniques to improve software fault tolerance.

There may also be other factors, depending on market pressures. Jim McCarthy (1995) of Microsoft described the situation well in an interesting book. McCarthy was the project manager for release 1.0 of the Visual C++ product. Because of market competition, it was considered imperative to ship updates, including major system upgrades, at six-month intervals. Keeping these updates on schedule was considered more important than including certain technical improvements, particularly if these improvements were not of highest priority. For example, a decision was made to delay the incorporation of C++ templates into version 1.0 of Microsoft's product. This decision was made in order to have a more robust (fault-free) implementation of the compilation semantics and an assessment of the importance of including templates. For a long time, updates of Visual C++ were produced at even shorter intervals.

This is typical of the influence of market factors in much of modern software development. Certain desirable requirements are often postponed until later releases of software. You should note that this assessment might mean that the requirements will have to be rewritten. This is standard practice and is no cause for alarm. As we have seen, it is cheaper to fix problems in the requirements phase than later.

In some cases, the assessment may indicate that the system is too costly to build. At this point, there are two choices: look for cost savings or scrap the project. The primary sources for cost savings are revision of requirements and the collection of existing, reusable software that can be used in the system.

Requirements can often be contradictory. For example, a system may have a requirement for an elegant user interface and also may have a requirement for real-time performance of certain operations. These may be contradictory and it is the responsibility of the requirements process to determine such conflicts.

The requirements assessment should pay particular attention to any system or subsystem that is safety-critical, in the sense that loss of life or major destruction to property can occur if the software system does not function properly.

3.8 USABILITY REQUIREMENTS

One of the goals of software engineering given in Chapter 1 was to produce software that is usable. The beginning software engineer is generally not aware that the usability of software can be measured, at least to a first approximation. Indeed, even experienced software engineers and managers frequently are not aware of the techniques of evaluating user interfaces.

This area is often called human–computer interaction, or HCI. It has been the subject of a considerable amount of research for many years. There is a special interest group, SIGCHI, of the Association for Computing Machinery (ACM) that is devoted to the study of this issue. (The acronym SIGCHI indicates the Special Interest Group on Computer–Human Interaction.) There are several annual conferences on this topic and a considerable amount of interdisciplinary research, often involving researchers and practitioners from psychology, fine arts, communications (for multimedia applications), and, of course, computer science.

The user interface is an integral part of a computer system and, thus, it must also be subjected to requirements engineering. The requirements for a user interface are often phrased in terminology that is far less precise than the terminology used to describe the actions of software that is used for process control, database management, or even the major software project that will be introduced later in this chapter. The requirements documents for a user interface are described as being nonbehavioral by Davis (1990) in his book on software requirements. See also Stone et al. (2005).

One of the first steps in evaluating a human–computer interface is to model the typical user. Users can range from complete novices unfamiliar with any aspect of computers, to experienced computer users who may be unfamiliar with the particular software, to experienced users of particular software packages. Clearly, the needs and desires of such types of users will be different. The designer of a user interface must have target users in mind when he or she designs the interface.

Ideally, the interface is so simple that a novice user can achieve both reasonable success with little training, but sophisticated enough that an experienced user can accomplish more complex tasks with a minimal amount of effort. Anyone watching a small child use a smartphone or tablet knows the advantage of a simple, elegant interface. We will discuss some issues with usability requirements later in this section. For now, we will focus on general issues that will be more appropriate for displays with large amounts of screen real estate.

Designing user interfaces with this degree of flexibility is not an easy task. Operator error can be dangerous if the software controls an airplane, hospital patient monitor, or power plant. User interface design is serious business.

The relatively high-level guidelines in that document are intended to apply across a wide range of applications. The goal of a style guide is to encourage consistency in the "look and feel" of the application.

The process involved in developing a style guide includes the following steps, as recommended by Bleser (1994):

1. Identify relevant guidelines.
2. From the overall set of guidelines, select those that pertain to the application under development.
3. Narrow down the subset of pertinent guidelines.
4. Develop design rules from the guidelines.
5. Allow for reasonable exceptions.

If a guideline states that displays should be formatted consistently, for example, a set of design rules would be needed to specify the location of such display features as menu titles, icon labels, dialog boxes, and error messages. Design rules take the guidelines down to a concrete, highly specific level.

Because a particular guideline can be translated in numerous ways, translation requires designers to define interface components, application components, and constraints that must be met.

The subset of guidelines selected in the first step may include some that conflict. The choice of which guidelines to retain may be based on relative importance or impact, given constraints of time and budget. Resolve any conflict that arises by considering whether one or the other is more appropriate for their application. When the answer is not clear, however, the team can use a more formal decision-making process, according to the following steps:

1. Identify the attributes of user performance that may be affected by the conflicting guidelines (e.g., color discrimination, target detection, speed of response).
2. Weight the importance of those attributes for overall system performance. These weightings are likely to vary from project to project.
3. Using a numeric scale, rate the conflicting guidelines for their expected effect on each performance outcome.
4. Multiply ratings by weights and sum the products. Select the guideline with the higher total.

Here is an example of the usability guidelines. When the same buttons are used for different windows, consistently place them in the same location and keep related buttons together. Some issues are:

- Which buttons are involved?
- Are there any related buttons?
- Where should these buttons be placed in this application?

Design rules should be specific enough that different developers will produce exactly the same features when applying them. For this reason, they should be pretested to ensure that developers will agree in their interpretation. There should be little room for a variety of interpretations.

Consider the two alternative screen designs indicated in Figures 3.6 and 3.7. Note the differences in both placement and the amount of space devoted to textual information. Which screen enables a user more easily to select the proper button in an emergency? We will give a partial answer to this question in the next few paragraphs.

Note that the upper left-hand window in Figure 3.6 contains much more textual information that the corresponding window in Figure 3.7. Note also that there is a button in Figure 3.7 that indicates that all valves are to be closed with a single action from the software user. There is no such facility in Figure 3.6.

Specification of color changes in the graph in the upper right-hand corner can increase the usability of the software, as can having the computer make an audible sound such as an alarm. Such interface specification should be part of the system requirements. Analysis of other issues, such as the amount of text displayed in a window, might be included in the requirements analysis, although it is more likely to be left to system design.

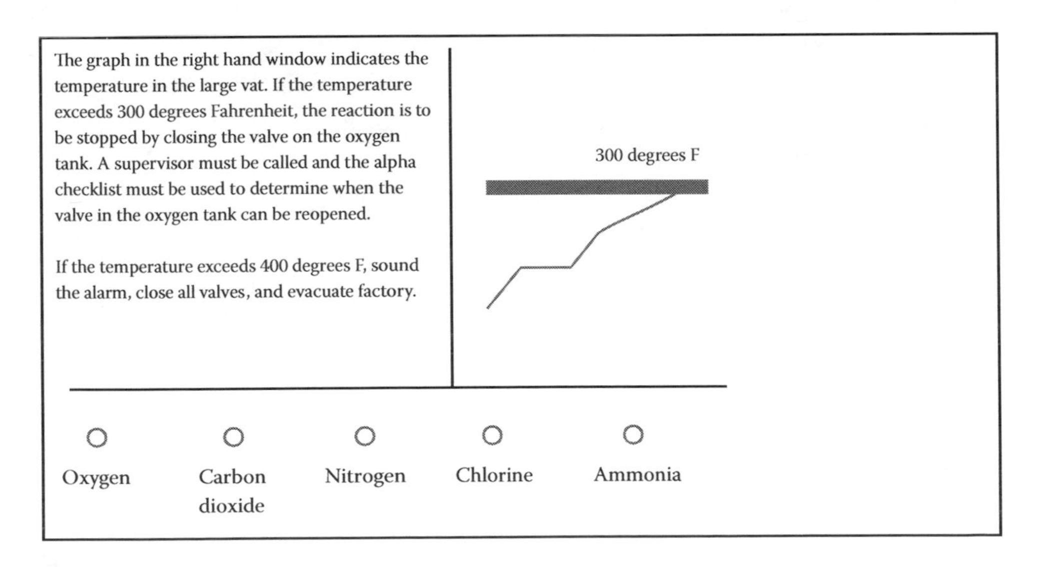

FIGURE 3.6 A screen design.

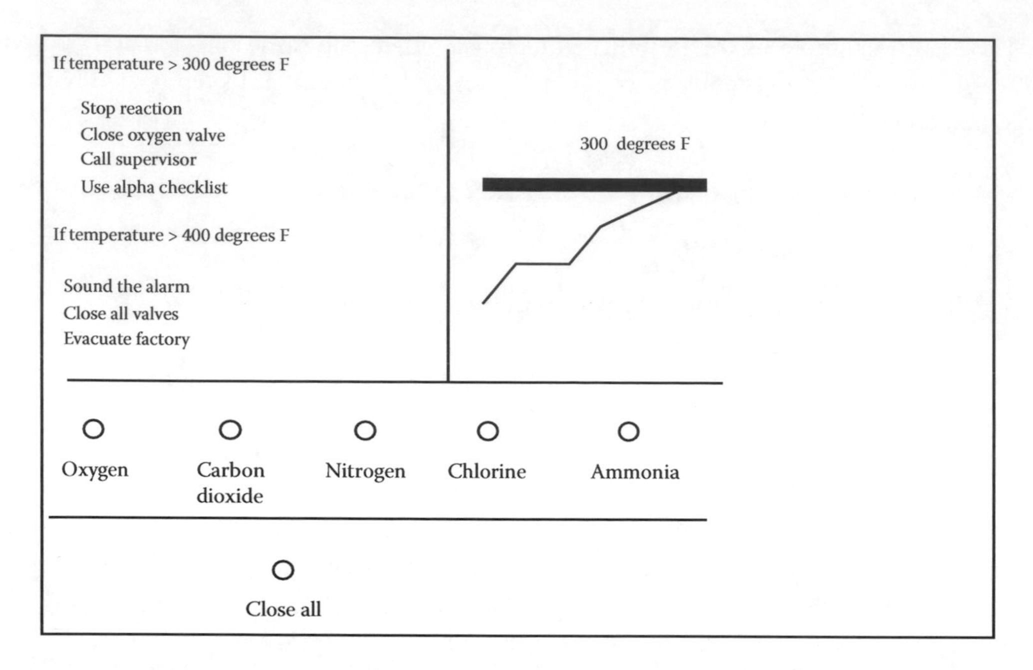

FIGURE 3.7 An alternative screen design.

A more detailed model of a user interface based on an object–entity relationship is given in the remainder of this section. The model incorporates attributes of the user, computer hardware, and input/output (I/O) devices. It also includes real-time and environmental constraints, as well as the underlying applications software.

The model presents a high level view of the interface and includes formal evaluation of the interface. Both human factors and user interface management systems research ideas are incorporated.

In this model, all the features of a user interface are abstracted as objects with attributes that indicate the properties that the objects may have. Any instance of an object has values for each of its attributes; these values indicate the degree to which the software possesses the attributes. The relationships between the objects making up a user interface are also important and are included in our information model.

The model includes seven objects that make up a user interface:

- Constraint
- Environment
- Hardware
- Input
- Output
- Software
- User

The relationship between these seven basic objects is shown in Figure 3.8. A brief description of each object in Figure 3.8 together with some of the more important attributes is provided next. The sets of attributes of different objects are relatively complex and complete lists of attributes for some of the objects are given in the exercises.

A constraint object is important for real-time systems in which a user/operator of the system must react to certain situations within prescribed time limits. This is the type of requirement that might apply to a control system for a chemical treatment plant or a medical monitoring device. Constraint objects embody the idea that the user's response is critical and that the timing constraints must be reflected in the model of the user interface. The critical attributes of a constraint object are called TIME_CONSTRAINT, TIME_LIMIT, and CRITICAL; these attributes reflect timing demands that might be present in the interface to a real-time system.

A hardware object involves those portions of the system that are not essentially I/O devices and are primarily hardware oriented. Included here might be portable versus nonportable computers, presence or absence of networks for remote data access, and so on. Typical attributes include OPERATING_SYSTEM, HARDWARE_VERSION, RESPONSE_TIME, HARDWARE_ERRORS, and PORTABLE.

Input and output objects are distinct from one another and from hardware objects. Input objects specifically include keyboards, mice, touch tablets, smartphones, and light pens. LCDs, printers, speakers, and strip chart recorders are output objects. Typical attributes of input objects are DEVICE_TYPE and NUM_INPUT_TYPES. Some typical attributes of output objects are DEVICE_TYPE, NUM_DEVICES, HORIZ_DIMENSION, and VERT_DIMENSION.

Software objects include all of the features of the software being interfaced to, whereas a user object includes all of the features of a user during learning or becoming expert with the software. Of course, the value of certain attributes of a user object will change as the user learns more about the system. A software object is important because of the inherent differences between software, such as word processing or image processing.

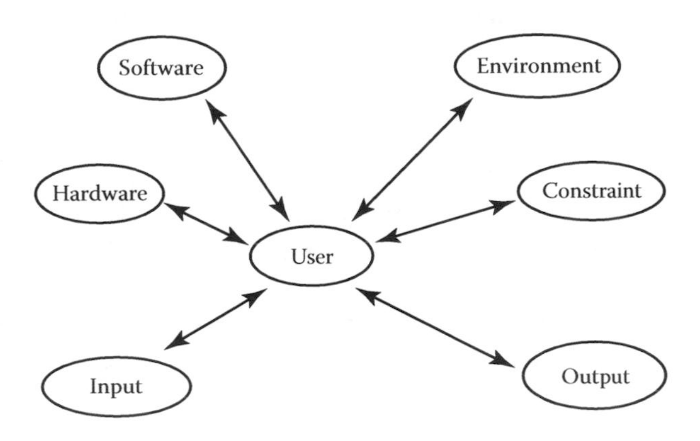

FIGURE 3.8 The relationship between objects in a user interface management system (UIMS).

Following are some attributes of alphanumeric portions of displays:

TEXT_DENSITY

TEXT_DENSITY_PER_WINDOW

NUMERICAL_DENSITY

NUMERICAL_DENSITY_PER_WINDOW

TEXT_COLUMN_DENSITY

TEXT_COLUMN_DENSITY_PER_WINDOW

NUMERICAL_COLUMN_DENSITY

NUMERICAL_COLUMN_DENSITY_PER_WINDOW

TEXT_ROW_DENSITY

TEXT_ROW_DENSITY_PER_WINDOW

NUMERICAL_ROW_DENSITY

NUMERICAL_ROW_DENSITY_PER_WINDOW

We now consider how this model can be used in the evaluation of an interface. Attributes can be classified into three classes:

1. Naming attributes give the name of an attribute.
2. Descriptive attributes give the description of the attribute.
3. Referential attributes refer to an attribute of some other object.

Consider the attributes of a user object, which is an abstraction of the actual human user of a computer system. Naming, descriptive, and referential attributes are denoted by N, D, and R, respectively.

NAME: text, the name of the user of the user object. (N)

EXPERIENCED_WITH_HARDWARE: Boolean, TRUE if the user has experience with the hardware previously, FALSE otherwise. This attribute indicates the user's experience with this particular computer and may be FALSE if the user has used different models of the same brand and operating system. For example, the computer to be used is a portable one with different key placement from an office system used before. (R)

EXPERIENCED_WITH_INPUT_DEVICE: Boolean, TRUE if the user has experience with the input device previously, FALSE otherwise. This attribute indicates the user's experience with this particular device and may be FALSE if the user has used similar but different models of the device. For example, the user may have used a three-button mouse previously but now needs to use a one- or two-button device. Obviously, this should be set to TRUE for most smartphone software. (R)

EXPERIENCED_WITH_OUTPUT_DEVICE: Boolean, TRUE if the user has experience with the output device previously, FALSE otherwise. This attribute indicates the user's experience with this particular device and may be FALSE if the user has used similar but different models of the device such as a non-scrollable or a scrollable CRT terminal. (R)

SOFTWARE_EXPERIENCE: integer valued in the range 1 to 10, with 1 indicating little experience with the particular software and 10 indicating being relatively experienced. (R)

COMPUTER_EXPERIENCE: Boolean, TRUE if the user is familiar with computers, FALSE otherwise. (D)

The numerical values can be used as part of a model of the user interface. It is very clear that detailed evaluation of user interfaces is much more complex than just looking at a site on the World Wide Web and pronouncing it "cool." (Many sites that supposedly have cool graphics and animation have poor designs that require users to either wait long periods for desired information to appear on their screens or else not get the desired information at all.) The effective use of computer resources for user interfaces is not trivial.

User interfaces for smartphones and tablets have some additional issues that should be addressed in the system's requirements. There are huge differences in connection speed, depending on the strength of Wi-Fi or cellular connections. The choice to use, say, accelerometers as part of the system interface requirements might have some ramifications for software portability, since not all smartphone and tablet families use them the same way. This difference means either increased development effort or time to deploy on some platforms, or not deploying whatsoever. For safety-critical systems, deciding on what must always be displayed on a screen is an important task.

You might feel that some of the points made in this section appear to be more appropriate for the chapter on software design rather than a chapter devoted to requirements. However, you should recall the goals as stated in the first paragraph of this section: to provide a systematic approach to developing requirements for the nonbehavioral portion of the software.

3.9 SPECIFYING REQUIREMENTS USING STATE DIAGRAMS AND DECISION TABLES

Until now, we have discussed software requirements using natural language. We have used a relatively structured pseudocode that can serve as the basis for later designs of the software to be developed. This is acceptable for projects that are relatively small. If the requirements are sufficiently small and unambiguous, we can manage with textual descriptions that are augmented by graphics as necessary.

However, in many larger software systems, the inherent ambiguity of natural language descriptions will make the project's requirements unsuitable for textual requirements. Even if the requirements could be written solely in a natural language, the resulting document might be too large to be of any use. Detailed documents that are several thousand pages long are rarely read, much less understood!

In many cases we are lead to the use of more formal and precise methods of specifying system requirements. Two such methods will be described in this section: finite state machines and decision tables. The descriptions are brief and are intended to complement your previous knowledge of these and related concepts in a course in discrete mathematics or automata theory.

A finite state machine is a model of computation in which the behavior of the system is dependent upon which of a finite set of "states" the system is in and which of a set of "events" is observed. One particular state is used to represent the initial state of the system. For a software system that controls a process in a chemical power plant, the states might be those shown in Table 3.6.

The software to control the chemical process would use these nine states to determine which actions would be appropriate at different times. For example, the software should never allow the contents of the vat to be poured into molds until the reaction is complete. The reagent should never be poured into the vat unless the vat's temperature has been reduced by refrigeration.

If the number of states is not too large, then a finite state machine can be described visually as a state diagram according to the following informal conventions:

- Each state is represented as a bubble.
- The flow of control of the software is illustrated by a set of directed arcs between the bubbles.
- Each arc is labeled by the action that occurs.
- Transitions from one bubble to another depend only on the state from which it emanates and the label on the arc.
- The process has one or more preferred states in which it would ideally terminate.

TABLE 3.6 Examples of States in a Chemical Process

1. Vat empty
2. Base chemical placed in vat
3. Temperature of vat lowered by refrigeration
4. Reagent added
5. Reaction occurs
6. Temperature of vat raised by heating
7. Reaction complete (pressure in vat is normal)
8. Compound in vat poured into molds
9. Failure; chemical plant must be evacuated

We can use these informal rules to develop a state diagram for the chemical process described in this section. The diagram is given in Figure 3.9.

The movement from state to state is controlled completely by the transition function. Since the transition function is mathematically precise, both the allowable states and the transitions must be completely specified.

The rules describing the creation of state diagrams can be mathematically formulated in a precise definition. Formally, a finite state machine is a quintuple consisting of the following:

- A set of allowable symbols for the inputs
- A nonempty, finite set of “states”
- A single distinguished state called the “start state”
- A (possibly empty) set of states called “final states”
- A transition function that takes as input a state and an allowable symbol and produces as output a single state

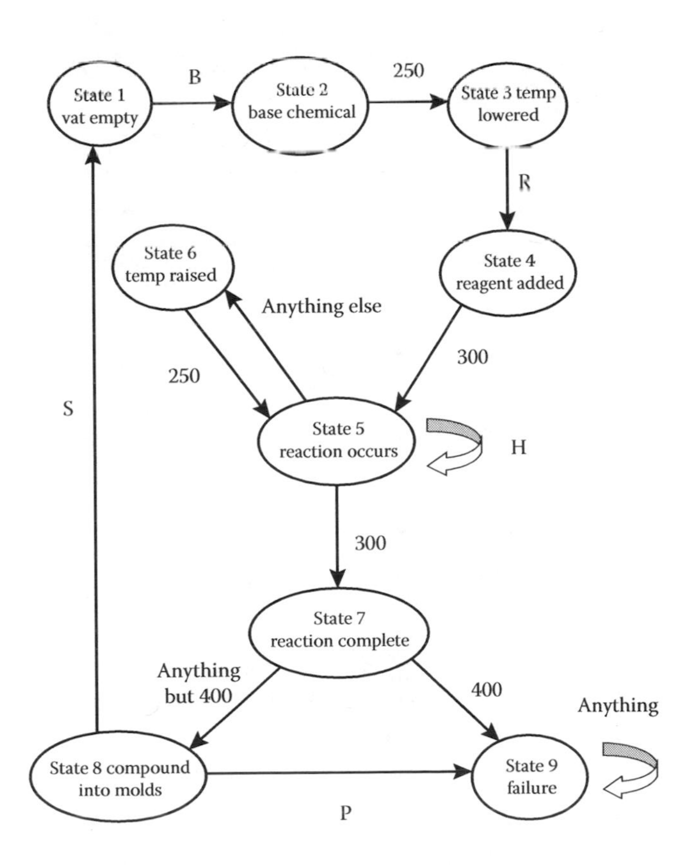

FIGURE 3.9 A state diagram for a simple chemical reaction.

For our example, the states are those shown in Table 3.6. The allowable inputs and their associated meanings might be the following:

B (indicating base chemical)

R (reagent)

S (solvent)

250 (low temperature limit)

300 (high temperature limit)

400 (temperature failure)

H (heat)

C (cool)

M (ready for molds)

The transition function for the diagram in Figure 3.9 is given in Table 3.7, and the decision table for the state diagram is shown in Table 3.8.

There is an alternative representation that can be used to describe this system: decision tables. A decision table is essentially a description of the choices made at certain states. A decision table for this system might be something like the illustration given in Table 3.8.

Decision tables have one advantage over the state diagram approach: they can indicate the influence of a range of values, rather than single inputs. Even though we do not illustrate it here in detail, you should be aware that decision tables can allow the associated state changes to reflect several ranges of inputs.

TABLE 3.7 State Transition Function for the State Diagram Shown in Figure 3.9

Current State	Input	New State
1	B	2
2	250	3
3	R	4
4	300	5
5	H	5
5	300	7
5	anything else	6
6	250	5
7	anything but 400	8
7	400	9
8	P	9
8	S	1
9	Anything	9

TABLE 3.8 Decision Table for the State Diagram Shown in Figure 3.9

Current State	Input	New State
1	B	2
2	if temp < = 250	3
3	R	4
4	if temp > = 300	5
5	H	5
5	if temp > = 300	7
5	if temp > = 400	9
5	anything else	6
6	if temp < = 250	5
7	anything but temp > = 400	8
7	if temp > = 400	9
8	P	9
8	S	1
9	Anything	9

3.10 SPECIFYING REQUIREMENTS USING PETRI NETS

Many software systems must handle concurrent events, with or without synchronization. Several approaches have been developed to specify how concurrency is to be represented, which processes must be synchronized, and which processes may execute concurrently, without synchronization. One such technique is the Petri net (Petri, 1975).

A Petri net is a graph in which the nodes are called "places" and the arcs are called "transitions." Each place in the graph is allowed to contain 0 or more "tokens," which indicate that a computation has reached the place. A transition is said to "be enabled," or "to fire," if there is at least one token in each of its input places. When a transition fires, one token is removed from each of the input places, and new tokens are placed in each of the output places. There is no requirement that the number of input tokens be equal to the number of output tokens. This approach enforces software systems adhering to the "happens before relationship."

Examples of Petri nets are given in Figures 3.10 through 3.12. Note that we have used a line between each pair of places to indicate transition. When a transition fires, tokens are put in each of its output places, as illustrated in Figure 3.12.

There are two advantages to the use of a Petri net to describe the operation of concurrent software: it allows a graphical representation and is automatable, so that consistency checks can be made. For noncurrent software that executes sequentially in a single process, Petri nets have essentially no advantage over flowcharts in their representative power.

3.11 ETHICAL ISSUES

There is always a lot of pressure during the requirements portion of the software development process. Every software project manager wants his or her project to produce a high-quality system that is completed on time and within budget. Every commonly used approach to software development depends on obtaining a number of correct requirements

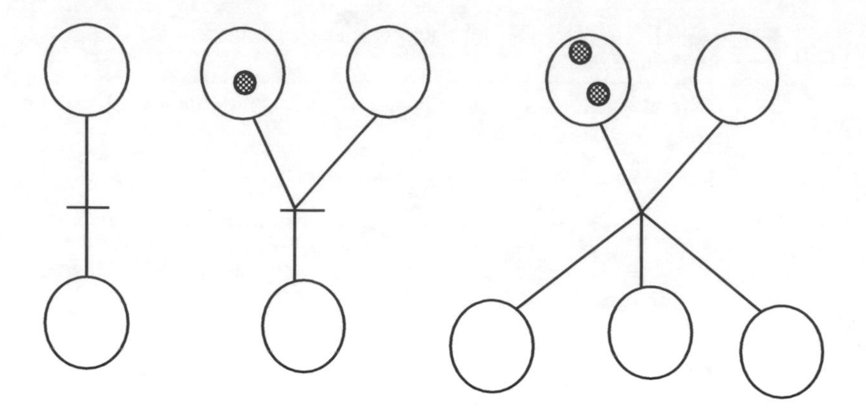

FIGURE 3.10 Some Petri nets with none of their transitions enabled.

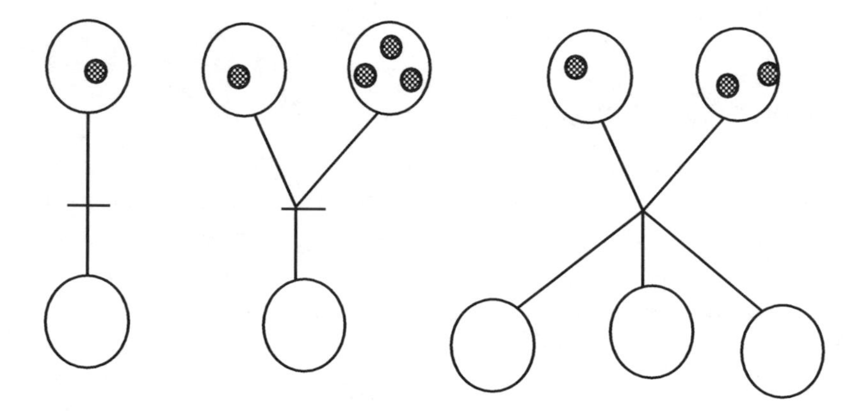

FIGURE 3.11 Some Petri nets with all their transitions enabled.

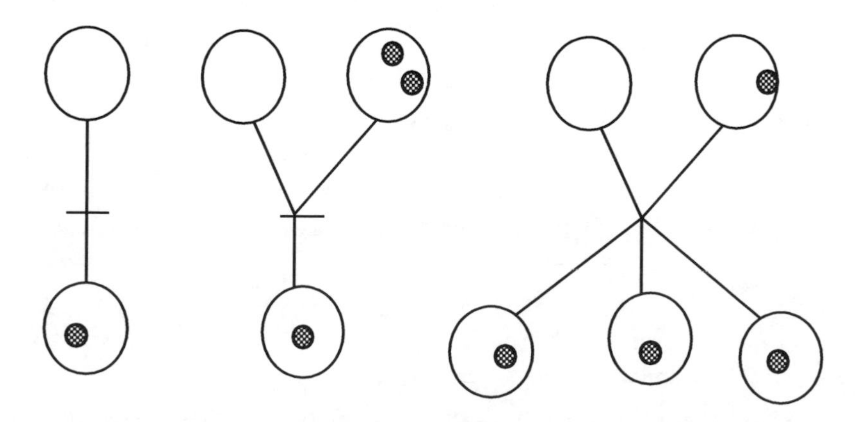

FIGURE 3.12 The Petri nets in Figure 3.11 after transitions are fired.

that comprise at least a relatively large subset of the final requirements. Without this, the software development is certain to be inefficient at best.

What does this mean for the requirements-gathering process? It is tempting to rush through requirements (and the later design process) in order to begin coding the software. In iterative approaches, this can be acceptable, provided that few of the requirements are actually incorrect. It is less satisfactory for software development that is based on the classical waterfall method. In short, incomplete requirements are acceptable as long as there is a general understanding that the requirements are incomplete, and that they can be made relatively complete with reasonable expenditures of effort.

What is clearly not acceptable is presenting a set of requirements as being correct when they are known to be incorrect. The incorrectness can arise from any of the following:

- Some computations are theoretically impossible.
- Some interactions with different subsystems are inconsistent within the software itself.
- Some interactions are inconsistent with software systems with which the software is intended to be interoperable.
- There may be impossible performance demands on the software's execution time or space requirements.

These types of errors in system requirements are a serious matter. Consider the problem of determining a user's password from its encrypted version. On the Linux and UNIX operating systems, only an encrypted version of the user's password is stored in a password file. The algorithm for encrypting the password uses what are called "one-way functions." Such functions do not have inverses and, therefore, the password cannot be recovered from the encrypted version. When a user attempts to log on, he or she is prompted for a password. The purported password is run through the one-way function for encryption and the result is matched against the previously encrypted version in the password file. If the two encrypted password versions agree, the user is allowed to log on. Otherwise, the login fails and the potential user is notified of that failure. On modern Linux and UNIX systems, the password file is not readable by ordinary users and, thus, the passwords are theoretically completely secure. Clearly, a requirement to recover a user's password from the encrypted version in a Linux or UNIX password file is theoretically impossible.

(Technically, the conclusion of the previous paragraph is not correct. A user with essentially unlimited time and unrestricted access could try all one character passwords, then all two character passwords, and so on until he or she either discovered the password or died of old age. A word to the wise: use long passwords with embedded non-alphabetical characters and few recognizable names, dates, or initials as part of your password.)

As another example, consider the problem of developing software for an emergency telephone system that will be used to track the phone numbers and addresses when emergency calls are made. The 911 telephone system in the United States is a typical example. In every area code, there is a maximum of 10^7 possible seven-digit telephone numbers. It is easy to trace the signal of a caller to the emergency telephone service. Indeed, such calls are never terminated, even if the calling party hangs up or is disconnected.

Determining the location of the telephone is another matter. Here is a description of the process for landlines that I encountered when reactivating a landline to a condominium in a resort community that had no obvious street addresses. Matching an address to a telephone number is effectively a sequential search, and this is not feasible if the system is to respond within proper time limits. Thus the matching of telephone numbers to addresses is generally done one time, before 911 emergency telephone service is installed. In fact, this information has been available for many years in what is known as a "cross directory." The preexistence of such a cross directory, or creation of one if it does not exist already, should be part of the requirements for an emergency telephone system. Otherwise, the requirements are not feasible because of the time needed to search the pairs of telephone numbers and addresses.

For cell phones, smart or otherwise, the position is usually found by triangulation, somewhat similar to the way that a GPS functions.

Preventing internal software incompatibility is the responsibility of the project team, and no incompatibility problems should be set by the requirements team. An example of this type of problem is requiring two different components of the system to write to memory buffers at the same location without control. Using the address of the start of a printer's buffer queue for storage of intermediate results in a desktop calculator program on a personal computer is an example of this problem.

External compatibility problems with interoperable software are much harder to detect. The user of COTS software often has to know which ports are being written to or which names are used for binding socket descriptors to locations. A typical example of this unwanted interference between software subsystems might be an existing system that runs on a Linux version and assumed that all relevant files were on the /usr file system. This can cause problems for another software application that used some of the same files and expected that the common files would be on the /var or /usr2 file systems.

A recent examination of the computer listings in the help wanted section of the *Washington Post* showed several jobs in the area of COTS. Industry is beginning to recognize the importance of avoiding such conflicts. Of course, an in-depth knowledge of both the COTS product and the software's application domain are necessary before one can predict with any confidence that no such conflicts occur. Clearly there are ethical considerations in assessing the appropriateness of COTS products as solutions to a system's requirements.

The IEEE Code of Ethics, IEEE Policy Number 7.8, is very informative in this regard. It is posted at the website http://www.ieee.org/about/corporate/governance/p7-8.html. For completeness, this code is listed next. Note the guidelines for taking responsibility when making technical decisions described in item 1 and the need for constant retraining described in item 6.

We, the members of the IEEE, in recognition of the importance of technologies in affecting the quality of life throughout the world, and in accepting a personal obligation to our profession, its members and the communities we serve, do hereby commit ourselves to the highest ethical and professional conduct and agree:

1. to accept responsibility in making engineering decisions consistent with the safety, health, and welfare of the public, and to disclose promptly factors that might endanger the public or the environment;
2. to avoid real or perceived conflicts of interest whenever possible, and to disclose them to affected parties when they exist;
3. to be honest and realistic in stating claims or estimates based on available data;
4. to reject bribery in all its forms;
5. to improve the understanding of technology, its appropriate application, and potential consequences;
6. to maintain and improve our technical competence and to undertake technological tasks for others only if qualified by training or experience, or after full disclosure of pertinent limitations;
7. to seek, accept, and offer honest criticism of technical work, to acknowledge and correct errors, and to credit promptly the contributions of others;
8. to treat fairly all persons regardless of such factors as race, religion, gender, disability, age, or national origin;
9. to avoid injuring others, their property, reputation, or employment by false or malicious action;
10. to assist colleagues and co-workers in their professional development and to support them in following this code of ethics.

The rest of this section is the author's personal opinion of the ethical issues involved with the software requirements process.

What should you do if you are involved with the requirements-gathering team and the requirements include one of the potentially fatal flaws? The first step is to check your analysis and make sure that you have not made a mistake. Do not rely too much on hazy recollections of meetings or on courses dimly recalled from college. Being human, students and professors make mistakes. (So, too, do authors of software engineering books, but at least there is the possibility of errors being caught by one or more technical reviewers, as well as by an editor. Authors of books are human, as are some editors and reviewers.)

If you have not made a mistake, consider discussing your concerns informally with a trusted coworker. Be careful about this, because you do not want your ideas presented to someone else as their own. Keep your own detailed notes about the perceived problem.

Perhaps a more direct approach is to make an appointment with your manager to explain the problem and your analysis. Remember that managers hate unpleasant surprises. Do not spring your objection on him or her during the middle of a meeting or during a review with the client or your manager's upper level management.

What should you do if your manager does not seem to be willing to accept your analysis? Go through the technical reasons with him or her again, allowing for the possibility that you were wrong or that you did not explain it well. Pursue the point, perhaps in future public or private meetings, until one of three things happens:

1. The manager does not agree with your analysis.
2. The manager agrees with your analysis and will ask the requirements team to fix the problem.
3. The manager seems to be ignoring the problem, allowing a potentially serious problem to occur.

In the first scenario, you have to consider the possibility that your analysis is wrong, even though you cannot see why. You need to keep a good relationship with this manager, even though he or she probably has some reservations about your judgment. Keep your technical skills up to date and begin to consider the possibility of taking another job.

In the second scenario, continue to be prepared. A bonus or promotion may be in your future.

In the third scenario, there is a serious ethical dilemma. In an operational environment in which human lives are at stake, such as in a computerized monitoring system in a hospital, control of a railroad, or almost anything in a nuclear power plant, there really is no choice: you must take all necessary actions to prevent what could be a disaster.

The choice is almost as clear if a major financial disaster would occur as a result of using software that is implemented to meet these erroneous requirements. You cannot argue that affecting an unknown person or corporation financially is not sufficient grounds for not adhering to the highest level of personal professional conduct.

In essence, the only way to justify taking little or no further action is to claim that the project is a throwaway research prototype with no possibility of becoming operational. If this is not the case, then you must do what Watts Humphrey (1989) recommends in his classic book *Managing the Software Process* when an experienced project manager is asked to complete a project with insufficient resources or within too short a time period: resign.

Of course, you must do more. The customer must be made aware of the problem. Unfortunately, whistle-blowers rarely have successful, remunerative careers. Be sure that you are right before taking any drastic action.

Ethical problems will be discussed further in Chapter 6 on software testing.

3.12 SOME METRICS FOR REQUIREMENTS

Many organizations attempt to predict project cost at the requirements phase. This means that they attempt to measure the size of the system that is to be produced. One way to estimate the size of (and resources needed to create) a software system is to determine the size of the system's requirements. There are several commonly used approaches to measuring the size of a set of requirements:

- Count the number of distinct requirements. This method is probably too simplistic, since it does not consider system complexity. In addition, this metric is influenced by the separation of multiple clauses in a single requirements sentence into separate sentences, each representing an individual requirement.
- Compute the sum over all requirements of the "functionality" expressed by each requirement. This approach requires describing the functionality that is associated with each requirement. The functionality is matched to a scale with a set of weighting factors.
- Match the requirements to those of existing systems that are to be reused. In this method, only new software is counted in any measurement of the size of a system's requirements.
- Compute the "size" of the interfaces to other software systems to which the desired software will be interoperable.

We will now discuss each of these measurement methods in turn.

Computing the number of distinct requirements is the simplest method of estimating the eventual size of the system, since it requires no special analysis of the individual requirements. The number can be computed by examining the number of rows in the requirements traceability matrix.

Unfortunately, simply counting the number of requirements does not take into account the difficulty of implementing different individual requirements. Thus, this approach does not distinguish the complexity of implementing a requirement such as

> The software will begin execution when the user enters the single UNIX command "go."

Or from the complexity of implementing the requirement:

> The system will be capable of processing 10,000 distinct transactions per minute, with no transaction suspended in a waiting queue for more than one second before it is processed.

In spite of the obvious imprecision, the number of requirements does provide at least a hint as to the amount of difficulty of implementing an entire software system. We can get more accurate information by examining the requirements in detail. The most common technique of estimating the functionality of a set of requirements is called "function point analysis." The term was coined by Albrecht, who first developed the concept at IBM in 1979 (Albrecht, 1979). Jones (1993) describes the use of function points in project estimation.

The function point estimation technique is based on an assessment of the functionality needed for each requirement based on the following:

- The number and type of user interactions
- The number and type of interfaces to internal files
- The number and type of interfaces to external files
- The general perception of the difficulty of programming in the project's application domain

We will now describe a formal procedure for computing function points. All function point metrics are based on the number and types of interfaces in the program. Interfaces can be either internal or external. The total number and sizes of interfaces are used as the basis for the collection of function points. Once this total is obtained, it is modified by several weighting factors, some of which are subjective and some of which are objective. The subjective factors include an assessment of the complexity of the interfaces and the overall complexity of the system.

More precisely, weights are assigned according to the rules that are given in Table 3.9. After the weights are assigned, a complexity factor is determined in the range from 0 (representing no influence) to 5 (representing strong influence) for each of the factors given in Table 3.10.

For each software system to be analyzed, the sum of the "complexity adjustment values" f_i is computed using the values determined from Table 3.10. The final result is the total number of function points for the system according to the equation:

$$\mathbf{FP = count\_total * (0.65 + sumf_i/100)}$$

As you can see, counting the number of function points in a proposed software system is somewhat subjective in the sense that there is no well-defined definition of terms such as "average," "simple," or "complex." Other terms in the "complexity analysis" list clearly indicate a lack of precision in their definition. Therefore, it is highly probable that different people will produce very different estimates of the number of function points for the same software system. There are two general approaches to the ambiguity that is associated with the function point analysis process.

TABLE 3.9 Use of Weights for Function Point Analysis

	Simple	Average	Complex
External input or files	3	4	6
External outputs or reports	4	5	7
External queries or responses	3	4	6
External interfaces to the systems	7	10	15
Internal files	5	7	10

TABLE 3.10 Use of Complexity Factors and Weights for Function Point Analysis

Complexity Factor	**Weight (0–5)**
Reliable system backup and recovery	
Data communications	
Distributed functions	
Performance	
Heavily used system	
Online data entry	
Ease of operation	
Online updates	
Complex interfaces	
Complex processing	
Reusability	
Ease of installation	
Operation at multiple sites	
Modifiability	

The first approach is to ignore minor differences, working under the assumption that many software measurements have errors and that minor differences can be ignored because they are not statistically significant. This makes it difficult to determine the cause of any variability between function point data that is gathered from several different software development projects. It might not be clear if any correlation between the number of function points and the amount of effort expended (as measured in person-months) is due to actual relationships or just to differences in counting techniques.

The second approach is to attempt to make the collection of function point data more rigorous. This is the approach taken by the various function point users' groups.

We note that there have been several attempts at establishing standards for function point analysis (see Leach, 1998c for a detailed example). Rather than describe these standards in detail, consult the electronic mailing list for the function point users group at a site hosted by the Computer Research Institute of Montreal: function.point.list-request@crim.ca.

Regardless of the approach used to standardize the collection of function points, the objectively determined weights are obtained by determining if the program will be interactive or will execute without user interaction. The effect of the program executing concurrently with other applications is also included in these weights.

You should be aware that the function point metric is not directly associated with programming effort or system size. An organization using the function point metrics must define some of the subjective terms carefully, collect data, compare the data with other metrics that are directly computed using correlation and other methods, and then calibrate any models that use this metric. Only then can this metric be used with confidence in a particular environment.

Because of the inherent subjectivity of function point analysis, there are several distinct standards for it. Unfortunately, there is no single international standard that can be used to determine some of the definitions needed for consistent application of the function point method. Thus, there is little reason to expect exact correlations of your organization's experience using function points for cost and schedule estimation with another organization's use of function point data.

3.13 THE REQUIREMENTS REVIEW

Once the requirements team has determined a set of requirements for the system, it is generally necessary to have a requirements review. The purpose of this review is to allow the customer to react to the requirements produced by the requirements team. If no customer is known at the time of the review, then members of the marketing staff and other personnel are often used instead.

A requirements review is the one thing that is common to all software development environments that have more than one software engineer or programmer. A requirements review is a check to ensure that the requirements as set by the requirements team actually meet the needs of the organization for which the software is intended. The basic principle is that it is pointless to proceed if the software to be developed does not meet the needs of its users. As such, a requirements review provides a much-needed "sanity check."

There may be several different requirements reviews, depending on the organization's software development approach. In the classical waterfall software development process, there usually will be at least two scheduled requirements reviews: a preliminary and a functional review. An iterative software development process may use a single requirements review if the resulting requirements are considered to be satisfactory and complete. It is more likely, however, that there will be many reviews in organizations that use an iterative approach. Note that reviews in iterative software development processes often include designs, source code, and requirements. We will discuss some issues in design reviews as they can be used for an agile software development process at the end of this section.

We will now consider requirements reviews in more detail. Depending on the nature of the organization developing the software, a requirements review can range from a relatively informal presentation where a few people talk while sitting around a table, to a highly scripted, well-rehearsed formal performance with overhead transparencies, slides, multimedia computer-generated presentations, and similar levels of communications aids. In general, the formality and complexity of the requirements review increases as least as fast as the complexity of the software being developed and the size of the organization.

The requirements review can be a fundamental milestone in an organization that uses a process based on the classical waterfall model of software development. As such, its importance is obvious. If the organization uses an iterative software development process, the requirements review might appear to be slightly less important, since a deficiency in the software requirements will be corrected in the next iteration of the software. In fact, one might almost think that requirements reviews are not important at all, since any deficiencies can be repaired in a later iteration.

However, this is not the case. A good requirements review indicates if the requirements are close to being satisfactory, if they are correct but incomplete, or if they are incorrect. (In actuality, all three of these things frequently occur simultaneously in the same software project. However, it is conceptually simpler for our discussion to consider requirements where only one of these things happens at a time.) Even if the software's requirements are expected to undergo one or more additional iterations, the process is certain to be much more efficient if any incorrect requirements are detected early, before too much effort goes into developing software based on these erroneous requirements.

Notice what this view of requirements suggests about software testing. In an academic setting, a student can argue with an instructor that his or her program is correct because it met the requirements as set by the instructor. Often such conversations are exacerbated by an ambiguity in the project specifications as set by the instructor. Only those test cases that are consistent with the instructor's specifications are considered to be fair.

Unfortunately, obtaining good requirements is more essential than this purely academic view of software development would indicate. If the requirements are not satisfactory because of being incomplete or not being what the customer really wanted, then the eventual agreement of the software with the (incorrect) initial requirements does not mean that the software will be correct or that it will be a success in the marketplace. Think of building a thirty-story building on top of a set of thin wooden boards that rest directly on top of a sandy beach. The rest of the building may match the wooden boards precisely, but the entire structure is inherently unstable.

Clearly, requirements reviews are important, even if they are expected to be repeated as part of an iterative software development process. Even in a relatively noniterative process such as the classical waterfall process, there may be multiple requirements reviews. The first and last of such reviews are often called preliminary and critical requirements reviews, respectively.

There is another way for you to understand the importance of a professionally run requirements review. Let us begin by estimating the cost of holding the requirements review meeting itself. The typical work year consists of 52 weeks of 40 hours each, or approximately 2,000 hours. The typical "loaded cost" of a software engineer includes the following costs:

- His or her yearly salary
- Fringe benefits (approximately 25 to 30 percent)
- The average prorated cost of technical writing, secretarial, custodial, and human resources (personnel) support staff
- The average prorated cost of computer maintenance, network manager, and other technical support staff
- The prorated cost of equipment, utilities, and other physical costs

This is frequently considered to be well in excess of $100,000 per year, or roughly $50 per hour, even for very junior personnel. Therefore, a four-hour meeting with five people presents costs of least a thousand dollars, and probably a whole lot more, assuming that there was no preparation time needed for the meeting, no coordination of schedules was necessary to schedule the meeting, and that no time was needed to travel to the meeting.

A more realistic estimate for a system of the same size is eight hours of meetings, ten people attending the meeting, with two days of preparation for the meeting. Typically, at least three people are part of the requirements presentation and, thus, they are involved with the preparation time including rehearsals; preparation of the documents, slides, transparencies, or computer-generated multimedia presentation; and proofreading the documents to be handed out at the meeting. Another fifteen minutes should be allowed for travel each way, even if the meetings are scheduled in the conference room down the hall. Because of limits to the attention span of human beings, such meetings are often spread out over several days, due to the number of breaks that are usually necessary. This conservative estimate of cost is more than $6,500.

The expense is much larger in many environments. Note that there is no special expense listed separately for food, entertainment of customers, or travel to a customer's site. Clearly, reviews must be taken seriously. Wasting the customer's time or that of your upper-level management is not recommended.

Now that we have established the importance of requirements reviews, it is time to describe how they should be conducted. We will describe two types of requirements reviews: (1) a typical relatively formal review process performed in a meeting and (2) an inspection of the requirements, usually done by one or more individuals but without formal presentations. Use the discussion here as a guide to how a requirements review should be conducted within the artificial confines of a software engineering class. In an actual software requirements review in industry or government, there may be company policies that supersede the suggestions here. If so, use the company policy.

We first consider a requirements review that is performed during a formal meeting. The most effective reviews have a person responsible for taking notes and providing a transcript of the meeting. The purpose of the note taker is to provide a written document that indicates any changes made to the requirements as part of the meeting. Such written documentation is provided to all meeting participants after the meeting is over. This allows the organization to have a complete record of all major actions on the project. Reviews are often recorded using audio recorders, with some organizations also using video or electronic storage of data passed by means of conferencing systems.

Generally, one person is responsible for the organization of the presentation. The person may be called the requirements team manager. In some organizations, he or she may also be the project manager. In other organizations, he or she may have no other project administrative responsibility. In any case, the person will introduce the members of the requirements presentation team and coordinate the rest of the discussion.

A checklist for a requirements review meeting is given in Table 3.11.

TABLE 3.11 Checklist for a Requirements Review Meeting

Presentation rehearsed.
Time of presentation within allotted limits.
Sufficient time is available for questions.
People who can answer technical questions are in attendance.
Slides, transparencies, and multimedia materials checked for accuracy.
Slides and multimedia materials checked for spelling.
Paper or digital copies of slides and multimedia materials are available to all participants.
Meeting room has all necessary equipment: microphones, overhead projector, slide projector, videodisk/videotape player and monitor, tablet/projector hookup, computer-monitor hookup, etc. Networks and connectivity needed to control presentation from devices such as smartphones, including Bluetooth, infrared, and ad hoc networks, are available.
An attendance list with names, addresses, phone numbers, and affiliations is passed around, and all attendees sign it.

The details given at the requirements review will depend on the nature of the review. At a preliminary requirements review, the requirements may be given at a very high level. This is especially true for software development that follows the classical waterfall model. Most other requirements reviews are much more detailed.

The organization of a requirements review presentation is generally guided by the organization of the software. A typical review presentation has the following topics:

- Introduction
- System Overview
- Subsystem Requirements
- Subsystem 1 Requirements
- Subsystem 2 Requirements
- Subsystem 3 Requirements
- Subsystem 4 Requirements
- Subsystem 1 Interfaces
- Subsystem 2 Interfaces
- Subsystem 3 Interfaces
- Subsystem 4 Interfaces
- Proposed Resources
- Proposed Schedule
- Potential Problems
- Summary

Sufficient time should be allotted for questions during each major section of the requirements review. The goal is to have a clear, comprehensive document.

The next type of requirements review to consider may be characterized as an inspection. The typical set of (oversimplified) inspection guidelines is described next.

The system requirements must answer the following:

- Which hardware platform will be used and which operating system versions will be supported?
- Will the different versions of software run on different computers and operating systems? If so, which ones?
- Will the file formats be consistent across operating systems?
- Will the software have a graphical user interface?
- Will the software's graphical user interface be consistent with that of the operating system and with other applications running in the same environment?
- Will there be multiple implementations, such as a minimal size for laptops with small amounts of memory and disk space, and larger implementations with more support files and features, to be used in computers with fewer limitations on memory and size?
- What are the minimal system requirements, in terms of memory size and available hard disk space?
- Will the program's existence in memory cause problems for other applications that are also running on the computer?
- Are there any other software packages, such as word processors, spreadsheet programs, database managers, or drawing tools, with which the program must be able to share data?
- Will online help be provided within the software?
- How will the software be delivered?
- Will the software be provided in compressed or uncompressed form?
- Will installation and setup software be provided?
- Will training be required?
- Are there any real-time constraints, such as completion of certain operations within specified limits?

The requirements-gathering process has two complementary principles (Berry, 1995):

1. The requirements for the system must be modified and refined until the system can be developed by software designers and coders who are familiar with the fundamental algorithms of the application domain but who are ignorant about the requirements

engineering process. The eventual requirements will be so complete and unambiguous that the design will be clear and the coding can be done easily by implementing standard, well-known algorithms.

2. The requirements for the system must be modified and refined until the system requirements can be understood by requirements engineers who are knowledgeable about requirements engineering but who know nothing about the application domain. The eventual requirements will be so complete and unambiguous that the requirements engineer will understand them completely, even without knowledge of the application domain.

The requirements for the system must be iterated until each of these two principles is true for each of the requirements.

The following tasks must be completed in order to develop specific verifiable requirements:

1. Number each of the requirements.
2. For each requirement, develop a specific, unambiguous test case that will determine if the requirement is met.
3. If you are unable to develop a specific, unambiguous test case that will determine if the requirement is met, either rewrite the requirement or break it down into smaller requirements until you can develop a specific, unambiguous test case that will determine if the new requirement is met. If new requirements are created as a result of breaking down existing requirements, number them using dotted notations showing the origin of the requirements. (Thus, new requirements obtained by breaking down requirement 3 will be numbered 3.1, 3.2, and so on.)
4. Repeat the process in steps 1 through 3 until you can develop a specific, unambiguous test case that will determine if the requirement is met.

Once you have developed a specific, unambiguous test case that will determine if each system requirement is met, it is time to determine the feasibility of implementing the requirements.

For each requirement developed for the system, perform the following tasks:

1. Develop a high-level design that will be used to implement the specific requirement.
2. Iterate that design until the design is sufficiently detailed to bring to a programmer.
3. If any requirement cannot be designed so that it can be implemented as code, mark the requirement as ambiguous or unimplementable and then review the requirements. (Some of the requirements, although individually testable, may be inconsistent with other requirements.)
4. If any of the requirements are inconsistent, revise the set of requirements to remove the inconsistency.

Once requirements have been developed, the customer must be given an opportunity to interact with them. Determine if you will present the system to the customer as a series of prototypes or as a single delivered system.

If the customer will be given a series of prototypes, begin development of the initial prototype. In this case, the customer is allowed to interact with the development process and to make changes up to the last possible minute. If the customer will only be given a single-system delivery, iterate with the customer to determine if the project team is building the right system.

You can see from the aforementioned guidelines that requirements reviews are serious affairs and demand considerable resources from an organization. Experience indicates that they are worth it.

It is clear that development of requirements is a major part of software engineering. An assessment by the quality assurance (QA) team is definitely appropriate for any requirements document produced. The QA assessment can take place before, during, or after a requirements review, depending on the organizational structure.

As promised earlier in this section, we now discuss some issues in requirements reviews as they can be used for an agile software development process. A quick examination of the online literature on requirements reviews for agile processes suggests that, especially in the relatively early years of the adoption of agile processes, say, before 2008, there was considerable difficulty in getting design reviews to work well in agile environments. The consensus appears that the difficulty was the tendency to try to force all features of, say, reviews in the classical waterfall software life cycle into an agile review (Gothelf, 2012).

The common thread from successful agile projects is that reviews should be limited in frequency and duration, and should include designers and customer representatives who are highly knowledgeable experts in the particular application domain. The requirements review has to result in testable requirements, because the agile team will be under severe pressure to complete their projects within strict deadlines. You will see that these issues were an integral part of the self-organizing teams discussed in the case study in Section 3.15.

3.14 A MANAGEMENT VIEWPOINT

In many very large software projects, the overall project manager may have a separate manager for the requirements, design coding, and testing teams. It is perhaps more common for the project manager to provide guidance to each of these teams, while remaining in overall control. For simplicity, we will have to ignore the extra level of management and assume that one manager is responsible for the entire project. In this section we will describe what this manager expects from the requirements phase. We describe both the product that he or she expects to be delivered and the process by which it is created.

The goal of the requirements process is to obtain a clear, unambiguous, consistent, and complete set of requirements. The manager wants these in a timely manner because he or she has deadlines that are usually imposed by higher authorities within the organization. He or she will expect frequent updates as to the requirements-gathering process. These updates may be informal. More likely, there will be weekly or monthly written status reports, perhaps with requirements reviews.

At most meetings, a manager wishes to show off the progress of his or her project. Often the most important thing is to avoid giving the manager an unpleasant surprise. Generally, a manager prefers to receive a report that is incomplete (and which he or she knows is incomplete before the meeting begins) than one that is inconsistent. The manager expects to see a lot of "action items" to perform specific tasks with roughly accurate estimates of when the tasks will be completed.

The manager will expect his or her people to attend briefings and reviews in a manner that is consistent with the organization's culture and the people present. In an internal meeting of a casual organization, there may not be any special dress code. In a meeting with customers or in a more formal environment, coats and ties may be expected of men, with equivalently formal work clothes for women. The rule of thumb is do not embarrass the manager, the customer, or the organization. When in doubt, ask.

There will always be a final presentation of the requirements. The manager expects the presentation to be carefully rehearsed, with media aids that include overhead transparencies, slides, videotapes, or computer-generated slide shows, as appropriate. Sufficient copies of all relevant documentation should be handed out to all participants. Copies of slides and overhead transparencies for all attendees are a must.

The manager also expects that all documents intended for internal use be given to him or her at the appropriate time and place. Generally the customer does not need to see the requirements traceability matrix document. (This is not correct if the process must be certified by QA, as might be the case for medical software products needing approval by the Food and Drug Administration.) Internal memorandums should also be kept private.

The manager expects to have cost and staffing estimates for the system. If such estimates already exist, then the manager needs to understand any changes in the estimates so that he or she can make a case for more resources for the proposal.

The manager also wants to be knowledgeable about any unanticipated problems and how they were resolved. The basic reason for this information is that the manager wants to be prepared for similar situations in the future. Managers hate unpleasant surprises, especially at the end of a software life cycle phase activity.

The manager also must be certain that all essential project documents are shared in an appropriate project archive. Hopefully the organization has prepared against computer crashes and other disasters by having multiple copies of all essential documents stored in separate locations safe from fire, flood, and theft.

Essentially, the manager wants one of two states to occur, depending on the organization's software development practices:

- If the organization uses the standard waterfall approach, then the requirements should be complete, consistent, and free of any ambiguities. The software design process should be ready to begin.
- If the organization uses an iterative approach to software development, then the next iteration should be able to continue without delay. This means that the requirements should be consistent and free of any ambiguities. Any incompleteness in the

requirements should be resolvable in the next iteration of the software's development. The requirements should be more complete than they were at the end of the previous iteration of development, if there was one.

A software project manager wants his or her project to produce software to be developed efficiently and the final product to be of high quality. Some inefficiency or detours during the setting of requirements and the system's design are probably unavoidable. What a manager hates is unexpected surprises.

Some typical unpleasant surprises include:

- Incoherent requirements review presentations
- Missing documentation
- Obvious omissions
- Anger on the part of the customer because technical and other issues raised at an earlier meeting were ignored
- Missing logistical infrastructure, such as proper audiovisual equipment or a room that is much too small

The old adage is still valid: You only get one chance to make a first impression. Keep this in mind when planning a requirements review. Act as if both your future and your company's are at stake. They are.

Keep in mind, however, that requirements are likely to change, even in the classical waterfall process. A software manager expects changes in the system requirements and will have a systematic plan for determining which to treat first and which to defer for future action. Change management is a vital part of the software process.

The project management plans will have been developed earlier and will have included development of requirements and planning for the requirements review. Once the requirements have been relatively set, the project plan can be described in more detail. In general, a project manager will expect to have a relatively detailed project management plan in place after the requirements review is complete.

As with all phases of any software development life cycle, there must be a mechanism for configuration management in order to properly treat the changes that are bound to occur in any project. In some projects, if sophisticated configuration management tools are not available, it may be useful to have all changes listed in the requirements traceability matrix, with revisions listed by number and accessed via hyperlinks to appropriate files.

3.15 CASE STUDY OF A MANAGEMENT PERSPECTIVE ON REQUIREMENTS IN AGILE DEVELOPMENT

This section may be confusing to students studying software engineering encountering a discussion of agile development at this stage of their educational development. At this point in your career, you are well beyond the perhaps apocryphal description of the student who is able to "create 300-line programs from scratch." Many readers of this book have had substantial work experience with writing computer software as part of a full-time job in industry or government, and may be attending school on a part-time basis. Perhaps you have participated in a co-op program and gained valuable work experience. You may have worked in a campus computer laboratory, which can often be a wonderful learning experience. Even if you are a full-time student of "traditional college age," age eighteen to twenty-two, you are likely to have had a part-time job in the computer industry, or at least a summer job or internship. Even with all this possible experience, you are unlikely to have had any experience working as part of an agile development team on the particular part of an agile team working on requirements.

Here's why. A critical part of the agile development process is the ability to create systems without doing a large amount of coding. This forces the agile team to reuse already deployed software components with as much functionality as possible as part of the software that the team will build. This can only be done if the team is aware of the existing software components and how their functionalities match up with the functionality desired in the requirements of the end product. Frequently, the functionalities guide the general setting of requirements. In most situations, this awareness of what is available in the application domain is only obtained after considerable experience in that application domain.

It is the combination of this domain knowledge, with intelligence, technical knowledge and excellent interpersonal skills that is most highly prized by employers.

You are probably not at that stage of career development. In forty-one years of teaching, I only had one undergraduate student hired as a project manager immediately after her graduation at age twenty-two. She was obviously exceptional, even when compared to many, many talented graduates I have taught. In most cases, the detailed domain knowledge comes much, much later.

Here are some specifics about the requirements process as it was used in our case study of agile development.

Team members worked toward a broadly determined goal with a high degree of freedom. At the same time they each knew where the project had to go and thus had to be self-motivated and self-directed. A self-regulating process with as few obstacles as possible was essential. When conflicts arose, the team lead would work to resolve issues. In many cases, where resolution could not happen quickly, prototypes would help the process. Compared to previous projects, less time was spent discussing problems and more time doing actual work.

We note in passing that the effect of goals on the behavior of software teams where the goals are set by the teams, by management, or even if no goals are set at all, has been studied by Babar et al. (2004). This is, of course, related to the self-organizing factor (SOF) discussed in Chapter 2, Section 2.10.

Another key component was an environment that allowed for errors. In this way, ideas and small experiments could be carried out when team members could not resolve a conflict that would provide good data for the team to creatively resolve conflicts. This required a higher degree of trust in one's teammates to accomplish their share of the load. The collective inquiry process ran counter to the traditional internal and external competition modes of the past.

Work was not conducted from a detailed set of requirements or design as was the tradition for older projects. Rather, each member was driven by some broad guidelines and perhaps some fuzzy requirements. In this way, everyone spent the majority of his or her time doing work on the product rather than creating detailed documents. Because the fuzzy requirements did not force the "Not Invented Here Syndrome," more possibilities were opened to the team. The attitude projected was more of "I can't tell you what I want, but I'll let you know when you get there!"

This broad attitude encouraged creative grouping of functions in the different phases of a satellite mission and the application of nontraditional software to solve problems. For example, the team used LabView, a relatively inexpensive software package typically used to test engineering bench instruments controlling experiments, to service functions that would normally be performed by three distinct groups of people. Recall that we have used the common term COTS (commercial off-the-shelf) software for the use of such commercial products within applications.

I hope it is clear that this use of the relatively inexpensive LabView software within this application would have not been possible if the agile development team did not understand the software's capabilities and how these capabilities could be applied to nontraditional applications in this project. Keep in mind that the team would have to know precisely what kinds of data interfaces the LabView software had in order to integrate its input and output with other large-scale software components. This illustrates the need for members of agile development teams to be well versed in the application domain. This knowledge of the data interfaces would be a critical step in the management of the project. Such a description would be in some form of an ICD (interface control document).

It is natural to ask what the requirements looked like at this point in the agile development process. After all, the systems developed are highly likely to have a long lifetime of producing useful scientific data.

The starting point was a general description of what the software was likely to do. Many of the details were deliberately left fuzzy at the beginning, awaiting clarification during the agile development process.

The clarification process is unnerving, especially to senior-level managers who may have limited familiarity with agile processes. The functional capabilities of the software components become part of the requirements of the desired system! Here is how this works, at least in the case study discussed here.

LabView, which was originally designed for testing engineering bench instruments that ran experiments, could receive and output data at certain rates, and these rates of data transmission would be used as part of the requirements. A major part of requirements that were created in the traditional method of requirements development would be needed to provide information on the status of the spacecraft's health and safety, and these requirements could now be provided much more quickly.

Once the system is built, the requirements would be placed into an archive where they would be made available to others if and when future changes would be needed. (Recall

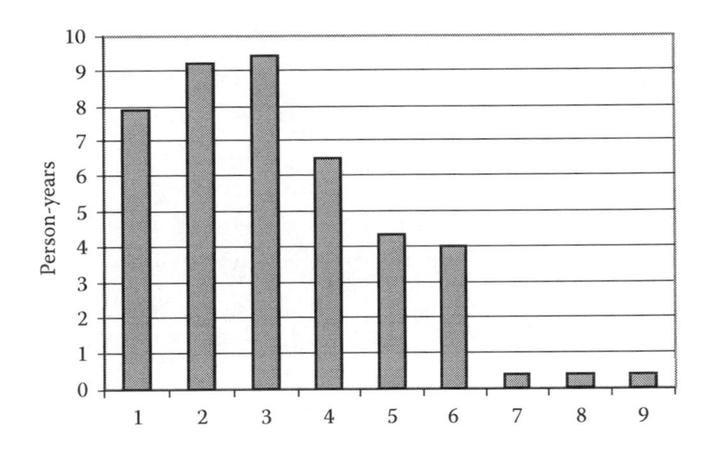

FIGURE 3.13 Decreasing time to gather requirements as a function of the team's increased experience in the use of agile methods.

that such changes are considered to be part of software maintenance, which is the primary topic discussed in Chapter 8.) It is highly unlikely that the same agile development team used to create the software would be used if these maintenance changes were necessary, so documentation of requirements (and design and so forth) is absolutely necessary.

Figure 3.13 shows the pattern of improvement of the requirements-gathering process for a series of software projects using agile methods. The later projects, largely developed after the team became more adept at employing agile development processes, are shown at the right-hand side of the graph.

You might have noticed that there is a potential divergence of interests that might occur in this case study. The LabView software is sold in many places and the relatively small number of seat licenses purchased by NASA is not likely to affect decisions by LabView's manufacturer to change the project greatly or even discontinue it. What happens to systems that use this software as a COTS product within a larger system? (We will discuss this issue in Chapter 8 in more detail when we describe the evolution of systems.)

An important recent paper by Jacobson et al. (2014) suggests that an approach named SEMAT (Software Engineering Methodology and Theory) can improve the requirements process by, among other things, detecting systemic problems and taking appropriate action. Checklist information can be found at http://www.ivarjacobson.com/resource.aspx?id=1225. Be sure to read the blogs, especially the one created on March 24, 2014.

3.16 THE MAJOR PROJECT: PROBLEM STATEMENT

So far, we have discussed the requirements process at a relatively abstract level. It is now time to illustrate the process by considering a concrete example of moderate size. As was indicated in both the Preface and Chapter 1, this example will be discussed throughout the remaining chapters of this book. The only thing unrealistic about the running example is that it is necessarily small enough in order to be discussed completely within the context of this book.

We now describe the setting of the problem to be solved. A major difficulty faced by software managers is the lack of good tools for measurement of software and for determining if an implementation of source code for a project is consistent with the original software design.

Imagine two software professionals meeting over an informal meal at a local restaurant. The two of them are old friends, with one of them being a high-level employee whose many responsibilities include managing the other. We will call them the Manager and the Software Engineer, respectively. The Software Engineer is also a manager, but only manages a small team. Being ambitious, the Software Engineer wishes to change this. Career advancement is a major goal of the Software Engineer.

The Manager is complaining about the difficulty in obtaining good information about the size of the projects that are under her direct or indirect control. The Manager also complains about having to monitor the organization's software development in three different languages: C, C++, and Java. The Manager also complains about the lack of software tools that enable maintainers of code to examine the program's control flow structure. In addition, she does not have any good way of tracking the software problems reported against systems at a lower level to the particular functions or source code files involved. There is a lot of data that has been collected about software errors, but the Manager just cannot use it efficiently.

The Software Engineer hears the problems of an old friend and would like to help. In addition, quickly providing a set of high-quality software tools can provide enormous visibility, aiding in career enhancement. The Software Engineer makes a mental note to consider this problem more fully. The conversation then turns to office gossip and politics.

While driving back to the office, the Software Engineer decides to work with a colleague to develop a set of requirements for what the Manager needs. The two of them will work together as a requirements team. They will produce a proposal to develop the software to meet the Manager's needs. If the proposal is accepted, resources will be allocated.

This scenario is not as far-fetched as it might appear. Many of the best software products had their inception at informal meetings such as the one described in this section. (Several agile software development projects get started this way.) Many other excellent ideas never became products because there was no follow-through. The Software Engineer is aware of this and is also aware of the many current demands on the time of the Manager, who will have the final say about the approval of the proposed project. The Software Engineer thus resolves to do the best job possible in developing the requirements, while making as few demands as possible on the time of the Manager.

3.17 THE MAJOR PROJECT: REQUIREMENTS ELICITATION

The next step in our discussion of requirements development for this example will take the form of an imaginary dialogue between the customer and the requirements team. The purpose of the dialogue is to illustrate the way that experienced requirements teams discern any omissions or hidden biases from the customer so that they can be used in the requirements as appropriate. Since not all of the dialogue will take place in the presence of the customer, we will use interludes to indicate the internal discussions that should be hidden away from the customer. We follow the approach of ignorance hiding that was discussed earlier in Section 3.4.1.

In our case, the requirements team is the Software Engineer and the unspecified colleague. To make the dialogue easier to understand, we will not distinguish which of them is speaking at any time. They meet in a small conference room with a large whiteboard, but no phones.

"I think we can produce something for the customer that can also help us move up within the company."
"What does the customer want?"
"I think there are real problems to solve. There is no way to figure out the size of software at present. Not being able to predict it, or even compute it for our systems, can get real expensive."
"What about our current techniques for counting code? We've been doing this for years. What's new?"
"Well, we now develop code in C, C++, and Java. Three different languages, three different approaches, three different counting standards."
"OK, let's write a new standard for code counting."
"We'd have to write lots of new utilities. If we have to do that, why not do it right?"
"OK, I'm convinced we need some changes. But what about the other stuff? What should we do there?"
"Well, I think she wants some kind of graphical tool that shows the program's structure. Maybe the call graph."
"Remind me. The call graph is the listing of which functions call other functions, right?"
"That's right. And if a function isn't in the call graph, then it's dead code and we can ignore it when we try to understand the code."
"Should we actually draw a graph, or just determine the functions that call other functions?"
"I don't know."
"What about different tools for different programming languages?"
"That might be a problem. I don't know about command-line arguments. Maybe a GUI should be used to avoid the command line problem."
"I don't know. It might be easier to use the tool if there were no GUI, if the tool could be used in batch mode."
"What about our existing data on software errors? Can we interface to them?"
"Who knows? Some of that data may not be in a consistent electronic form."
"We have a lot of questions to be answered. Let's meet Tuesday at 10."
"OK."

After thinking about the problem for a few days, they meet again. They have prepared for the meeting and have brought in both a statement of the problem and lists of potential requirements for the software solution to the problem. Because each of them is familiar with the company's business strategy, they have already made the decision to include the requirements that were given previously in Table 3.3.

The requirements team members provide each other with copies of their prepared materials and resume their dialogue.

"OK, we agree on the problem statement."

"Yes."

"What do we do about the multiple languages? Are there separate tools?"

"We want a common output so that only a small amount of new training is necessary for users."

"What about a common input?"

"Yes, as much as possible. Let's keep a GUI out of it until the last possible moment. Perhaps we should delay GUI introduction to the second generation of the tool."

"I don't agree. But let's keep the user interface simple."

"Let's do a demo of the simplest prototype."

"We'll handle one source code input file at a time."

"What about multiple languages used for source? That's the way we develop source code now, multiple languages for software systems. We use Motif and Builder Xcessory for UNIX software development, spreadsheet software for PC applications, and SQL or MySQL everywhere. Many of our systems get ported to other environments as part of the company's reuse program. We often have to interface software components written in different languages."

"Let's simplify. We'll only consider software subsystems written in a single language."

"Huh?"

"OK, that was vague. Just restrict the requirements to have only one source code language treated at a time. That is, the system must be invoked one time to handle C source code files, a second time for C++, and a third time for Java."

"Do we prompt the user or allow the type of input files to be specified as a command-line argument?"

"Having command-line arguments usually force coding in C or C++. Do we want to make this decision yet?"

"Probably not. Let's defer action on this item until the customer decides."

"OK."

"What about displaying the call graph?"

"That's a more fundamental problem. If we only analyze portions of a program at a time, say the portion written in C, then how do we treat calls to functions not included in the source code files we analyze?"

"It's the same problem if the entire program being analyzed was written in C, but we only analyzed some of the source files in the program at a time. We would simply flag any functions that were called, but for which the source code was not provided to be analyzed."

"I see. It's the same way we would treat any library functions."

"Do we want to use the IDE of our CASE tool to do the graphs?"

"Maybe. We already have that capability. Still, it might be better to find something open source that does that. I hear we might be changing our CASE tool anyway, to save some costs."

"Right. Anything else?"

"No. Let's write these up. I think we are ready to talk to management. Let me set up a meeting next week."

"OK. Let me know a few days early so that we can prepare. Don't want to look bad in front of management. See you next week."

A use case, such as we described earlier, would be extremely helpful in explaining some of the interactions between a user and the data collection, analysis, and storage.

Before meeting with the Manager, who will act as the customer in this example, the requirements team will prepare for a formal presentation. There will be a presentation in either PowerPoint or Keynote, and paper documents that will be made available to the customer. The documents comprise a formal proposal to develop the software. They include an initial assessment of cost.

We now turn to the hypothetical meeting between the requirements team and the customer. Because the two members of the requirements team should speak as a unit, we will identify the participants in the dialogue as Requirements Team and Customer, abbreviated R and C, respectively.

R: Thank you for meeting with us today.

C: You're welcome. What's on your mind?

R: Two weeks ago you talked about a problem in software tools. You mentioned a need for a good tool to measure software size.

C: Sure. But that's not my only problem in the software tools arena. I need some other things, too.

R: We think we know some of your needs. We've developed a set of requirements for a simple software system to meet these needs.

C: Go ahead.

R: The first thing is a consistent way to measure the size of software. Since the maintenance and testing groups provide their test data according to functions in which the problems occur, you need to compute the size of each function included in the source code. The size of individual functions must then be totaled into the size of each source code file. This must be repeated to compute the total size of a software system by adding the sizes of each of the component source code files. Of course, these totals should be computed automatically.

C: Sounds good. Please go on.

R: We want a standard way of doing this computation, since the company develops software in three different languages: C, C++, and Java.

C: Yes. It's a terrible problem for us. We can't get good numbers.

R: We want to have a simple, batch-oriented system with as simple a user interface as possible. We don't want the software to be too complicated.

C: I want a standard definition of a line of code. I've heard too many horror stories about how programmers organize their code to look more productive. They bump up the lines of code measurements artificially to make themselves look better.

R: We've thought about that. Here's an example.

The Requirements Team then shows the Customer the well-known C program that is listed next. (This example was discussed previously in Chapter 1.) The Requirements Team and the Customer then discuss the size of this C program.

```
1. #include <stdio.h>
2. main(int argc, char * argv[])
3. {
4.   int i;
5.   for (i = 1; i < 10; i++)
6.     {
7.     printf("%d\n", i);
8.     }
9. }
```

R: How many lines of code are there in this program?

C: Looks like nine to me.

R: Look at lines 6 and 8. They really aren't necessary. The program would have the same meaning and be shorter if we remove those two lines.

C: I agree. There's also a difference if we join all the statements together on the same line, or if we count all the function prototypes in the header file. What does this have to do with your tool?

R: Just that we will develop a precise measurement that is consistent with different language standard practices, such as the JDK style guide for Java. Our requirements will specify standards for code counting measurements.

C: Sounds good. I need this tool right away. When can I get it and how much will it cost?

R: We have some cost numbers and a tentative schedule for you. It might be better to wait until we discuss all the tools we have developed requirements for.

C: OK.

R: We also want to develop a tool that will provide a description of the program's call graph.

C: What's that?

R: The call graph tells which functions call which functions. It provides a view of the program's structure and organization. The maintenance people really want this.

C: I can see how it is important. You said "graph." Do you actually have a graphical output?

R: Not in the first prototype. However, we do have it in our requirements for the second iteration of our tools.

C: Is there a single tool or a set of programs forming a software tool kit?

R: There is a single system. It may be made up of several software tools that can be used either as stand-alone utilities or as part of an integrated system.

C: I like this. It seems like a flexible system.

R: It should be. At least, we hope so.

Note that a use case, such as those shown in Figures 3.4 and 3.5, can help with the understanding of what the project is supposed to do. (A misuse case would not usually be discussed at this point of the requirements process.) Let us return to our imaginary dialogue.

C: This means that I can get something useful out of this effort, even if it doesn't produce all that I want.
R: That's right. We can do prototyping and show you interim progress frequently.
C: OK. How are you going to handle multiple languages? Many of the company's systems use a combination of programming languages.
R: We are going to measure the C, C++, and Java codes separately. We'll make the user responsible for determining the input files and making sure that they are all the same type of source code language.
C: I'd really prefer to see a system that took all kinds of source code as input, regardless of the language.
R: We've thought about that. There are too many options with this, too much error checking. We'll try this in a later prototype.
C: I hope there's a good reason for this.
R: It's easier this way. We're going to specify a software interface that will allow some flexibility.
C: Anything else I should know?
R: Yes. We've decided to write the front end of the tool in either C or C++ in order to be able to use command-line arguments.
C: What's the point?
R: We can use the tool for batch processing. It can work overnight if it has a lot of files to analyze.
C: Will your tool work on our systems? I don't want to buy any more equipment or software that we can't charge to a project.
R: Yes. We'll use the minimal interface and make as little use of platform-specific software as we can.
C: What about file names? Can wild cards be used?
R: We don't know. We'll get back to you on that.
C: It seems to be well thought out. The tool fits our needs. When can I get it and how much will it cost?

At this point in an actual project, the Requirements Team would defer to a person who had produced an initial cost estimate for the project. The schedule would also be developed by a person who was likely to be the project manager. Often, the detailed cost estimate is not presented to the customer at this point because charging the correct amount for software may be the difference between a company thriving or going bankrupt. Cost estimates should be done with extreme care, as should project schedules.

In our example, this was not necessary, because the project was to be done internally, a preliminary cost estimate and schedule had been produced already, and the requirements were not changed appreciably during the meetings with the potential customer.

This hypothetical discussion illustrates some of the ways that a dialogue might take place between members of a requirements gathering team and potential customers or their surrogates. In the next section, we will provide a complete set of requirements for the system. The rest of the discussion will follow a more formal approach. Always keep in mind, however, that some sort of informal elicitation of the initial system requirements is almost always necessary.

3.18 THE MAJOR SOFTWARE PROJECT: REQUIREMENTS ANALYSIS

We will now consider the requirements for the software development project that we will discuss throughout the remainder of this book. Our starting point will be the set of requirements that was elicited by the requirements team. The goal is to take these informally obtained initial requirements and transform them into requirements that are precise enough for us to tell if the software system that we eventually develop will have actually met the desired requirements.

We will assume that the initial attempts at a requirements document would look something like the following.

Problem statement for the major software project

- Determine the size of software source code written in any one of three languages: C, C++, and Java.
- Develop a common standard for code counting for programs written in C, C++, or Java.
- There should be as simple a front-end user interface as possible.
- There should be a common output interface.
- Functions in other languages should be treated as library functions.
- Develop a tool that shows the program's call graph.
- Integrate the code counting and the call graph display into a single software system.

Let us go through the list, applying the ignorance hiding technique of Berry that was presented in Section 3.4.1. Which of these words and phrases are vague and untestable?

- "standard"
- "As simple a user interface as possible"
- "flexible enough"
- "should have a common output interface"
- "treated as library functions"

TABLE 3.12 Requirements Traceability Matrix for the Major Software Project

Requirement	Design	Code	Test
1. Intel-based			
2. Windows 8.1			
3. Windows 8.1 UI			
4. Consistent with Excel 14.0			
5. System one size only			
6. 200 MB system			
7. 500 MB disk space			
8. One CD			
9. Includes installation			
10. No decompression utility			
11. One input file at a time			
12. Develop standard for C, C++, Java			
13. Size of each function			
14. Size of each file			
15. Size of system			
16. Compute totals			
17. Front end in C or C++			
18. Batch-oriented system			
19. Precisely define LOC			
20. Measure separately			
21. No error checking of input			
22. File names limited to 32 characters			
23. Wild cards can be used			
24. Dead code ignored			
25. No special compilers needed			

The list could certainly go on. The term *front end* almost certainly means *user interface* to a user of the system, but might mean the initial component of the software: the one that interfaces with the operating system and software such as spreadsheets and databases. We will not belabor the point, but will simply indicate a revision of the initial requirements list.

The requirements traceability matrix is given in Table 3.12. We have included some of our generic requirements into this matrix.

Initial requirements list for the major software project

- The project must develop a standard way of computing lines of code for C, C++, and Java.
- The system is to be batch-oriented.
- The system is to have as simple a user interface as possible, since we have it in our requirements for the second iteration of our tools.
- All input is assumed to be syntactically and semantically valid. No error checking will be provided.

- No special compilers or application programs will be needed.
- The system will determine the functions that call other functions. It will be flexible enough to be extended to the drawing of a graph in a future release.
- The system will be flexible enough to be extended to interfacing with existing maintenance and testing data on software errors in a future release.
- Dead code can be ignored.
- The system should have a common output interface for multiple languages.
- The system should have a common input for multiple languages.
- It will be assumed that each input software subsystem is written in a single language.
- The front end is to be coded in C or C++ to use command-line arguments.
- Calls to functions not included in the source code files being analyzed are to be flagged and treated as library functions.
- The software must compute the size of each function included in the source code.
- The software must compute the size of each source code file from the sizes of individual functions.
- The software must compute the total size of a software system by adding the sizes of each of the component source code files.

As we indicated earlier in this chapter, when the essential requirements have been determined, the project manager must complete the details of a project plan for developing a system that meets these requirements. (We did not create a preliminary project plan until now because we had not introduced the problem statement previously.)

You may have some objections to the list of requirements (and the associated requirements traceability matrix) given above. We have not attempted to improve the software architecture to make it more modular. Certainly we have not considered the possibility of reusing any existing software at this time.

However, we have some flexibility. The back end of the system appears to be interfaced to the Microsoft Excel spreadsheet. So we are at least using a COTS product for any additional analysis of the data that our new software tools will collect. This approach to software development emphasizes reuse at the highest levels and is becoming very common in industry.

This section will close with the following changes to the requirements for the major project. Can you guess what they are?

The hypothetical discussion we presented earlier is based on command lines and batch processing, not a modern networked-based environment. The customer now wants a web-based user interface, with opening and closing screens such as the ones shown in Figures 3.14 and 3.15, respectively.

The second change is that the customer now wants to use a cloud for data storage instead of one that is PC-based.

Obviously, some changes need to be made to the requirements. You will be asked to modify the requirements in the exercises. Keep in mind what changes need to be made to the requirements process and what changes need to be made to the requirements themselves.

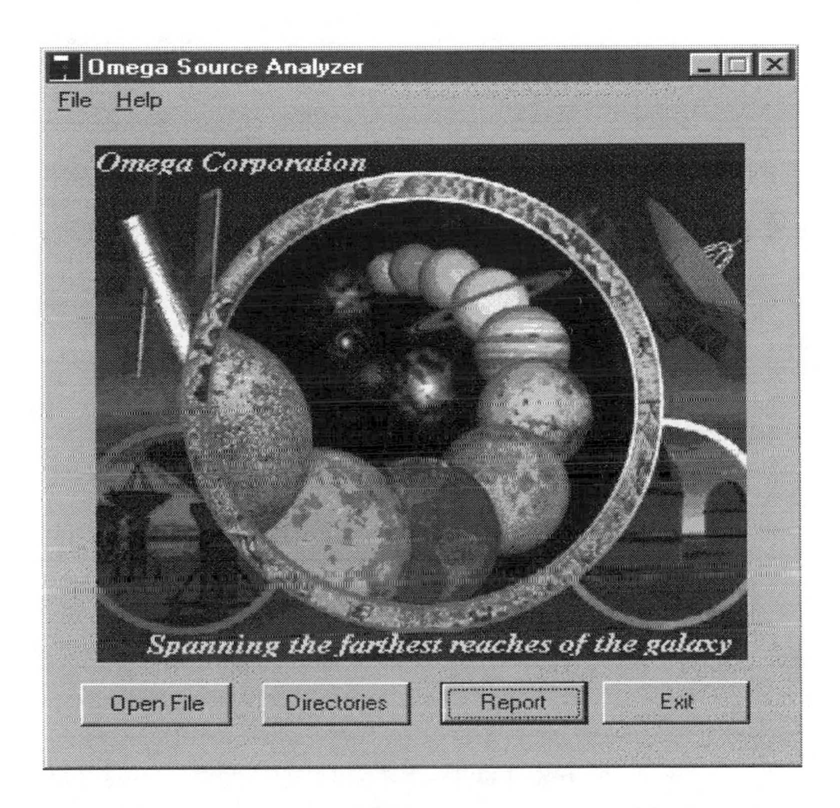

FIGURE 3.14 A sample opening screen. (Software system prepared by Sean Armstrong, Aaron Rogers, and Melvin Henry.)

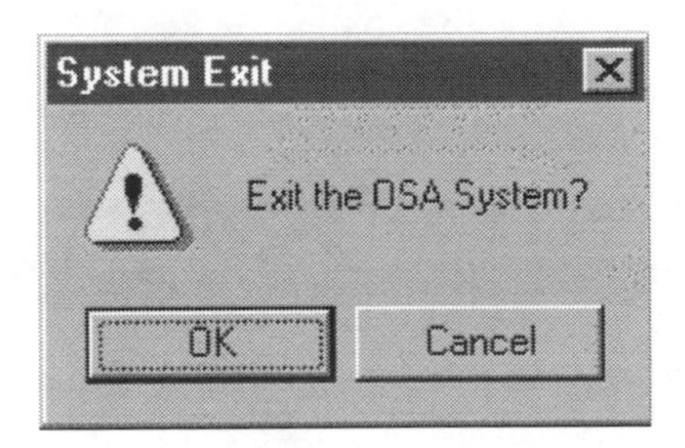

FIGURE 3.15 A sample closing screen. (Software system prepared by Sean Armstrong, Aaron Rogers, and Melvin Henry.)

SUMMARY

The development of a proper set of requirements for a software system is the key to the success of the project. Without complete, consistent, and clear requirements, the project is doomed to failure. This is true regardless of whether the software is developed according to the classical waterfall, rapid prototyping, or spiral model, or if the development process is based on open source software, or even if an agile development process is used. Eventually, the requirements must be determined correctly, even in an iterative process.

There are several techniques that can aid in requirements organization:

1. Use data abstraction and information hiding to produce a complete description of the system's functionality.
2. Regroup requirements to be consistent with the requirements and design of both existing and planned systems.
3. Reuse requirements in order to be able to reuse existing designs and source code.
4. Automate the requirements engineering process.

The requirements must be clear and unambiguous. In order to trace the requirements throughout the software development process, a requirements traceability matrix must be created. This matrix lists the requirement's number, each of the requirements, and entries for checking off that the requirement was met in the design, code, and testing phases.

Once developed, requirements should be assessed for their efficiency, potential cost, and attention to safety-critical issues. A study of current technology to determine feasibility might be appropriate at this point. Ideally, the organization's software development process includes metrics to estimate project size and cost.

Requirements represent a major portion of software development effort. As such, they will be emphasized in project schedules and will be subjected to one or more requirements reviews. The purpose of a review is to ensure that the product being evaluated (in this case, requirements) is of sufficiently high quality. A quality assurance evaluation is often used with requirements reviews.

Software requirements often change, even if the development process was not explicitly iterative. Management of change includes prioritizing tasks, reallocating resources, and making sure that the effects of the change make the most efficient use of existing requirements.

Agile development processes are different, primarily because they are best used in situations where the application domain is well understood by the agile development team, and there are many available software components to be employed to speed up the development process. Using a COTS product such as LabView, as was described in our case study on management of agile processes when evaluating requirements in Section 3.15, is an excellent example, because the choice could not have been made unless the agile team knew the essential requirements and could foresee how LabView would help meet them.

KEYWORDS AND PHRASES

Requirements engineering, requirements elicitation, function points, COTS, software architecture, data abstraction, information hiding, testable requirements, function point, function point metrics, requirements review

FURTHER READING

There are few books devoted exclusively to the requirements generation process. Perhaps the books *Software Requirements: Analysis and Specification* by Alan Davis (1990) and the second edition of *Software Requirements* by Richard Thayer and Merlin Dorfman (1992) are the most accessible. One of the major advantages of the book by Davis is that it includes an annotated bibliography of 598 references in software requirements and related research. The book by Thayer and Dorfman includes many of the most important papers on software requirements and provides an excellent overview of the state of the art in the late 1990s in this active area of research.

It may be easiest for a person new to the area of software requirements engineering to examine either proceedings of conferences devoted to requirements or to read some books that are more focused on specific areas of requirements engineering. An important website for the proceedings of annual conferences in the area of requirements engineering can be found at requirements-engineering.org.

Two papers that appeared in proceedings of this conference are of special note, having been listed as being among the most influential papers presented at these conferences, and especially accessible to readers of this book. The 1994 paper by Orlena C. Z. Gotel and Anthony C. W. Finkelstein (1994), "An Analysis of the Requirements Traceability Problem" focuses on the traceability issue. A paper by Barry Boehm and Hoh In (1996), based on their extensive industrial experience, "Identifying Quality-Requirement Conflicts" appeared in 1996. The awards for most influential paper were awarded ten years after the papers appeared, in 2004 and 2006, respectively.

The paper by Daniel Berry (1995) in the *Journal of Systems and Software* is essential reading in requirements engineering.

It is often hard to find information on the use of agile processes that provides enough information to be of use to a would-be user of agile requirements processes. A relatively recent book by Dean Leffingwell (2001) provided a detailed perspective on the interaction between requirements engineering and agile software development processes. A comprehensive article by Carol Woody (2013) provides an excellent research perspective.

Albrecht (1979) introduced the notion of function points as a measurement of program size. He noted that function points could be applied at many places in the software life cycle, including requirements engineering. The book by Jones (1993) describes the use of function points in project size estimation.

User interface research has been concentrated in two major areas: (1) psychological experiment to determine efficiency and usability of I/O devices or screen organizations, and (2) development of software models and associated tools to assist an interface designer. See articles by Card (1982) and Miller (1968) for examples of some early psychologically based work on user interfaces. Green (1985) and Olsen (1986) describe some projects in user interface management systems (UIMS) that are tools for the rapid development of user interfaces. An article by Hartson and Hix (1989) in *ACM Computing Surveys* has a large bibliography. Shneiderman (1980) and Nielsen (1994) have excellent introductory texts on this subject. The work of Bleser (1994) and Campbell (1994) provides additional information on guidelines for the design of classical user interfaces.

Of course, the user interface community has been profoundly affected by the advent of mobile devices such as smartphones and tablets, with their touch-screen interfaces with capacitative multitouch capabilities. Perhaps the most accessible information on user design that takes into account these devices can be found on the web pages of the Nielsen Norman Group, http://www.nngroup.com, which emphasizes the use of evidence in evaluating user interfaces. We note that Jakob Nielsen was one of the original user interface designers for Apple. On the website useit.com, which was the website for his previous consulting company, he indicated a price for an on-site seminar as $70,000 per day, plus expenses. (The consulting rate is impressively high. Mine is lower.)

EXERCISES

1. In our initial discussion of software engineering, we presented in Table 3.3 a simple example of the decisions that had to be made before marketing software for personal computers. This is an ideal opportunity for automation. Write a set of requirements for a software system, called RGS, that will interact with a user to determine the eleven requirements with minimal input from a requirements engineer. (The acronym RGS stands for requirements generation system.)

2. Complete the analysis of the remaining requirements in Table 3.4 using the ignorance hiding approach of Daniel Berry that was discussed in Section 3.4.

3. Provide a state diagram for the requirements given in Table 3.4.

4. Provide a decision table for the requirements given in Table 3.4.

5. Provide a Petri net for the requirements given in Table 3.4.

6. Estimate the size of the requirements given in Table 3.4, using the function point approach.

7. This exercise concerns the model of a user interface that was discussed in this chapter. Use the detailed descriptions of the object to develop a quantifiable evaluation of the displays shown in Figures 3.4 and 3.5. Apply the model to your favorite word processing software.

8. A system will have two concurrently executing processes, A and B. Consider the requirements:

    a. Processes A and B will execute on different machines on a network.

    b. Processes A and B will communicate as little as possible.

    Which of these two requirements is most consistent with the system having no single point of failure? Which is most consistent with a real-time performance requirement? Explain your answers.

9. Develop a Petri net for synchronization of the following tasks, which may occur concurrently:

    Task 1: Put on sock

    Task 2: Put on sock

    Task 3: Put on right shoe

    Task 4: Put on left shoe

    The synchronization requirement is that at least one of tasks 1 and 2 must be complete before either task 3 or 4 can happen.

10. Same as question 9, except that both tasks 1 and 2 must be complete before either task 3 or 4 can happen.

11. Select a software package with which you are familiar. Use the software package to perform some operation that you have done many times before. While you are doing this, note the changes in the user interface, including any new screens or messages that might appear. Now write down a list of requirements that this software must have had. (Determination of a system's requirements from its performance without any existing requirements documents is part of a process called "reverse engineering.")

12. Consider the state diagram given next. It is related to, but not identical with, the diagram given in Example 3.9. How can you determine if these requirements are inconsistent?

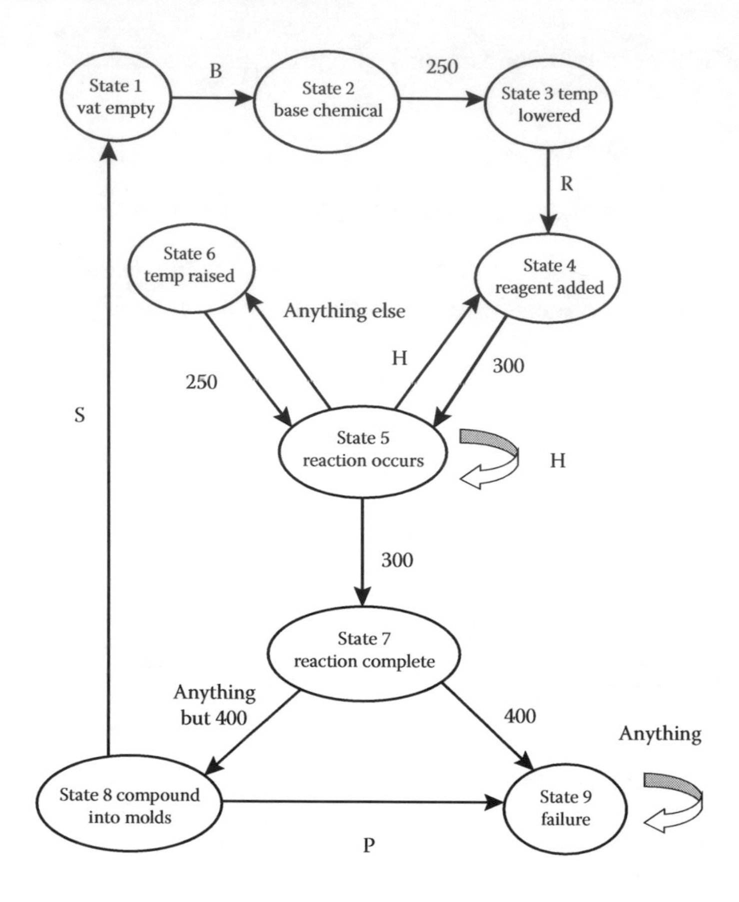

EXERCISES FOR THE MAJOR SOFTWARE PROJECT

1. Write a state diagram for the requirements of our continuing software project as they were developed in Section 3.18 and Summary of this chapter.

2. Write a decision table for the requirements of our continuing software project as they were developed in Section 3.18 and Summary of this chapter.

3. Write a Petri net for the requirements of our continuing software project as they were developed in Section 3.18 and Summary of this chapter.

4. Estimate the size of the requirements of our continuing software project as they were developed in Section 3.18 and Summary of this chapter, using the function point approach.

5. Consider the requirements for our continuing software project as they were developed in Section 3.18 and Summary of this chapter. Apply the suggestions in this chapter to reorganize the requirements to make the software architecture simpler.

6. Write a state diagram for the requirements of our continuing software project as they were developed in Section 3.18 and Summary of this chapter as they were modified in Exercise 5.

7. Write a decision table for the requirements of our continuing software project as they were developed in Section 3.18 and Summary of this chapter as they were modified in Exercise 5.
8. Write a Petri net for the requirements of our continuing software project as they were developed in Section 3.18 and Summary of this chapter as they were modified in Exercise 5.
9. Write a set of use cases for the requirements of our continuing software project as they were developed in Section 3.18 and Summary of this chapter.
10. Change the requirements to allow the software to have a web-based interface. Apply the ignorance hiding technique of Section 3.4.1 to analyze these new requirements. What sort of technology assessment must take place? Which existing requirements must be changed?
11. Develop a detailed project plan for the major software project we introduced in this chapter.
12. What attributes would you expect to have for a project that is attempting to complete our major project using an agile software process? Does your project team possess these attributes at this point in this course? Begin your answer by assessing your team's knowledge of the application domain.
13. This question considers the ramifications of a considerable change to the requirements. The first change is that the customer now wants a web-based user interface instead of one that is PC-based. Determine the changes need to be made to the requirements to have a different type of user interface, with different issues in the possible configurations (multiple operating system versions on the PC, multiple browsers, and multiple versions of HTML and Java). What changes, if any, need to be made to the software development life cycle model or the requirements elicitation process?
14. This question considers the ramifications of a considerable change to the requirements. The second change is that the customer now wants to use a cloud for data storage instead of one that is PC-based. Determine the changes that need to be made to the requirements to have cloud storage for the data, with different issues in scalability and the size of the cloud that is needed for the project. Also, determine the changes that need to be made to the requirements in order to have cloud storage for the source code modules being analyzed, with different issues in scalability and the size of the cloud that is needed for the project. What changes, if any, need to be made to the software development life cycle model or the requirements elicitation process?
15. This question considers the ramifications of a considerable change to the requirements. This change is that the customer now wants to use open source software and the Linux environment as much as possible instead of one that is based on the Microsoft Windows environment. Determine the changes that need to be made to the requirements to do this. What changes, if any, need to be made to the software development life cycle model or the requirements elicitation process?

CHAPTER 4

Software Design

4.1 INTRODUCTION

It is essential to develop a model of a system before writing any software that will be used to control the system or to interact with it. Modeling the software is probably the hardest part of software design. Producing good software designs requires a considerable amount of practice, with a great deal of feedback on the quality of the designs with the resulting software products influencing the design of the next software system.

We view software design as an iterative process, with the goal being design improvement at each iteration. The purpose of this chapter is to discuss some basic principles that can guide the software design process. These principles will be illustrated by several examples.

You are undoubtedly aware that a fundamental change is taking place in the software industry. With the advent of cloud computing, software is not as tethered to particular servers as before. The notion of a computer itself is changing, moving from a powerful desktop or laptop that connects to a server as necessary, to much more ubiquitous devices such as smartphones and tablets, with cloud computing often added.

There also is heavy emphasis on object-oriented technology and software reuse. Some aspects of the object-oriented methodology are so different from corresponding issues in the traditional, still-common procedural approach to software design that they should be discussed separately. This is especially true in software design.

However, a considerable amount of current software engineering effort is still far more focused on requirements analysis, design, implementation, testing, and maintenance of programs that have been developed under the more familiar procedural approach rather than object-oriented techniques. Therefore, a software engineer must be familiar with designing both types of software systems. This presents a problem for the writer of a book on software engineering.

Our solution, and the one most appropriate for both the current software engineering world and that in the foreseeable future, is to use a hybrid approach that combines components that are written in either procedural or object-oriented languages. After all, an existing software module that performs a specific requirement well should not be rejected for potential reuse just because a new application is object oriented. The use of even larger-scale software components, such as the use of LabView that we discussed in Section 3.15 during our case study of agile processes, is a prime example. If we use such commercial off-the-shelf (COTS) products or any large-scale components as is, the main consideration is their applicability to our particular project. This is, of course, especially important for any agile software development process. Even if the software component is open source, we are still primarily concerned with its functionality and its quality, not whether the internal design of the component is object-oriented.

We note that this hybrid approach is common in the software engineering community and has been formalized by several organizations. Here's a historical example. Consider the experience of the U.S. Department of Defense, which spends an enormous amount of money on software development and maintenance. A survey in the late 1970s showed that software was being developed or maintained in more than 1600 different computer languages and dialects. This was the impetus for the development of the Ada programming language, which was designed to make use of the lessons learned in many years of software development (Ada, 1983; Ichbiah, 1986). Some of these lessons were information hiding, separate compilation of software interfaces and software internals using package specifications and bodies, support for polymorphism and abstraction through "generics," and strong typing. Major revisions, including increased object-oriented technology occurred in Ada in 1995, 2005, and 2012.

As this chapter is being written, there is considerable public attention directed toward a technology named "positive train control" that many experts believe might have prevented a recent train derailment when a train exceeded the 50 miles per hour speed limit on a tight curve north of Philadelphia's 30th Street Station. Measurements suggest that the train was moving at an amazing 106 miles per hour around this curve. Many people not familiar with the technology suggest simply measuring a train's speed using a GPS and slowing it automatically if it is determined to be too fast. Unfortunately, that approach is simply not up to the complexity of the problem.

Positive train control makes use of the information provided by local tracking devices and uses these to provide a real-time control system with a high degree of fault tolerance. On a personal note, I had an interesting discussion with a technical manager for the Federal Railway Administration at a time (the early 1990s) when the speed controls and collision prevention software was written in a specialized language known originally as Tandem Application Language, which was later renamed Tandem Transaction Application Language. Even then, a purely object-oriented programming language did not seem adequate for the task. A hybrid approach is being deployed.

While the number of programming languages currently in use is far less than 1600, it is large and expanding, including languages for statistical data analysis, for scripting, for visual programming, for textual analysis, and smartphone applications development. You should expect to learn many languages, software development kits, and development environments during your career.

The rest of this chapter will be organized as follows: First, we will present some basic, general-purpose design principles. Some of the design principles will be based on the basic technique of software pattern matching. The goal is to reuse software components at as high a level as possible. This discussion will be followed by some representations that are commonly used for the design of procedurally oriented systems, with examples of each representation. After completing our initial discussion of the design of procedurally oriented systems, we will discuss some design representations for them, again with examples. The same steps will be followed for object-oriented software. This will complete our initial discussion of software design techniques. We will also discuss our agile case study.

The last three sections of this chapter will be devoted primarily to the design of the example begun earlier in this book. This example will be discussed in detail. Special emphasis will be placed on those issues that arise because of the necessary combination of both procedurally and object-oriented designs.

4.2 SOFTWARE DESIGN PATTERNS

One possible way of designing software is to attempt to match the problem to be solved with a preexisting software system that solves the same type of problem. This approach reduces at least a portion of the software design to pattern matching. The effectiveness depends upon a set of previously developed software modules and some way of recognizing different patterns.

Following are some types of patterns that seem to repeatedly occur in software development. Our patterns are relatively high level, as opposed to the patterns described in what is commonly called the "Gang of Four Book" after its four authors, Erich Gamma, Richard Helm, Ralph Johnson, and John Vlissides (Gamma et al., 1995). These high-level patterns are

1. A menu-driven system, where the user must pass through several steps in a hierarchy in order to perform his or her work. The menus may be of the pull-down type such as is common on personal computers or may be entirely text based.

2. An event-driven system, where the user must select steps in order to perform his or her work. The steps need not be taken in a hierarchical order. This pattern is most commonly with control of concurrent processes where actions may be repeated indefinitely, in contrast to pattern 3. It is also typical of a user interface that is guided by selection of options from both menus and combinations of keystrokes, such as in menus that may pop-up in response to user requests. In any case, there is less of a hierarchy than in the previous pattern.

3. A system in which the actions taken depend on one of a small number of "states" and a small set of optional actions that can be taken for each state. The optional action taken depends on both the state and the value of an input "token." In this pattern, the tokens are usually presented as a stream. Once a token is processed, it is removed from the input stream.

4. A system in which a sequence of input tokens (usually in text format) is processed, one token at a time. This pattern differs from the previous pattern in that the decision about which action to take may depend on more information than is available from just the pair consisting of the state and the input token. In this pattern, the tokens may still remain in the input stream after being processed.

5. A system in which a large amount of information is searched for one or more specific pieces of information. The searches may occur once or many times.
6. A system that can be used in a variety of applications but needs adjustments to work properly in new settings.
7. A system in which everything is primarily guided by an algorithm, rather than depending primarily on data.
8. A system that is distributed, with many relatively independent computational actions taking place. Some of the computational actions may communicate with other computational actions.

How many of these patterns are familiar to you? The first two describe human–computer interfaces, with the first one being a rigid, menu-driven system, and the other being controlled by actions such as moving a mouse or other pointer and pressing a button. Much of these two patterns is still relevant in an age of tablets and smartphones. Note that pattern 1 seems procedural in nature, whereas pattern 2 fits well with objects communicating by passing messages in an object-oriented system.

Software pattern 3 describes what is often known as a "finite state machine" or "deterministic finite automaton." This pattern is especially useful in controlling some processes in a manufacturing plant such as was shown in Figure 3.9. It is probably easiest to imagine it organized as a procedurally oriented design, at least at top level.

Pattern 4 occurs often in utility software. It is the basis for lexical analyzers and the parsing actions of compilers. Many software systems have some sort of parser to process input commands.

Obviously, software pattern 5 refers to a database searching problem.

Pattern 6 is quite general but suggests that there is a general-purpose system that must be specially configured for each application setting. The installation of a printer for a personal computer (or a network of computers) follows this model.

Software patterns 7 and 8 are also very general in nature. However, pattern 7 appears to suggest that the solution will be procedural in nature, whereas pattern 8 might be better suited to a solution using object-oriented methods. Pattern 8 suggests cloud computing and the Software as a Service (SaaS) model.

These software patterns are not absolute rules. Neither are they intended as a complete classification of software. Nonetheless, they can guide us in the design and implementation portions of the software development process.

We note that these high-level patterns can be further broken into three groups: creational patterns, structural patterns, and behavioral patterns. These groups can be broken further into twenty-three patterns as follows.

Creational patterns

1. Abstract factory—This pattern provides an interface for creating families of related objects without specifying their classes.

2. Builder—This pattern separates the construction of a complex object from its representation.
3. Factory method—This pattern defines an interface for creating a single object and allows subclasses to decide which class to instantiate.
4. Prototype—This pattern specifies the kinds of objects to create using a prototypical instance and creates new objects by copying this prototype. In theory, this can aid in the development of projects that use a rapid prototyping life cycle.
5. Singleton—This pattern makes sure that a class has only a single instance.

Structural patterns

6. Adapter—This pattern converts the interface of a class into another interface that the clients of the class expect. This function is called bridgeware or glueware in other contexts.
7. Bridge—This pattern ensures that an abstraction is separate from details of its implementation.
8. Composite—This pattern allows objects to be grouped into tree structures where both individual objects and compositions of objects can be accessed.
9. Decorator—This pattern attaches additional responsibilities to an object dynamically while keeping the same interface.
10. Façade—This pattern provides a unified interface to a set of interfaces in a subsystem.
11. Flyweight—This rare pattern uses a form of sharing to treat large numbers of similar objects efficiently.
12. Proxy—This pattern provides a placeholder for another object to control access to it.

Behavioral patterns

13. Chain of responsibility—This pattern allows giving more than one object a chance to handle a request by the request's sender. This can be highly useful if there are alternative responders, including fallback operations.
14. Command—This rare pattern encapsulates a request as an object, thereby allowing clients to be given different requests.
15. Interpreter—This large-scale pattern allows designers to use a representation of a language grammar for parsing. This can be helpful for developing code metrics.
16. Iterator—This pattern provides a way to access the elements of an aggregate object sequentially without exposing its underlying representation, which need not be an array.

17. Mediator—This pattern defines an object that encapsulates how a set of objects interact, preventing explicit references.

18. Memento—This pattern allows objects to capture and externalize an object's internal state, while allowing the object to be restored. This can be very useful in creating rollback states in the event of an unforeseen system error.

19. Observer—This useful pattern defines a one-to-many dependency between objects where a state change in one object results in all dependents being notified and updated.

20. State—This pattern allows an object to alter its behavior when its internal state changes. This can be useful for rollbacks in the event of system errors.

21. Strategy—This pattern defines a family of algorithms, encapsulates each one, and makes them interchangeable. Strategy lets the algorithm vary independently from clients that use it. This can be used to improve system fault tolerance when incorporated into an "N-version programming" or "multiversion" scheme where correctness is evaluated by majority vote on a set of independently implemented program versions.

22. Template method—This pattern, which is familiar to those who have taken a data structures course, defines the skeleton of an algorithm in an operation, deferring some steps to subclasses.

23. Visitor—This pattern is used to represent an operation to be performed on the elements of an object structure without changing the classes of the elements on which it operates.

I refer the reader to the original Gang of Four book (Gamma et al., 1995) for more detail on these twenty-three patterns. The reader can find many online examples of implementations of these examples in a variety of places. Of course, a 1995 book as important as this one has spawned many research efforts. There are at least forty-eight patterns described in the literature, far too many to discuss here.

It is important to note that the original twenty-three design patterns were given in C++, a language that is not as object-oriented as Java, Objective C, Swift, Ruby, and the like. Hence, some of the aforementioned patterns that were intended to solve issues specific to C++ may not apply to other, later object-oriented programming languages.

We note explicitly that is not necessary to have the entire software system under consideration match one of the aforementioned patterns. It is perfectly reasonable to have different portions of the same software system match several different patterns. This will be the case when we discuss our large software engineering example later in this chapter.

Separation of a software system into several subsystems is common during software design. Indeed, some sort of stepwise refinement of designs is essential for an efficient software design process. Decomposition of system designs will be a recurring theme during this chapter. For now, we will ignore system decomposition and concentrate on software patterns.

Let us consider how these patterns can be used as part of a design process. Suppose that we recognize that a particular pattern is applicable to a software system that we are to design. Where do we go from there?

The first step is to see if we can make our work simpler. Specifically, we should search through any software available to see if, in fact, we do already have a full or partial solution to our problem. This situation is an example of software reuse, which is the most efficient way to produce software. If the preexisting software solves the problem exactly, then we are done. The needed modifications should be compared with the estimated cost of entirely new software to see if we should attempt to reuse a partial match. We will illustrate the application of software reuse to our continuing discussion of our major software engineering example in Section 4.20 of this chapter.

4.3 INTRODUCTION TO SOFTWARE DESIGN REPRESENTATIONS

Any notations, techniques, or tools that can help to understand systems or describe them should receive serious consideration from the person modeling the system. We will focus our attention in this section on techniques for modeling and will briefly discuss design notations.

Suppose that we cannot find any existing software that either solves our problem directly, or else is a candidate solution after it is modified. In this case, we are forced to design new software to solve the problem. How do we do this?

There are several ways to describe a software system:

- The software can be described by the flow of control through the system.
- The software can be described by the flow of data through the system.
- The software can be described by the actions performed by the system.
- The software can be described by the objects that are acted on by the system.

Each of these system descriptions, in turn, leads us to one or more design representations that have been found to be useful for describing and understanding these types of systems.

The first type of description of a software system leads us to the concept of a flow graph or flowchart. The earliest popular graphical design representations were called "flowcharts," which were control-flow oriented. The term *control flow* is a method of describing a system by means of the major blocks of code that control its operation. In the 1950s and 1960s, a flowchart was generally drawn by hand using a graphical notation in which control of the program was represented as edges in a directed graph that described the program. Plastic templates were used for consistency of notation.

The nodes of a control flow graph are boxes whose shape and orientation provided additional information about the program. For example, a rectangular box with sides either horizontal or vertical means that a computational process occurs at this step in the program. A diamond-shaped box, with its sides at a 45-degree angle with respect to the horizontal direction, is known as a "decision box." A decision box represents a branch in the control flow of a program. Other symbols are used to represent commonly occurring situations in program behavior. An example of a flow chart for a hypothetical program is given in Figure 4.1. More information on flowcharts is given in Appendix C.

The second method of describing designs is appropriate for a data flow representation of the software. As was mentioned in Chapter 1, data flow representations of systems were developed somewhat later than control flow descriptions. The books by Yourdon, one by him and the other coauthored with Constantine, are probably the most accessible basic sources for information on data flow design (Yourdon and Constantine, 1979; Yourdon, 1989). Most software engineering books contain examples of the use of data flow diagrams in the design of software systems.

Since different data can move along different paths in the program, it is traditional for data flow design descriptions to include the name of the data along the arrows indicating the direction of data movement.

Data flow designs also depend on particular notations to represent different aspects of a system. Here, the arrows indicate a data movement. There are different notations used for different types of data treatment. For example, a node of the graph representing a transformation of input data into output data according to some rule might be represented by a rectangular box. A source of an input data stream such as an interactive terminal input would be represented by another notation, indicating that it is a "data source." On the other hand, a repository from which data can never be recalled, such as a terminal screen, is described by another symbol, indicating that this is a "data sink." See Appendix C.

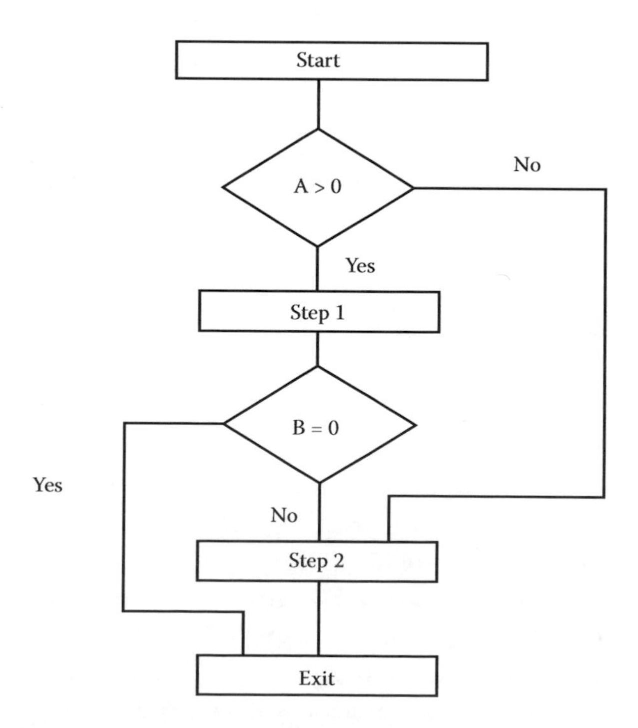

FIGURE 4.1 A flowchart description of a hypothetical program.

Since different data can move along different paths in the program, it is traditional for data flow design descriptions to include the name of the data along the arrows indicating the direction of data movement.

Typical data flow descriptions of systems use several diagrams at different "levels." Each level of a data flow diagram represents a more detailed view of a portion of the system at a previously described, higher level. A very high-level view of the preliminary analysis of the data flow for a hypothetical program is shown in Figure 4.2. This simple diagram would probably be called a level 1 data flow diagram, with level 0 data flow diagrams simply representing input and output.

The third method of representing a software system's design is clearly most appropriate for a procedurally oriented view of the system. It may include either control flow or data flow, or even combined descriptions of the software. The notations used for this hybrid approach are not standard. Such a design may be as simple as having a link from, for example, the box labeled "Step 1" in Figure 4.2 to the flow chart described in Figure 4.1.

Finally, the fourth method is clearly most appropriate for an object-oriented view of the system. This is obviously a different paradigm from the previous ones. We have chosen to use a modeling representation known as Unified Modeling Language, or UML. UML is an attempt to describe the relationships between objects within a system, but does not describe the flow of control or the transformation of data directly.

Note that each of the boxes shown in Figure 4.3 represents an object for which the inheritance structure is clearly indicated by the direction of the arrow: the object of type class 1

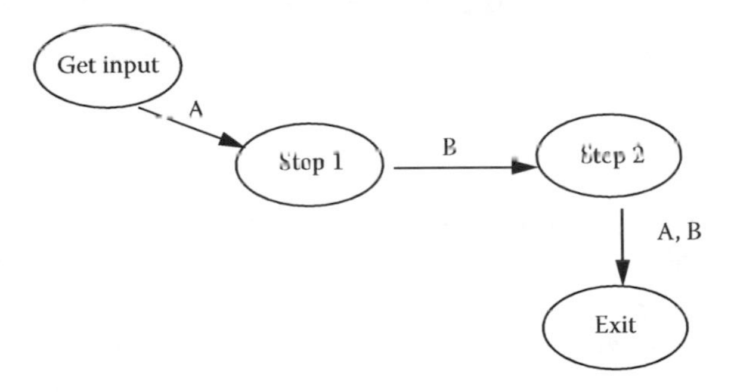

FIGURE 4.2 A level 1 data flow diagram (DFD) description of a hypothetical computer system.

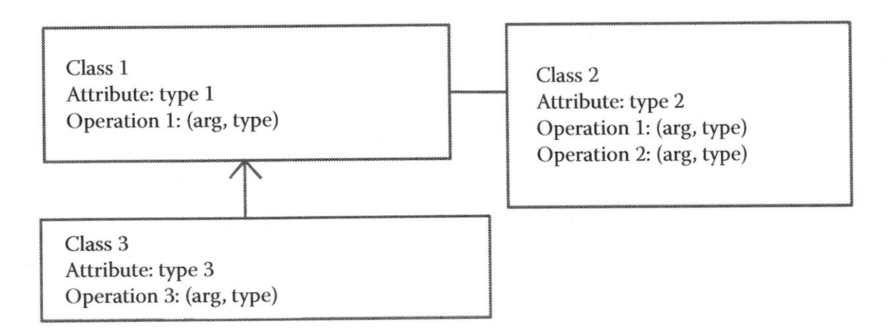

FIGURE 4.3 An oversimplified object-oriented description of a hypothetical computer system using an object model in UML notation.

is a superclass of the object of type class 3. In an actual design using the unified object model, the horizontal line would contain information about the type of relationship (one-to-one, many-to-one, one-to-many) and about the aggregation of multiple objects into a single object. We omit those details here.

It is often very difficult for a beginning software engineer to determine the objects that are appropriate for a software system's design. Although there is little comparative research to support the hypothesis that good object-oriented design is harder to create than design using traditional procedurally oriented approaches, we do believe that considerable training is necessary to appreciate the subtleties of the object-oriented approach. The student unfamiliar with the object-oriented approach is encouraged to read this and related discussions throughout this book several times after he or she has become more familiar with software development in order to understand why certain design decisions were made.

A helpful approach is to think of a description of the system's functionality in complete sentences. The verbs in the sentences are the actions in the system; the nouns represent the objects. We will use this simple approach when developing that portion of the design of our major software engineering project example for which the object-oriented approach makes sense.

It is natural for the beginning student of software engineering to ask why multiple design representation approaches are necessary. Of course some techniques evolved for historical reasons. However, many techniques and representations have survived because they provide useful, alternative views of software design.

The familiar problem of sorting an array of data provides a good illustration of the distinction among the four approaches. Consider the well-known quicksort algorithm developed by C. A. R. Hoare (1961).

The quicksort algorithm partitions an array into two subarrays of (perhaps) unequal size. The two subarrays are separated by a "pivot element," which is at least as large as all elements in one subarray and is also no larger than the elements in the other subarray. Each of the subarrays is then subjected to the same process. The process terminates with a completely sorted array. Note that this algorithm description is recursive in nature. Think of the quicksort algorithm as being used in a design to meet some requirements for an efficient sort of a data set of unknown size. (Students wishing a more detailed description of the quicksort algorithm are advised to consult any of the excellent books on data structures, or any introductory book on algorithms. Hoare's original paper (Hoare, 1961) is still well worth reading.)

A control flow description of the algorithm would lead to a design that would be heavily dependent on the logical decomposition of the program's logic. Think of implementing the recursive quicksort algorithm in a language such as FORTRAN or BASIC that does not support recursion. The logic of the design would be very important. On the other hand, too large a data set might mean too many levels of recursive function calls if a purely recursive algorithm is used. This control flow view might point out some limitations of the design of a simple quicksort algorithm.

A data flow description of the algorithm would emphasize the data movement between arrays and the two smaller subarrays that are created at each iteration of the quicksort algorithm. Attention to the details of setting aside storage locations for these subarrays might lead to a consideration of what to do if there is no room in memory. Most courses in data structures ignore the effects of data sets that are too large to fit into memory. However, such sets exist often in applications and efficient sorting programs must treat them carefully.

A third view can be obtained by considering a purely procedural solution. In this case, a standard library function can be used. For example, the standard C library has a function called qsort(), which takes an array argument and produces a sorted array in the same location. That is, the input array is overwritten by the output array.

A user of the qsort() function in the standard C library must provide a comparison function to do comparisons between array elements. The function prototype for this comparison function, compare(), is

```
int *compare(*element1, *element2);
```

This user-defined comparison function takes two arguments (which represent arbitrary array elements) and returns 0 if the two arguments are the same. The comparison function must return –1 if the first argument is "less than" the second, and must return 1 otherwise. The number of elements in the array and the size of an array element must also be provided.

The qsort() function in the standard C library is accessed by including the header file stdlib.h within a C program. This function has the syntax

```
void qsort(
 const void *base,
 size_t num_elts,
 size_t elt_size,
 int (*compare(const void *, const void *)
 );
```

and the typical usage is

```
ptr = qsort(arr,num_elts,elt_size, compare);
```

Finally, an object-oriented approach would most likely use a general class for the array of elements and invoke a standard member function for sorting. Such a function typically would be found in a class library for some abstract class. An example of this can be found in the class libraries that are provided with most C++ software development systems.

A general class in an object-oriented system will contain the methods that can be applied to the object represented by the class. In C++, the methods that are associated with an object are called the "member functions" of the class. In the case of C++, a class description is likely to contain what are called "templates," or general classes in which the specific type of a member of the class is only relevant when the class is used. Thus a template class can refer to an abstract array of integers, character strings, or any other relevant type for which the class operations can make sense.

Provided that the methods of the template class can be implemented for the particular type of object in the class, the general method known as a sorting method can be invoked by simply calling a member function of an object as in

```
A.sort();
```

Here, the type of the object A is compatible with the template type and the sort() member function for that object is used.

Even for the simple case of a sorting algorithm, we have seen several different approaches that can be used for software system design. Each has advantages and disadvantages. We have not examined any of the disadvantages in detail. In particular, we have never considered the question of the efficiency of software written according to any of the designs given previously. For example, a slow sorting algorithm, or one that uses recursion, would be completely inadequate for many software applications.

There is one other commonly used method for describing a system. This involves pseudocode that provides an English-like (at least in English-speaking software development environments) description of the system. The pseudocode is successively refined until the implementation of the source code is straightforward, at least in theory. Note that pseudocode is a nongraphical notation. An example of pseudocode is shown in Example 4.1. The pseudocode describes a portion of an authentication system for password protection in a hypothetical computer system.

Example 4.1: A Pseudocode Description of a Hypothetical Computer System

```
GET login_id as input from keyboard
Compare Input to entries in login_id database
IF not a match THEN
   WAIT 10 seconds
   SET error_count to 1
   REPEAT
      PROMPT user for new login_id
      IF login_id matches database
               THEN PROMPT for password
      ELSE increment error_count
        WAIT 10 seconds
      END IF
   IF error_count > 3 EXIT password PROMPT
   END REPEAT
ELSE
   GET password as input from keyboard
   Compare password entries in password database
```

```
  IF error THEN
    EXIT
  ELSE
    BEGIN login process
END IF
```

Pseudocode representations have two major advantages over the graphical ones: they can be presented as textual information in ASCII files, and pseudocode descriptions of large systems are no more complicated to represent than those of small systems. Of course, pseudocode representations may be so long and have so many levels of nesting in their statement outline that they are extremely difficult to understand.

You should note one other advantage of pseudocode representations when used for designs: they can produce instant internal documentation of source code files. We will address this point in the exercises.

The design representations described in this section by no means exhaust the number of representations available.

A more complex design representation is available when using the Department of Defense Architectural Framework (DoDAF) processes. The framework consists of seven different "Viewpoints," each of which includes multiple views. The seven DoDAF Viewpoints (Department of Defense Architectural Framework, 2009) are

1. Capability Viewpoint
2. Data and Information Viewpoint
3. Operational Viewpoint
4. Project Viewpoint
5. Services Viewpoint
6. Standards Viewpoint
7. Systems Viewpoint

There is also a combined Viewpoint called "All Viewpoint." Each of these Viewpoints consists of several elements. For example, the Operational Viewpoint requires at least seven distinct model descriptions. Keeping these Viewpoints consistent for a DoDAF representation for any nontrivial system clearly requires the use of a high-quality CASE tool, such as *System Architect* from Telelogic AG. Such CASE tools are expensive and geared to very large-scale systems. They are unlikely to be available for general use at most colleges or universities for general instruction. Consult the references for more information.

4.4 PROCEDURALLY ORIENTED DESIGN REPRESENTATIONS

In this section we will provide some examples of the use of different procedurally oriented design representations for a single-system example. We will consider object-oriented design representations for the same system later in this chapter in Section 4.9.

Consider a common situation: software that controls a terminal concentrator. A terminal concentrator is a hardware device that allows many different terminal lines to access the same CPU. At one time, these were quite common, since the only means of access to computing power was through the use of relatively inexpensive "terminals" that were connected remotely to a mainframe or minicomputer. For reasons we will discuss in the next paragraph, this seemingly obsolescent technology is becoming more common in settings where the top-level domain is running out of fixed IP addresses, but the use of DHCP to extend network access is not always appropriate because of the need for some IP addresses to remain static to meet with the needs of specific applications.

Perhaps the most common example of this concept can occur in a hospital setting, where the output from a monitoring device is sent to a computer located at, say, a nurses' station in an intensive care unit. It is critical to know the patient's room in which the device is located, so dynamic IP is probably not appropriate to use. There is no need for all the data being monitored to be sent any further than the nurses' station, although there may be some secondary or tertiary data storage devices nearby because of the need for hospitals to be able to defend themselves against medical malpractice suits. Anyone who has either stayed in a room in an intensive care unit or visited a patient in one is well aware of the fact that these monitoring devices often need to be reset because of such things as inappropriate alarms, patients accidentally disconnecting the devices, or the plastic bags holding fluids used for IVs becoming empty.

Because this particular situation is likely to be more familiar to you than the classical terminal concentrator, we will use the term "medical device terminal concentrator" for the remainder of this discussion. We will reserve the word "terminal" to describe the devices located in patient rooms.

Input from any of the lines is associated with the medical device terminal concentrator to which the line is attached at one end. Corresponding output for a data collection or analysis program running on one of the terminals is sent to the appropriate medical device terminal concentrator's screen.

To keep the data being sent to and from different terminals from going to the wrong place, the signals are "multiplexed" by the medical device terminal concentrator. Multiplexing hardware and software allow the attachment of many terminals to the same port of the computer, thereby allowing for more simultaneous users (at a potential cost of reduced processing speed, which does not really matter in this application). All terminals on the same terminal concentrator share the same connection to the computer, as is shown in Figure 4.4.

Multiplexing means that a single set of wires is used for the connection from the terminal concentrator to the centralized computer. The decision about which terminal to communicate with along these wires can be made by using different frequencies for the signals or by attaching routing information to each packet of information sent.

The essential data structure in a multiplexed system is a queue. A user's data is sent from his or her terminal to the medical device terminal concentrator CPU in a stream in a first-in, first-out manner. The data passes through the multiplexing operation of the medical device terminal concentrator and is sent to the CPU when the process is scheduled

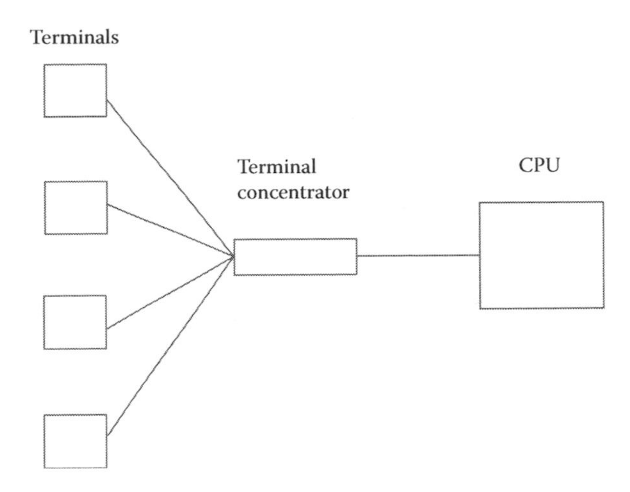

FIGURE 4.4 A medical device terminal concentrator. (From Leach, R. J., *Object-Oriented Design and Programming in C++*, Academic Press Professional, Boston, 1995. With permission.)

for execution. All other processes (and the input/output [I/O] from their terminals) wait for the process to relinquish control of the CPU. Output is then sent from the CPU to the appropriate terminal, also by means of the multiplexing operation of the medical device terminal concentrator.

Thus, there are queues for input, queues for output, and mechanisms for determining which data is attached to which terminal (for either input or output).

A small portion of a flowchart for the software that controls a terminal concentrator is shown in Figure 4.5. You should compare this with the example of pseudocode for the medical device terminal concentrator system (Example 4.2). Many of the steps indicated in Figure 4.5 can be decomposed into additional steps.

Example 4.2: A Pseudocode Representation for a Medical Device Terminal Concentrator

```
For each terminal
 {
 Repeat forever
  {
    When time for input to CPU
    {
    get input from medical device terminal concentrator's
        input queue
    place on medical device terminal concentrator input queue to
        CPU
    include information to identify terminal
    send input data from concentrator to medical device terminal
```

```
            concentrator CPU
        remove data from concentrator queue
        }
        When time for output from CPU
        {
            receive output data from CPU
            place on medical device terminal concentrator's output queue
            include information to identify terminal
            send output data to terminal queue
            remove data from medical device terminal concentrator queue
        }
    }
}
```

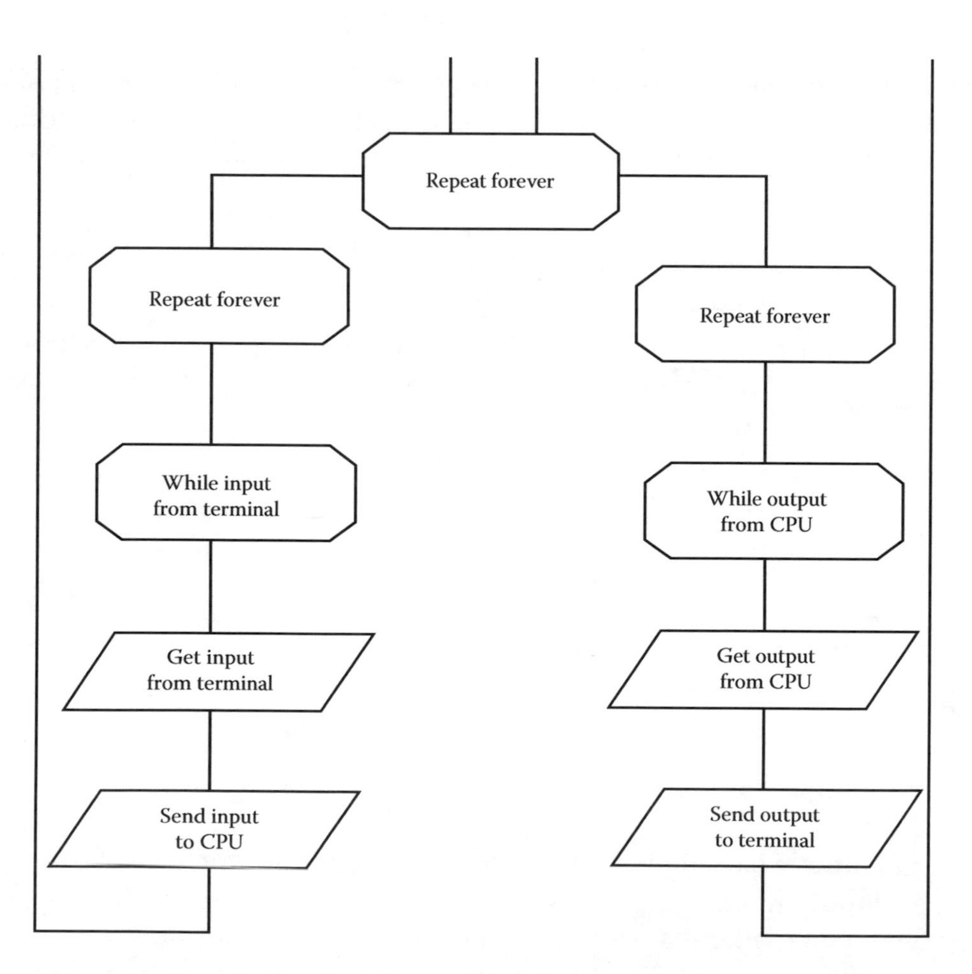

FIGURE 4.5 A portion of a flowchart for a medical device terminal concentrator. (From Leach, R. J., *Object-Oriented Design and Programming in C++*, Academic Press Professional, Boston, 1995. With permission.)

Of course, many different representations can be used to describe the same system; we are not restricted to flowcharts. The next design representation we discuss for this small example is the data flow diagram.

An example of the data flow description of a program in one graphical notation is given in Figure 4.6. The data flow diagram shown in this figure describes one direction of flow for the terminal concentrator example previously described. The medical device terminal concentrator system can be described completely by a more elaborate diagram. You will be asked to do this in the exercises.

In reality, changes in technology and rapid turnover in the software engineering industry mean that a software designer must be familiar with several different design representations.

The entity–relationship, or E-R, model is common in database design. Another is the information model that is frequently used in artificial intelligence and expert systems. We will discuss each of these models briefly in this section.

An E-R diagram represents a set of fundamental quantities, known as entities, and the relationships between them. The labeling of the arcs in the E-R diagram indicates the nature of the relationships between the different entities connected by each arc.

An E-R diagram can serve as a starting point for a preliminary set of objects. The diagrams' relationships often suggest some possible methods, or transformations, on objects in the system.

This method is only a first step in the development of a complete description of the objects and methods in a system because the E-R diagram generally lacks any self transformations

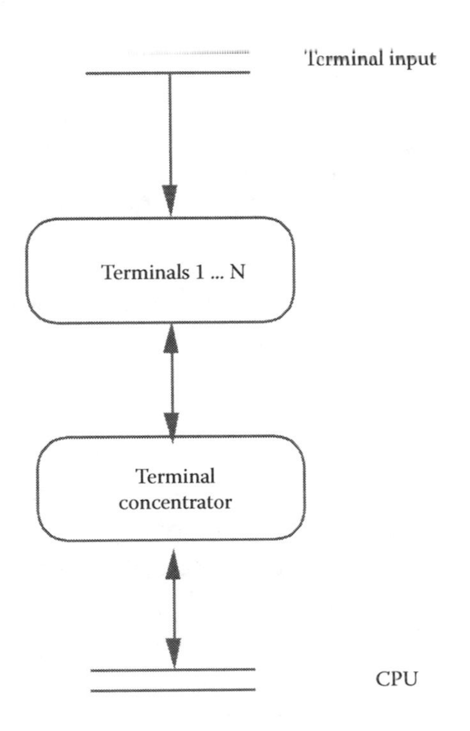

FIGURE 4.6 A data flow representation for a medical device terminal concentrator.

TABLE 4.1 A Methodology for Determination of Objects

1. Choose a candidate to be an object.
2. Determine a set of attributes and their possible sets of values. Use the has-a relation. List all relevant transformations on the object.
3. Develop an initial set of transformations on the object to serve as member functions. The list of attributes and their values provide an initial set of transformations by determining the value of, and assigning a value to, each attribute of an object. Constructor, destructor, and I/O functions should also be included.
4. Determine if there is more than one example of the object. If so, then place the proposed object in a set of potential objects. If not, discard it because it fails the multiple examples test.
5. Apply the is-a relation by considering all sentences of the form "object 1 is a object 2." Objects considered for this relation should include the object under development and any other objects believed to be related. (The class library and any SDK should be consulted during this step of the process.) Each valid sentence should lead to an inheritance relationship. Each inheritance relationship should be illustrated graphically.
6. Use polymorphism and overloading of operators (and functions) to check if we have described the objects in sufficient detail. Check the object description if no polymorphism or overloading is found.
7. Use the uses-a relation "object 1 uses object 2" to determine all instances of client–server or agent-based relationships. Use these relationships to determine issues of program design.
8. Review the object, its attributes, member functions, inheritance properties, polymorphism, overloading, and relationships to other objects to determine if the object is complete in the sense that no other functions, attributes, or relationships are necessary.
9. Repeat steps 2 through 8 for all combinations of relevant objects (triples, quadruples, and so on) until the object's role in any proposed system has been adequately described.

of an object. Thus, constructors, destructors, and initializers are not generally evident from E-R diagrams. Tests for equality are typically not clear. Many other common methods are not easily represented in E-R diagrams.

The observations of the utility of E-R diagrams in treating objects can be summarized as follows:

- If an E-R diagram already exists, use the entities as initial choices of objects and methods. Pay particular attention to the need for self-transforming methods such as constructors, destructors, and initializers.
- If no E-R diagram exists, do not bother writing one. Instead, proceed to the description of objects using the process in Table 4.1.

4.5 SOFTWARE ARCHITECTURES

The term *software architecture* has been used in many different, often conflicting, ways in the software engineering community. Regardless of the specific context, the term is commonly used to describe the organization of software systems.

We define it as follows: An *architecture* is the definition of the key elements that constitute a system, their relationships, and any constraints. We will view architectures as being composed of several types, each of which is also an architecture. We use the approach of Ezran, Morisio, and Tully in our description and classification of architecture types (Ezran et al., 1999).

Generally speaking, architectures of software-based systems can be classified into several categories:

- A business architecture describes the structure of the business tasks performed by the organization and the mechanisms by which the system supports these business tasks. The description of the business architecture is often created by a business analyst who is an expert in this particular business.
- A physical architecture describes the structure of the hardware platform(s) or network(s) on which the system will operate. Often, much of the physical architecture of a system is greatly simplified by the use of a cloud.
- A logical architecture describes the structure of the business and application objects. This architecture is often part of an object-oriented view of a system. Accordingly, the description of the logical architecture is often created by an object analyst who is well versed in object technology.
- A functional architecture describes the structure of the potential use cases and the requirements from a user's point of view. Again, this architecture is often part of an object-oriented view of a system.
- A software architecture describes the structure of the system into layers, such as the OSI (Open Systems Interconnection) seven-layer model of data communications or the layered architecture of the UNIX operating system. (See Figure 4.7 for a view of the layered architecture of UNIX.) A decision to decompose a system into a client and a server would be part of a software architecture.
- A technical architecture describes the structure of the major interfaces in the software architecture. Elements of a technical architecture include the application programming interfaces (APIs), middleware, database management systems, graphical user interfaces, and other glueware or bridgeware needed to interface components at various layers in the software architecture. Decisions to use CORBA, DCOM, Java RMI, or RPC or even cloud computing, with a protocol such as HTTP, SOAP, REST, or JSON, would be reflected in this type of system architecture. (The HTTP, SOAP, REST, and JSON protocols will be discussed in Section 4.6.)
- A system architecture describes the structure of the business, application, and technical objects and their relationships. The description of the system architecture is often created by a systems analyst.
- A deployment architecture describes the structure of the mapping of the system and technical architectures onto the physical architecture. The deployment architecture includes a static view of the basic files that will be necessary and a dynamic view of the concurrent processes and threads, and the mechanisms for their synchronization and communication. Decisions to place a thread on a particular computer or to have an autonomous agent performing computation on a processor that is idle would be reflected in this type of architecture, as would a decision to use some sort of cloud service.

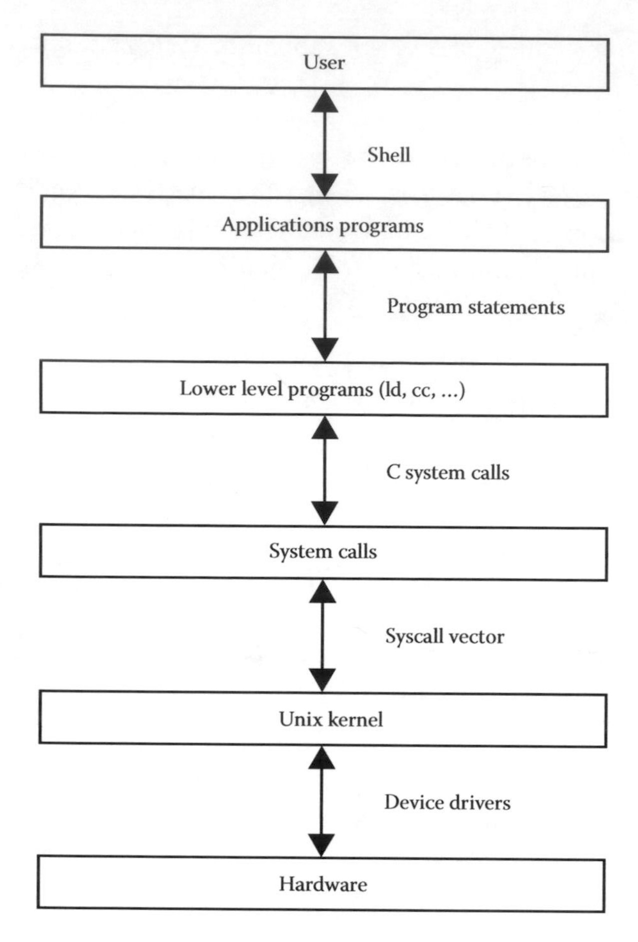

FIGURE 4.7 The layered software architecture of UNIX. (Leach, R. J.: *Advanced Topics in UNIX*. 1994. Copyright Wiley-VCH Verlag GmbH & Co. KGaA. Reproduced with permission.)

Why are architectures considered important? The success of the OSI model and the UNIX operating system show the power of the layered approach to software architecture. The other types of architectures are attempts to increase the level of reuse in systems by encouraging the description of systems at different levels of detail and abstraction.

The layered architecture of the UNIX operating system is shown in Figure 4.7. The architecture makes it clear that, for example, system calls interact with the UNIX kernel by means of the "syscall vector," that the UNIX kernel interacts with the hardware by means of device drivers, and so on.

It should be noted that the use of a notation such as UML can promote a consistency between different architectural views. Whether such consistency can be incorporated into computer-aided software engineering (CASE) tools and development processes in such a way as to improve efficiency remains to be seen, although preliminary efforts are promising.

At this point, we will be content with this high-level description of different architectural types. We will return to their descriptions when we illustrate the design of our major software project in Section 4.17.

4.6 SOFTWARE DESIGN PRINCIPLES FOR PROCEDURALLY ORIENTED PROGRAMS

Before we begin the study of software design representations in depth, we will introduce an element of reality into the discussion. Consider the case of a program of approximately 500 lines of code with perhaps seven or eight functions. A high-level view of the program's design is likely to be relatively small, regardless of the design technique or representation used. It is likely that a reasonably complete view of the design can fit on one page. Clearly any of the design representations described earlier in this chapter could be used if the systems were not too complicated.

We should not expect this to be the case for realistic programs that are used in industry or government. Clearly, the design of a program of 500,000 lines of code with over 2000 functions and complex control and data flow cannot be represented accurately on a single sheet of paper. Thus, every design representation technique will include some method of decomposing design representations into smaller units that are linked together.

Here is an example of this decomposition approach: the diagram in Figure 4.8 has been taken from a CASE tool called Technology for the Automatic Generation of Systems (TAGS)

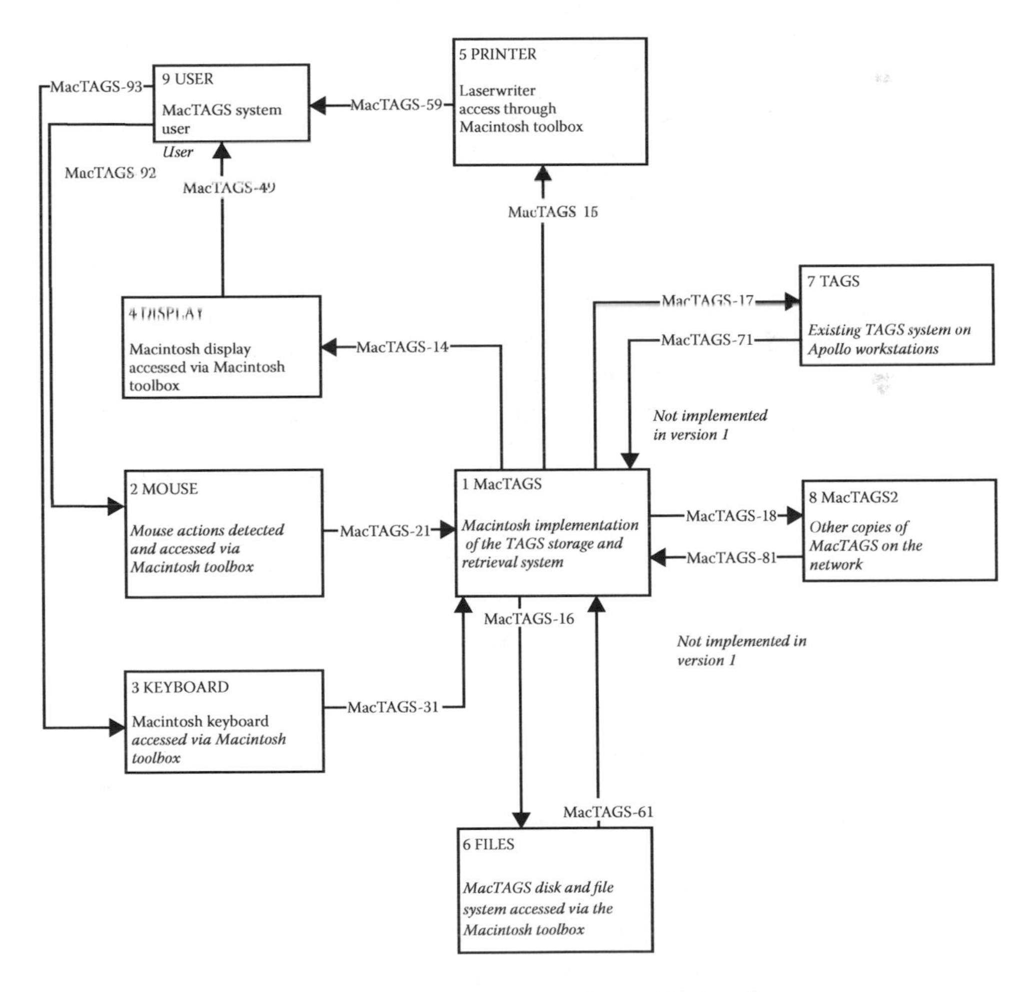

FIGURE 4.8 An example of a design representation using a CASE tool.

that is available from Teledyne Brown Engineering. The version presented here is a public domain version of the "front end" of the tool that was implemented for the Macintosh. This front end contained only the design representation portion of the full software. It used a special system description language called IORL (Input Output Requirements Language). Features of the complete CASE tool included a consistency check for the designs that were graphically represented and a code generation feature that allowed source code to be generated from the system's graphical description.

Only an example of the decomposition is represented in Figure 4.8. This figure describes the interaction of the Macintosh version of TAGS, called MacTAGS, with the Macintosh MacOS operating system and the Macintosh toolbox, which contains some of the routines for access to the Macintosh graphical utilities. (The diagram has been compressed to fit within standard page margins.)

It is not appropriate for us to discuss Figure 4.8 in detail. However, there are several things to note. The boxes are connected by well-drawn arrows. It is clear that this was done automatically, rather than by a simple drawing package. The arrows are labeled, providing more information about the interfaces of the boxes linked by the arrows. The labels on the arrows indicate the number of an interface.

The boxes in Figure 4.8 are also labeled, but for a different purpose. The numbers in the top left-hand corner of the boxes indicate other subsystems, which are also represented by similar diagrams. This is the mechanism for linking different graphical pages to the same system.

MacTAGS used the approach of having software made up of "systems," which in turn are composed of one or more sections. Sections are represented by "system block diagrams." The diagram in Figure 4.8 is system block diagram 0 and is denoted as section "SBD-0" of the system named "MacTAGS."

We note that this CASE tool is more than a drawing aid for providing a graphical representation of software. It also includes a "context editor" in the sense that double clicking on any interface arrow or labeled block provides access to additional detailed information such as the creator of the box or interface, the data flowing along the interface between different boxes, and so on. This very useful technique is similar to clicking on a Uniform Resource Locator (URL) in a mail document where the Eudora™ mailer software will automatically open a copy of an associated network browser. The same approach also works for other applications that have been associated with a particular type of file name, such as .htm, .html, .doc, .docx, .xls, .xlsx, .ppt., .pptx, and .exe.

The tool is even more advanced, as may be seen from the software's interconnection with other versions on a network. The advanced versions of the software for networks include an optional code generator so that a correct graphical design representation can also used to generate source code for the system being designed. Even if the source code that is generated by this tool is not complete or does not meet proper performance requirements, the source code produced can save considerable development time.

We now return to the study of design decomposition. The goal of design decomposition is to produce a detailed design from a very high-level one. How detailed should a detailed design document be? It should be sufficiently detailed to be able to have one or more software engineers implement source code on the basis of just this document. The source code should be implemented unambiguously, in the sense that minor variations in the implementation details should still be consistent with the design.

Unfortunately, it is not well understood how the choice of design representation leads to any particular decomposition of a high-level design into a more detailed design. For example, a decomposition of a system based on an analysis of the control flow would naturally lead to a detailed description of the actions taken by the system. On the other hand, a decomposition technique based on a data flow representation would lead to a detailed description of the movement of data that are made by the system. Transformations of data are easy to detect in a detailed, data flow-based design representation. Other design representations have different influences on the design decomposition approach. The effect of pseudocode, state diagrams, and Petri nets on designs will be discussed in the exercises.

Clearly, different design representations and decomposition strategies have a major influence on the eventual detailed design. Of course there is considerable variation between individual designers; designers working independently will also produce different detailed designs for the same system. Thus, it is often important for management to perform comparative analyses of designs.

The design decomposition step is one that must be examined to make certain that the resulting design is modular, efficient, and, above all, meets the requirements of the system. It goes without saying that the detailed design should be checked against the requirements traceability matrix to ensure that the requirements can be met by an implementation that is based on this design.

What about cloud computing and the related notion of SaaS? Design decisions to use these technologies demand adherence to specific protocols, in the same way that designing software for deployment on mobile devices requires adherence to the specified standards for APIs used by these devices' software development kits (SDKs).

Following is a list of the most typical APIs used in SaaS, together with some information on the status of the standardization of the protocols and where technical details can be accessed.

Hypertext Transfer Protocol (HTTP)—The Hypertext Transfer Protocol format was standardized by the W3 consortium and can be accessed at http://www.w3.org/protocols.

Representational State Transfer (REST)—The Representational State Transfer format was standardized by the W3 consortium (Erl et al., 2013) and can be accessed at http://www.w3.org/protocols.

Simple Object Access Protocol (SOAP)—SOAP was formerly known as the Simple Object Access Protocol, but is now simply known as SOAP. The SOAP format was standardized by the W3 consortium and can be accessed at http://www.w3.org/TR/soap.

JavaScript Object Notation (JSON)—The JSON Data Interchange Format was formalized in 2013 by Ecma and can be accessed at http://www.ecma-international.org.

We will return to cloud computing standards in Chapter 6, Section 6.10 when we discuss software integration in the "cloud."

4.7 WHAT IS AN OBJECT?

This section will give some guidelines for determining the objects that are present in the system we wish to design. The basic idea is that an object has attributes and that in any instance of an object, these attributes will have values. With proper abstraction, including careful design of the operations that can be performed on an object (the methods), the details of an object's implementation can be hidden from other objects communicating with it, thereby enabling reuse of an object's source code.

Unfortunately, there is often a great deal of confusion about the effect of object-oriented design and programming versus more procedurally oriented software methods on source code reuse. The confusion appears to occur because of misunderstanding of the general principle that reuse offers the greatest cost savings and development efficiency gains when done early in the software development life cycle, regardless of which life cycle model is used. Hence, any increase in code reuse due to object-oriented processes, while very useful, is not going to have the same cost reduction potential as reusing larger components and COTS products. The savings in cost and program development time when reusing and tailoring existing entire systems is one of the major advantages of an agile software development process, and is far greater than the advantages of reusing a single object class and its associated methods. With this point in mind software reuse will be emphasized as much as possible in the next few sections, hoping the suggestions encourage you to both reuse code and create code that is easier for your project colleagues to reuse.

Suppose that you have chosen a candidate to be an object in your system, and that you have determined a set of attributes and their possible sets of values. What informal tests can you apply to be sure that your objects are the most appropriate abstractions for your system?

The first test is the "multiple examples test." Simply stated, if there is only one example of an instance of an object, then the concept is not sufficiently general to have object-oriented techniques used to implement it.

The multiple examples test is consistent with the procedural programming dictum: "If you use it once, just write the code. If you use it twice, write a procedure or function."

There is a related perspective from the area of software reuse. Biggerstaff (1989) offered a rule of thumb "Don't attempt to reuse something unless you intend to use it at least three times."

Finally, this is also consistent with the advice commonly given to young mathematicians: "If a concept or theory does not have at least three different hard examples, it is not worth studying."

Once our candidate for an object passes the multiple examples test, it is placed in a set of potential objects. The attributes of the objects are now tested using the "has-a" relationship, which can help to formalize the set of attributes of the objects. We should write an initial set of member functions for this potential object based on reasonable transformations of the values of some of the attributes of the proposed object.

The next informal rule used for checking the appropriateness of objects concerns the class hierarchy in which new classes are designed from old ones. This relationship between the base class and the derived class is best described as the "is-a relationship." If the sentence "object 1 is a object 2" does not seem to make sense (ignoring the grammar), then the relationship is not of the form "base class, derived class" and, hence, we do not have an inheritance relationship. Why is an inheritance relationship important? Because we can avoid duplicating functionality if an object inherits any properties from another class (the base class). We can add new functionality to the derived class that is in addition to the functionality of the base class.

On the other hand, if the sentence does seem to make sense, then we have a candidate for such a relationship. This is called the "is-a relation test." In this case, we should draw a diagram of the object and any relationship between the object and other objects.

We should list the potential member functions and be alert for any examples of polymorphism. The appearance of polymorphism suggests that we have chosen our inheritance relationships properly. If there is no polymorphism, then we should be suspicious that we have not described the member functions correctly or at least not in sufficient detail.

Joseph Bergin, the author of a new book on polymorphism (Bergin, 2015), made the point (in a private email conversation) more clearly than most: "To build big things you need to build small things well." Keep his statement in mind when you develop objects. The set of potential objects and the descriptions of their member functions should be refined at each step.

There is one final relationship test that should be performed in order to incorporate the objects into a preliminary object-oriented design of a software system. The concern here is that the objects listed should form a complete set of the objects needed for the software system being designed. The relationship we are looking for is the "uses-a relation."

We use this relationship by asking if the sentence "object 1 uses object 2" makes sense for the pairs of objects considered. Every meaningful sentence suggests either a client–server or agent-based relationship and is to be considered as part of the program's design. If we cannot find any instances of this sentence making sense, then there are two possibilities: either the objects are insufficiently specified for us to be able to describe the entire system, or the natural description of the system is as a procedural program controlling objects.

Note that objects can be related to many other objects. Thus the previous steps should be repeated for groups of three objects, four objects, and so on, until the designer feels that the system's essential object-oriented features have been described.

We summarize the recommended steps for determining objects in Table 4.1 on page 174 in Section 4.4. Make sure you understand the steps used in this process of determining objects.

The importance of data abstraction and information hiding in objects should be clear to you. In addition, you should have some appreciation for the power of the concepts of operator overloading and inheritance. We now consider the development of larger object-oriented programs.

In the next few sections of this chapter, we will discuss some issues in object-oriented design and indicate how the object-oriented paradigm for software design differs from the procedurally oriented one. We will also indicate a methodology for the development of object-oriented systems that determines the fundamental objects (and appropriate methods) in these systems. This will be done in the context of the development of a class that describes strings.

We now return to the topic of object-oriented modeling of systems. The first fundamental question to address is: What are the objects in the system.

Once we have determined the objects, we must answer the second fundamental question: What is the set of transformations that can be applied to these objects?

A few definitions may be helpful. An abstract object is said to have *attributes*, which describe some property or aspect of the abstract object. For any instance of an abstract object, each attribute may have values.

The listing of attributes is helpful in assessing our candidate for an object. The attributes must be independent in the sense that a change in the value of one attribute should not affect the value of another attribute.

There is one other judgment that should be made from the attribute list. If there is a single attribute, then it is likely that the object is at a low level of abstraction. This should generally be avoided and, hence, we want to have at least two attributes per object.

We will illustrate these concepts by several examples in this and some of the following sections.

Consider the development of a class to describe the abstract concept of a string. (A string class is present in nearly all implementations of standard programming languages, including C++ and Java. Compare the discussion here and in the next few paragraphs to what is obviously the design of a class named "String" in your object-oriented programming language of choice.)

Some typical attributes of a string might include its length, its contents, its creation or destruction, and its display. These lead us to some potential values for these attributes: its length might be 42, its contents might be "I like this Software Engineering book" (not counting the delimiting quotes), it might be in a state called "created," and it might be displayed on the standard output stream cout. Alternate values of some attributes for a string object might be 80 for the length, "I like this Software Engineering book and wish to buy many copies," and a file named something like "outfile" for the "display" of the string.

The determination of the attributes of an object and the set of possible values for each attribute then suggests the functions and data types necessary to implement methods to be used with the class. For example, we must have a member function for the String class that is able to compute the length of the string. This length must be an int (since this is the appropriate elementary predefined data type in C++). Even more careful programming would declare the type to be an unsigned int, since strings cannot have negative length. We must have a constructor function to initialize the contents of the string and we must have a function to display the contents on the appropriate output stream.

Note what we have not determined yet. There has been no mention of the null byte \0 to be used as a termination byte to indicate the end of the string; hence, another question arises as to whether that termination byte is to be counted as part of the string, thereby increasing its length by one.

Indeed, there is no requirement that the string should be implemented as a set of contiguous bytes. We could choose to follow the lead of many word processing programs (including the one in which this book is written) and use linked lists to chain together contiguous arrays of characters (and other objects, such as graphs and images, in the case of the word processor). Linked lists are especially suited to insertions and deletions of text and other objects within a word processing document. Other data organizations are possible.

The important point is that none of the decisions about the implementation organization or details is relevant to the string object. What we need now is a determination of the attributes of the object and a set of possible values for each attribute. With such a determination, we can write the first attempt of a description of the class. Thus, the first attempt at the definition of a string object will probably have a definition something like the class string defined next. We will not worry about syntax at this point.

```
class String
Member functions:
   int length;
   int strlen( String s);
   void display( String s);
```

Some problems arise here because we have not precisely determined the interface. The constructor member function needs to have a size specified for the length of the string. Other member functions will have to have their interfaces determined, also. Thus, we might have to make a second iteration of the design of the class representing strings.

```
class String
Member functions:
   int length;
   String (char *arr); //terminated by \0
   int strlen();
   operator << ;
```

There are more iterations that need to be performed in order to have a complete definition of the interfaces of the String class. Of course, we still have to develop an implementation of the methods. We omit the details of the iteration for the class description at this point since this class is part of the standard library for the C++ language and nearly every other object-oriented programming language in common use.

4.8 OBJECT-ORIENTED DESIGN REPRESENTATIONS

In this section, we will present a simple object model for the medical device terminal concentrator software system described in Section 4.4. We will make no attempt to refine the object model and will keep it at the same high level that we used for the previous models using different design representations.

The representation is simple (at least at this point). We will use a rectangular box to describe an object, with diamond-shaped boxes and line segments used to indicate relationships between objects. The name of the class is given above a horizontal line inside the box. We will follow the convention of having class names begin with an uppercase letter.

The relationship between the objects is given a name in this representation. The relationship is called "is connected to" and behaves in a similar manner to the "uses" relationship that we will discuss later in this chapter.

Some of the attributes of an arbitrary object in the class are included in the box. We would include all attributes, but there were too many of them.

At the bottom of the diagram, we again list the classes that are in the upper part of the diagram, again using a graphical representation. The classes in this list have typical values for each of the attributes of an object in each class.

Note that this model makes no mention of the queue data structure that will probably be used to keep information going to and from the medical device terminal concentrator in buffers.

We can have two different views of an object-oriented system: the object model and the interface model. We will discuss these two views in order.

In the object model representation, we will use an extended E-R notation. We will use the convention that the name of a class will be given in upper case and the instances of a class will be given in lower case.

Our starting point for this discussion is the diagrams given in Section 4.4. Such diagrams provide a good high-level view of a system.

The object diagrams can be refined further by incorporating the cardinality of each relationship into the diagram. Figure 4.9 illustrates this. The numbers on the side or top of a relation indicate the number of items on each side having the relationship. The cardinality can be one of the following: a precise value, a range of values, the asterisk (*) denoting 0 or more, or the plus sign (+) denoting 1 or more. We have used the + and a range in Figure 4.9.

An object model should be expanded until it describes the essential abstractions of objects in the system. Unfortunately, the object model as indicated so far in this section is inadequate to fully describe the relationship between different objects. In view of this limitation, we will attempt to incorporate the interfaces between objects into our model. There are several methods of doing this.

One method is the use of state tables. The terms "state diagram," "state machine," and "finite state machine" are often used instead of "state table." This is one of the oldest methods for describing systems. It certainly predates any of the current efforts in object-oriented design.

A state table for the medical device terminal concentrator system might have six states, which we will call TIR, CIR, CPUIR, CPUOR, COR, and TOR. The acronyms stand for Terminal Input Ready, Concentrator Input Ready, CPU Input Ready, CPU Output Ready, Concentrator Output Ready, and Terminal Output Ready, respectively.

We illustrate the states for the medical device terminal concentrator system in Figure 4.10. The notation is slightly different from the most common one in that we have not specified the initial state where the inputs to the system arrive (from the keyboard) and the final state where the outputs leave the system for good (when they get displayed on the terminal screen).

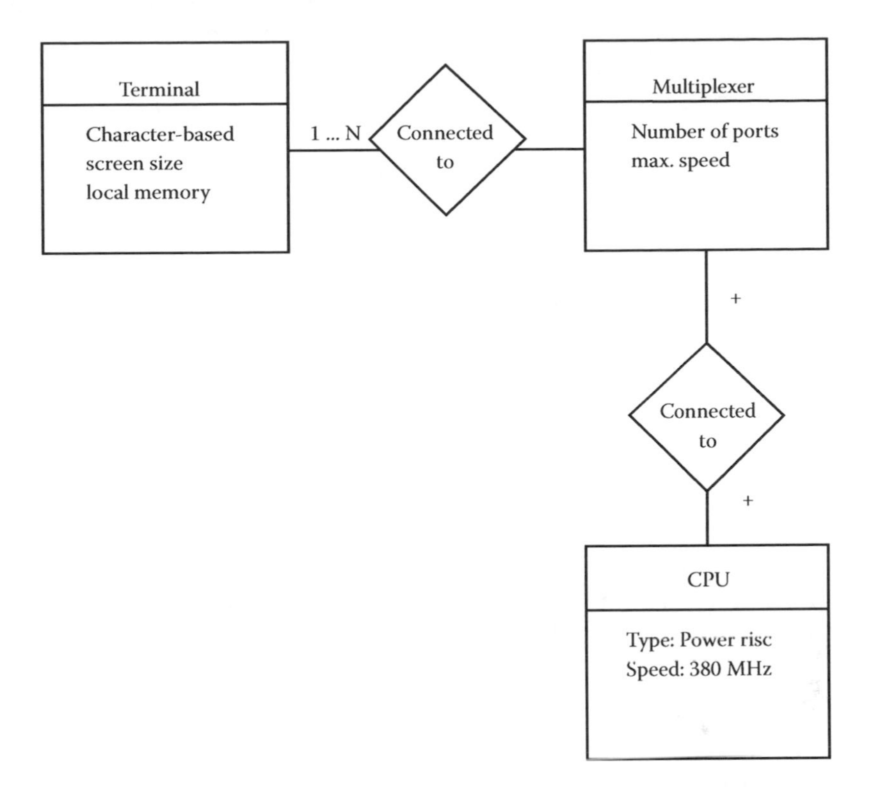

FIGURE 4.9 Addition of cardinality information to an object model notation. (From Leach, R. J., *Object-Oriented Design and Programming in C++*, Academic Press Professional, Boston, 1995. With permission.)

It is useful to know how to use existing representations for systems when attempting to describe them in object-oriented terms. We often have an object model for a system. Our suggestions for the use of existing information models and diagrams in object modeling can be summarized as follows:

- If an information model diagram exists, use it to determine initial relationships, especially those of aggregation.
- Examine attribute lists, for indication of data structures and a preliminary set of transformations to operate on the values of these attributes.
- If no information model is available, then proceed directly to a description of objects in the system.

For additional information on E-R diagrams and information modeling, consult the references.

4.9 SOFTWARE DESIGN PRINCIPLES FOR OBJECT-ORIENTED PROGRAMS

All the methods of design representation previously discussed were created to improve the quality of software designs and the efficiency of the processes used to create them.

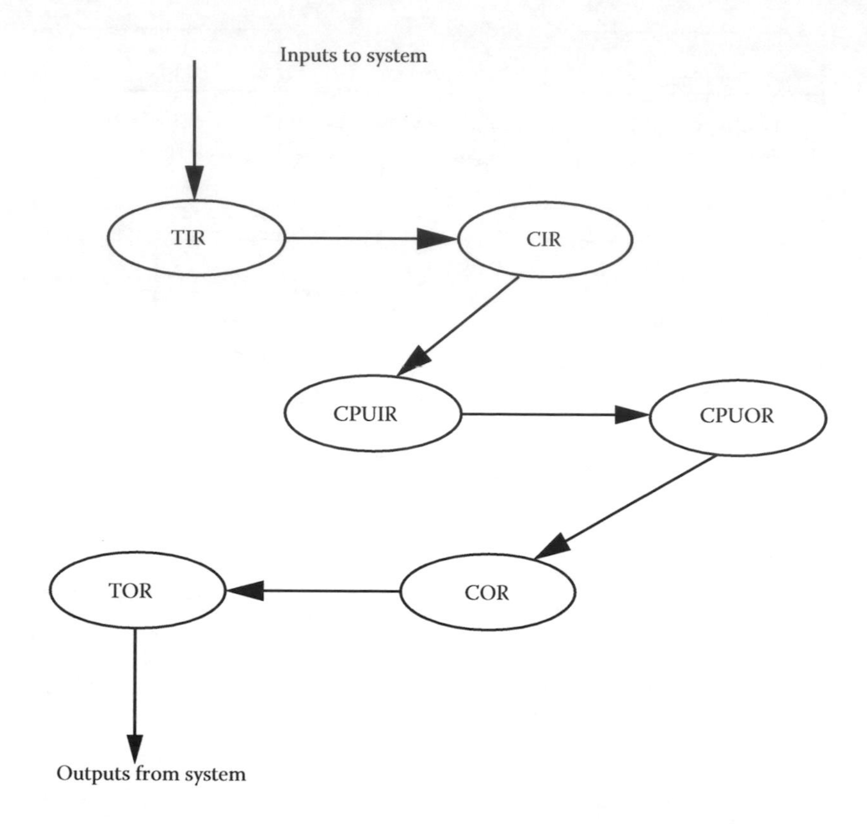

FIGURE 4.10 A state diagram for the medical device terminal concentrator system. (From Leach, R. J., *Object-Oriented Design and Programming in C++*, Academic Press Professional, Boston, 1995. With permission.)

Design representation techniques for traditional, procedurally organized systems are far more common than techniques for object-oriented systems. This is not surprising given the relative newness of object-oriented techniques. Other than UML, few of the earlier object-oriented design representations are still used.

As we saw in the previous section, decomposition was the major technique used for procedural software development once we had determined the high-level software patterns, if any, that described the system to be designed. The example that was previously constructed using the MacTAGS CASE tool emphasized this decomposition-based approach.

The complete installation of the TAGS system on powerful workstations also included a facility for automatic code generation. The availability of CASE tool enhancements that provide automatic code generators is not restricted to those CASE tools that support procedurally oriented designs. We note that some related features can be found in several readily available software packages used for object-oriented development. For example, the ROSE system originally developed by Rational Corporation is another popular system that supports object-oriented analysis with code frameworks. It has evolved into the Eclipse IDE, since the company that created this product is now a subsidiary of IBM.

Object-oriented design often focuses on the development of complete classes to describe objects. The idea is that all the necessary methods or member functions will be provided in the class description. Thus, the primary issue is the design of the class for completeness, so that the possible interactions are treated by having proper interfaces. Objects then interact by sending messages from one class to another and the individual classes react only to the messages that they receive. The objects are viewed as autonomous, with no interactions other than those specified in the object's interface.

The use of abstract classes can be especially useful if the software's requirements are not identified completely before design begins, as is certainly the case for the concurrently engineered, market driven software development process so common in the personal computer world.

In this approach, the interfaces for an object in the system are specified at an early stage. The implementation details for the methods can be developed later. If an object is only needed for a new project, it may not be necessary to fully implement all methods if a few methods are likely to be used. This frequently occurs in many iterative design processes that use integrated development environments (IDEs) and emulators for development of apps for smartphones and tablets. If new functionality is required, it is provided in the methods associated with an object, not in the object's interface. A pure virtual function (one that has no implementation details and is used as a base class with interfaces to be used by the classes that inherit from it) is often used for this purpose in C++.

Even in a worst-case scenario, development of an object's member functions can proceed concurrently with the object's design. If a new interaction is added to an object's interface, most of the existing methods do not have to be changed. Only new methods have to be added to implement the additional interface.

Here is an illustration of the effect of different design approaches to object-oriented systems. We consider the simple situation of describing a class to represent quadrilaterals, which are four-sided geometrical objects. We wish to develop a class structure to describe quadrilaterals, with squares and rectangles as special types of geometrical shapes.

There are several choices available to us. We can create a hierarchy with a primitive class called vertex. We can build a class called edge using the vertex class, and then use the vertex class to build a new class called figure. The figure class can be used to construct quadrilaterals as a subset.

This construction is illustrated in Figure 4.11. We present some C++ source code to implement this organization in Example 4.3.

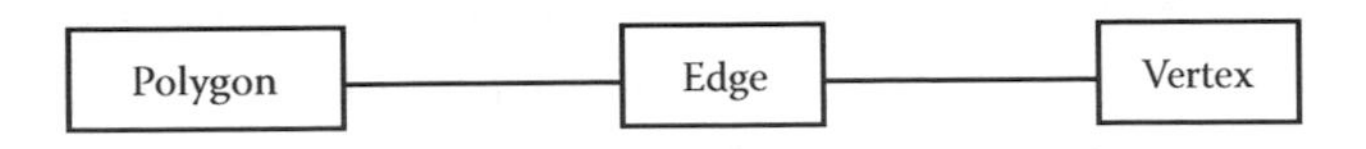

FIGURE 4.11 An illustration of a design organization of some geometric figures based on the vertex primitive.

Example 4.3: Some C++ Source Code for an Organization of Figures Based on the Vertex Primitive

```
// FILE: vertex.cpp
class Vertex
{
public:
  double x, y; // coordinates
  Vertex(double x1 = 0, double y1 = 0);
};

// FILE: edge.cpp
#include "vertex.cpp"
class Edge : public Vertex
{
public:
  Vertex first, second;
  Edge();
  void init(Vertex, Vertex);
  boolean is_adjacent_to(Edge E);
  double length;
 };

// FILE: figure.cpp
#include "edge.cpp"
class Figure : public Edge
{
public:
  List<Edge> set_of_edges;
  List<Vertex> set_of_vertices;
  void init(List<Edge> &,  List<Vertex> &);
  boolean is_adjacent_to(Figure F);
  int number_of_edges;
  int number_of_vertices;
};
```

Unfortunately, there is no obvious way to capture the notion of a geometric figure being a triangle, square, rectangle, rhombus, parallelogram, pentagon, and so on. We could modify the class given in the file figure.cpp, but this seems awkward.

There is one disadvantage to allowing this flexibility of concurrent requirements, design, and coding. Any method added to a class at the last minute before release will have minimal testing due to time pressures. It is highly unlikely that an object will be tested for all possible types of responses to methods, particularly if the interfaces are polymorphic or if any operators are overloaded. Thus, a user of the software may observe many system malfunctions, especially if the user operates the software in ways that were not expected by its designers. A potential reuser of code may find that the code cannot be reused without modifications.

In many software development organizations, treatment of such potential problems is postponed until the next release. Thus, corrective testing of software components of a previously released version may occur during development (requirements, design, coding, testing) of the next release. Determining when to release software, when to demand additional testing, and which features to incorporate in new releases are clearly essential to success in the market-driven software environment. We will return to this point in Chapter 6 when we study testing and again in Chapter 8 (and elsewhere) when we study configuration management.

The models we present in this chapter are intended for illustration only. Any realistic model of a complete system would be much more detailed than what is presented in the next few sections. The models given here only describe the system at its highest level. After reading this chapter, you should appreciate the expressive power of several different design representations for both object-oriented and procedurally oriented software systems.

4.10 CLASS DESIGN ISSUES

It is necessary to discuss some of the difficulties associated with the development of programs that use the class definition methods described in the previous section. The methodology described in the previous section assumes that the software system has been developed in a vacuum. That is, we have described some of the attributes that we believe are associated with a type of object and have described some potentially useful interfaces between this object and others (such as character arrays and I/O streams). We have not paid attention to any of the previously developed classes that might be related to our class and would encourage the use of similar interfaces.

If there is a set of related classes with similar standards for their interfaces, then we must at least consider these interfaces before we set in stone the interfaces that are part of the current description of the object.

This point cannot be emphasized enough. Development of an object must take place along the lines of defining its attributes and typical values for these attributes. This can be done by the usual method of stepwise refinement so familiar to software engineers. Development of a *useful* object (one that can be used in a variety of important situations) requires that one consider the interfaces to existing objects in the development stage. Otherwise, the objects that we develop will have very limited utility.

We illustrate the point by this analogy: Consider the current state of development of computer hardware. Unless a hardware designer is designing a special-purpose, high-performance supercomputer or a custom microprocessor for control of an embedded system, he or she will use previously developed components with well-defined and well-documented interfaces. These existing well-designed hardware components have predictable levels of performance in terms of clock speed, data transfer rate, reaction to specified interrupts, power usage, and so forth. A new piece of hardware that does not adhere to these standard interfaces is unlikely to be very useful outside a narrow range of applications. Only a revolutionary design with tremendous applicability or performance is likely to be useful; an unusual interface with only mundane applications or average performance is not likely to be very successful.

This is exactly the same situation that applies to developing classes that can be used as reusable software components. The development of a class cannot be done in a vacuum. It must take into account the other classes in the class library, SDK, or IDE in order to make efficient use of previously developed classes. The library catalog and any software tools for examining the different libraries relevant to specific application domains are extremely important when developing classes to interact with real systems. The advantage of deep knowledge of class libraries, SDKs, and IDEs is a major factor in hiring decisions of many employers. The same holds for knowledge of the HTTP, REST, SOAP, and JSON protocols.

Note that we must do one other thing to ensure that the classes we develop make efficient use of resources, especially programmer time. If there is a relationship between our class and one or more preexisting classes, it is possible that we can then make use of previously developed functions that were members of a preexisting class. This is easy to do if we can use inheritance.

We can also use functions associated with an object if the functions were originally declared as being C++ friend functions. (Recall that friend functions are allowed in C++ but not in pure object-oriented languages such as Java.) Even if the functions were not originally declared as friend functions, we might be able to change the definition of the preexisting class to make the required functions member functions and recompile the system.

In any event, an efficient software design process requires a check of available software resources.

Grady Booch is one of the most highly regarded experts in the object-oriented programming community. He observes that in most high-quality, object-oriented programs, many essential objects are clustered into several related classes, rather than being grouped solely by inheritance. He states that the only programs that have all objects related are relatively trivial ones.

For convenience, we repeat the 23 design patterns described by the Gang of Four (Gamma et al., 1995) that were discussed in Section 4.2. If your primary objects seem to fit one or more of these patterns, then use any of the readily available implementations as a starting point.

Creational patterns

1. Abstract factory
2. Builder
3. Factory method
4. Prototype
5. Singleton

Structural patterns

6. Adapter
7. Bridge
8. Composite
9. Decorator
10. Façade
11. Flyweight
12. Proxy

Behavioral patterns

13. Chain of responsibility
14. Command
15. Interpreter
16. Iterator
17. Mediator
18. Memento
19. Observer
20. State
21. Strategy
22. Template method
23. Visitor

Clearly, the determination of available classes in class libraries is only a starting point when designing object-oriented programs. However, it is an essential step in designing an object-oriented system.

4.11 USER INTERFACES

The marketplace success of many software products depends much more on the software's user interface or the set of visible features than on hidden features such as robustness. This is especially true in crowded markets such as applications software for personal computers in which there are many competing products.

In this section, we will present a brief overview of this essential topic. The intention is to convince you that there is more to user interface design than the use of multiple colors, flashing messages, clever icons, sound, or computer animation. There is both an art and a science to user interface design.

Most current software systems have user interfaces based on the window, interaction, mouse, pointer (WIMP) paradigm. The Macintosh operating system and the Microsoft Windows variants follow this paradigm, as does software written to use the X Windows and similar utilities for Linux and other UNIX-like environments. The books by Jakob Nielsen (1994) and Ben Shneiderman (1980) provide excellent overviews of human–computer interaction and their roles in good software design. Their discussion applies even today to most user interfaces of software systems that run on desktop or laptop computers with substantial screen real estate.

The WIMP paradigm has been extended by the user interfaces of tablets such as the iPad and smartphones such as the iPhone and Android, with their capacitative multitouch screens allowing one or more fingers to be used for both the selector and pointer actions of traditional mice. Jakob Nielsen and Raluca Budiu have written a book on usability of software on mobile devices (Nielsen and Budiu, 2012). A discussion of the problems in developing user interfaces for tablets and smartphones with their limited screen real estate is beyond the scope of this book.

The graphical user interfaces of high-quality software packages and web pages all have a consistency that is reflected in the simplicity of their menu design. For example, most graphical user interfaces for personal computer software have pull-down menus with file operations at the top of the screen on the left-hand side.

We will now distinguish between a pull-down and a pop-up menu. A pull-down menu is one in which the user selects an option from a set of options by moving a pointing device such as a mouse to a fixed position and selecting that position by pressing a mouse button or some other selection device. This sequence of operations makes a menu appear below the place where the pointing device was placed. The user then moves the pointing device down the list of menu options until he or she selects one of the options. (The user has the option of terminating this process by releasing the mouse button or moving the mouse to another position.) The initial set of options listed in a pull-down menu is always available to the user.

In contrast to the information in a pull-down menu, none of the information in a pop-up menu is visible to a user unless the user takes a specific action, such as pressing a mouse button or the equivalent combination of keystrokes.

The action of a typical pull-down menu before and after a mouse button is pressed is illustrated in Figures 4.12 and 4.13. Compare these menus to Figures 4.14 and 4.15, which illustrate the action of pop-up menus. Note that the pull-down menus appear at the top of the screen illustrated in Figure 4.12 as long as this application (Microsoft Word) is running and the window is displayed on the screen. Contrast this to the situation in Figure 4.13 in which the same pull-down menu is shown together with a new menu that is obtained by the selection of an option from this menu.

The wide range of options caused by the availability of pull-down and pop-up menus causes great problems for an interface designer. He or she has to be careful to not use these tempting features just because they are available. The following guidelines are helpful:

- Use a pull-down menu when the options should always be visible.
- Use a pull-down menu to supply a set of options for a novice user.
- Use a pop-up menu when the options need not always be visible.
- Use a pop-up menu to provide shortcuts for an experienced user.

FIGURE 4.12 An example of a pull-down menu before a selection is made.

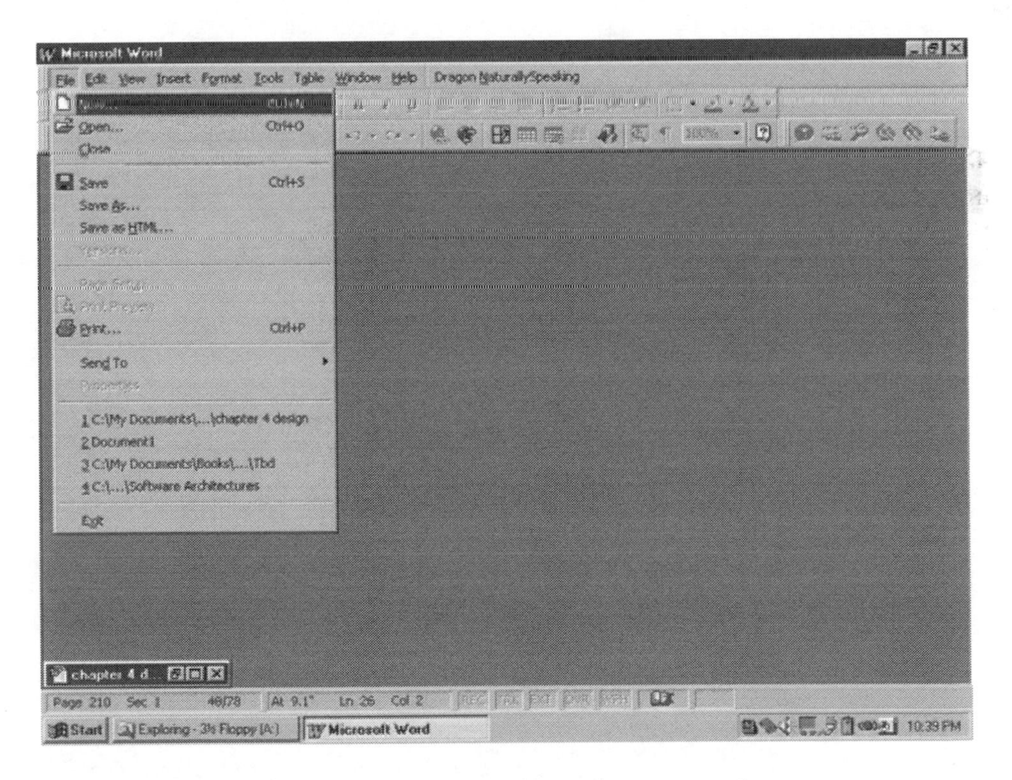

FIGURE 4.13 An example of a pull-down menu after a selection is made.

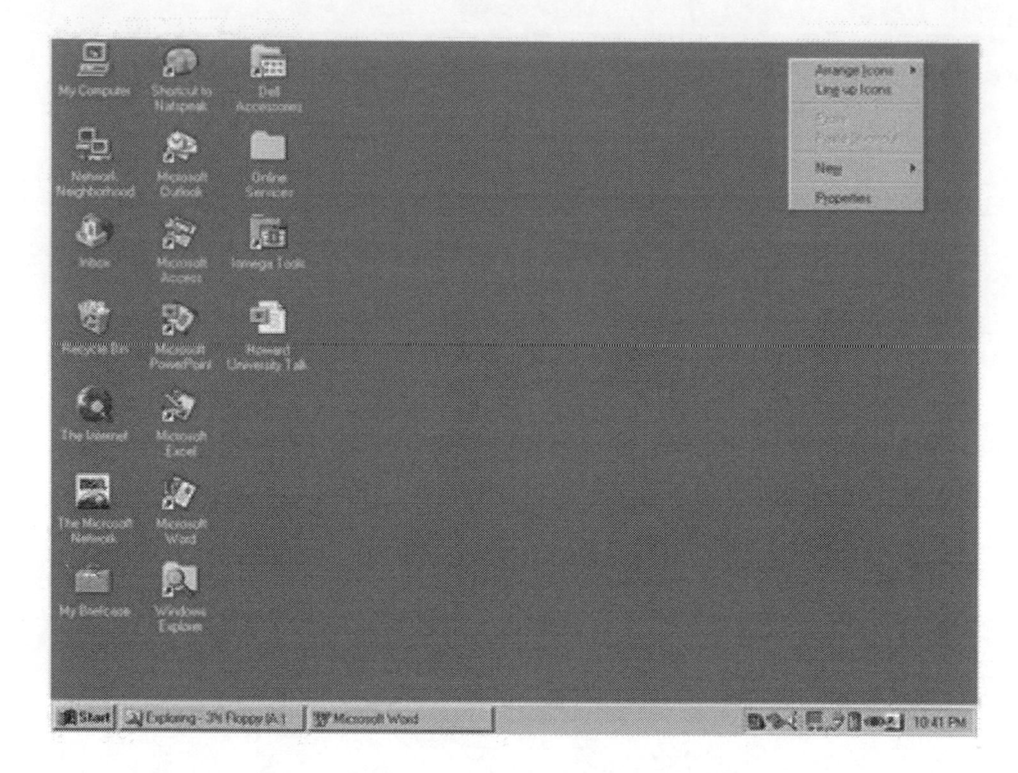

FIGURE 4.14 An example of a pop-up menu before an initial selection.

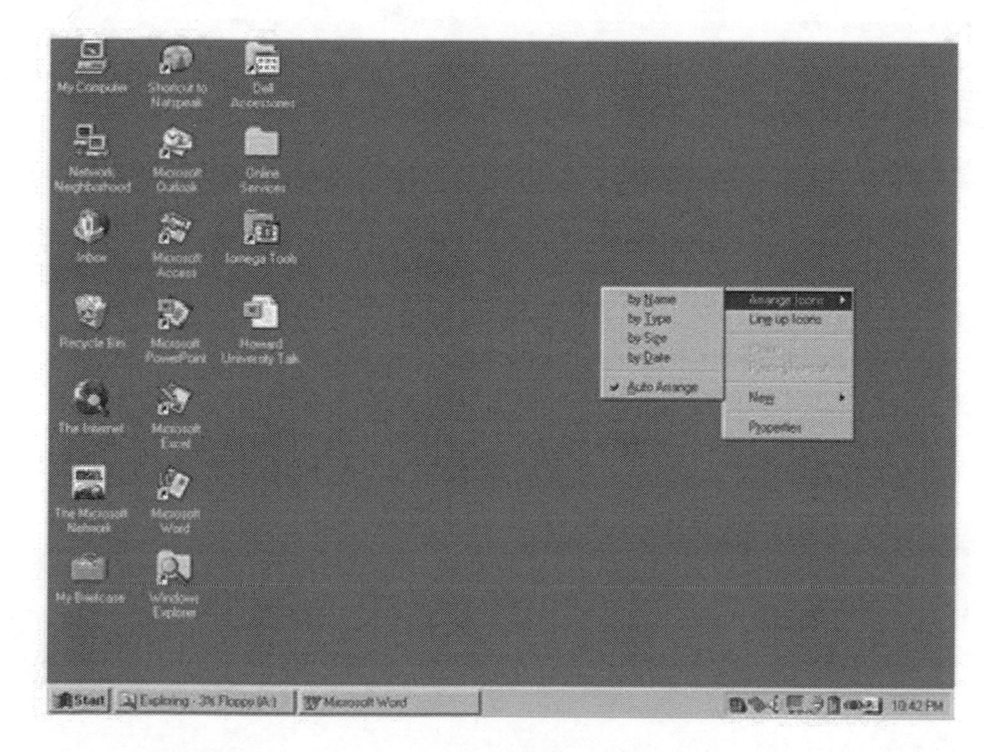

FIGURE 4.15 An example of a pop-up menu after a selection is made.

The guidelines are necessarily incomplete because of space limitations and because of the wide range of applications and potential users. Nonetheless, they illustrate some of the major problems for an interface designer: needing a consistent approach to the interface's design, while simultaneously providing the necessary help for a novice user and powerful time-saving shortcuts to make an experienced user more productive.

We note that in many situations, a graphical user interface is not appropriate. Consider the case of my colleague at Howard University, Peter Keiller, who collaborates with colleagues in software reliability research all over the world. He and his research assistant, Claude Charles, have developed a set of tools to analyze data files to assess the rate of growth in the reliability of software systems. The data files contain the times that software faults occurred during the testing process.

Because many of the systems Keiller considers are not connected to the Internet for security reasons, he often has to do this research without access to workstations and servers available at his home environment at Howard University. He thus developed a smaller version for his software tool set for a personal computer. The tool set must work in all environments and, thus, graphical user interfaces could not be used.

In Figure 4.16, we present another example of a menu-driven, nongraphical user interface for a system used to select a series of files to be analyzed by a software system. It was developed on a system with character-based terminals and considerable effort went into the development of text files that were printed to simulate graphical terminal screens. An example of such a menu for a system named SANAC used to measure coupling with programs is shown in Figure 4.16.

Menu items can be grouped in several ways. The most common organizations are by functionality or alphabetically. It is generally preferable to order different options by the ones most likely to be used. It is generally agreed that the exit option should be the last one available in each menu in which it appears.

Of course the worst possible mistake in user interface design is to expect the user to provide input without a prompt and a description of the input. What do you do in response to a user interface prompt such as that shown in Figure 4.17?

```
S   A   N   A   C     Main menu
1. Coupling measures
2. Coupling description
3. Help
4. Exit
```

FIGURE 4.16 An example of a main menu from a text-based, menu-driven system.

```
c:\
```

FIGURE 4.17 The worst possible user interface?

Now imagine that the executable version of the software is called "sanac." By setting up a simple command line argument such as in C or C++, we can reduce multiple menu selections in to the single command line

```
sanac file.a
```

which is much simpler and has far fewer keystrokes than the menu-driven system. The simplicity is appreciated most by an experienced user. Command-line interfaces, so essential in Linux systems programming, are discussed in Appendix B.

There is one final note about user interfaces. Use a spell checker in all text files, menu screens, or menu options. Follow this by proofreading carefully. If no spell checker is available, use a dictionary and proofread carefully. The point cannot be stressed enough. A sloppy user interface is a red flag to anyone evaluating the software. If the public face of the software is not correct, why should anyone have confidence in the software's internal correctness?

4.12 SOFTWARE INTERFACES

Some of the most interesting types of new software utilities for personal computers are known as "conflict resolvers." A conflict resolver examines software configuration files for possible inconsistencies among different applications, or even among different versions of the same application. These utilities attempt to solve the problems that can occur when software is interfaced to an environment with different versions of operating systems, compilers, utilities, or application programs. Conflict resolution detection programs are often run by users who are frustrated by apparently correct software that does not run correctly, crashes their computers, or even destroys data created by other application programs.

Of course, the same problems can occur when a software system is created from individual software components. The market-driven need to release software updates every few months places heavy emphasis on software reuse and higher-level software components, including complete subsystems. This, in turn, can lead to significant problems with interfaces between subsystems. Since users of the software often do not see these interfaces, the results of conflicts or inconsistencies are not obviously clear, especially to the casual or inexperienced user. Of course, configuration management is critical.

Many software development organizations create a formal document, usually known as an interface control document, or ICD, to help manage the interfaces between software components or subsystems. A typical ICD for a very small project will include the names of each of the major subsystems, together with a formal description of the interfaces between them. The interfaces can be as simple as the number, type, and usage for each argument to a function, the type and usage of any values returned, and any side effects of a function's operation, especially in the case of error conditions.

In more realistic applications, the ICD will be more elaborate. Frequently, the ICD is given as a matrix, where the rows and columns are labeled with the names of the major subsystems or components. The entries in the matrix are often names of documents that describe the actual interfaces in more detail than can be given in a small space.

For example, an ICD for a project with three subsystems named A, B, and C, might look something like this:

	System A	System B	System C
System A		Doc 2.0	Doc 2.3
System B			Doc 1.1
System C			

There is one point that needs to be mentioned. One of the purported benefits of object-oriented analysis and design is the separation of the interfaces of objects from their implementation. A software development project that determines the public interface of the major objects in a system at an early stage in the process will allow implementation of both the object whose interface is published and all objects that use that interface to proceed with little need for additional coordination. The disadvantage is that, although it is easy to determine the interfaces between small objects, it is hard to describe the interfaces between major subsystem components.

We note that software interfaces are easiest to manage when all software components are in the same application domain so there is control over component or subsystem interfaces. Established standards in the particular application domain contribute to reusability success. When there are enforced standards, the following good things can happen:

- The problem of reusing source code within a particular domain becomes more manageable, since the domain is more likely to be narrow.
- Components in the library have a higher probability of being reused than if no standards are used.
- There are few data types and the reusable parts are small and, therefore, more likely to be reused in multiple applications.
- The cost of development and maintenance is reduced, because there is less need to write filters or glueware to interface between different components.

The use of standard interfaces means that decisions made for one module or subsystem should not cause conflicts with other modules or subsystems.

Some standards are easier to determine than others. For example, storage of fundamental data types such as characters, integers, and floating point numbers might be different in different computer languages or even in different compilers for the same language. The movement toward internationalization of character sets can cause great difficulty for the interaction of software intended for the wide character set (the type w_char) of newer versions of C with the older versions in which the simple declaration (the type char) are sufficient. These differences can be found by searching the source code files for the presence of such words as w_char.

On the other hand, determining whether two software components handle dates properly might be harder. For example, one component might store the date as a set of three integers for month, day, and year. Another might conserve space by using only six characters with the form MMDDYY. The conversion routines, together with the limitations caused by the implicit assumption that two characters would always be sufficient to represent a year are not trivial, especially if the date is used in many different ways in the software components. Different countries often have different national standards for representation of dates, which can cause problems for software that are intended to be used internationally. Of course, limited space was the major difficulty that caused the Y2K problem.

The basic questions to ask about interfaces between software components are:

- Have the software components been developed according to coding standards?
- Have the necessary configuration and header files been documented?
- Are the software components of high quality?
- How much effort must be expended to incorporate the components into a system?

In the next section, we will address these points, keeping in mind the context of our large continuing software engineering project example.

4.13 SOME METRICS FOR DESIGN

The purpose of this section is to introduce you to the idea of measuring designs of software systems. Unfortunately, very few approaches are standard in this area and there is little definitive research as to the most effective metrics. Nonetheless, it seems clear that there will be increased emphasis on metrics that can provide some guidance to managers evaluating designs. Accordingly, we will provide a few suggestions for managers on this topic.

A major concern of many software managers at this point in the software development process is the modularity of their designs. Therefore, we suggest the following metrics for measuring the size of the interfaces between subsystems at the design stage:

1. One possible measure is a count of the number of functions, modules, subsystems, or components in the system.
2. Another possible measure is a count of the number of objects, functions, modules, subsystems, or components with which each component must interface. This is essentially a measure of complexity.
3. Another possible measure is a description of the size of the interfaces between distinct functions, modules, subsystems, or components. This measurement may include a count of the number of shared variables or may be enhanced to include an assessment of the complexity of the shared variables. As with item 2, this is a measure of complexity.
4. Still another possible measure is a count of the number of loops in a control flow graph used to describe a system's design.

5. There are some obvious metrics that a project manager may apply to the design produced by the design team. The function point metrics discussed in Chapter 3 can be used at this stage to provide an assessment of the difficulty in implementing the design.
6. A design that encourages high productivity by means of reuse should make certain that the different software components have standard interfaces. The degree of adherence to standards is also important and is measurable to some extent. Here the goal of the metrics is to produce as simple a design as possible, with the coupling between subsystems minimized. Metrics that compute the degree of interconnection among different subsystems or modules should be collected. System designs with broad interfaces should be reviewed to reduce integration and testing costs.
7. Cost metrics can be very important at the design phase. The cost estimates for the previous systems built using the design that is to be reused should be compared to the actual costs for implementation of the design, and any unusual deviations should be noted and explained.

You may be able to suggest some other metrics. Suggesting metrics for designs is easy; the hard task is collecting the metrics and analyzing the data collected to determine whether the metrics actually describe some aspect of a system. We discuss this issue briefly in Chapter 9.

Your software development project may develop other design metrics depending on the organization's experience and the preferred development environment.

4.14 DESIGN REVIEWS

Recall that a major goal of a software manager is to predict risk in a software development project. He or she would like to minimize risks entirely. Thus, there is heavy emphasis on those activities that the manager believes will reduce risk and allow the current software to be created to meet the customer's specifications and will ensure that the entire project will be completed on time and within budget.

Thus, the manager is likely to expect two things: a detailed design review and some sort of inspection. In many organizations there will be several design reviews, often with increasingly detailed views of the design being presented.

Design reviews have much in common with requirements reviews. They are extremely important to the success of a project. They require considerable planning, and a large expenditure of time and resources. For simplicity, we will only discuss design reviews that take place at a single meeting location with all stakeholders present. Coordinating meetings across geographically separated areas is, obviously, much more complicated.

There is at least one fundamental difference between requirements reviews and design reviews, however. In a properly conducted review of a detailed design, the design will often be examined in order to be certain of what could happen if the system was used in an unexpected way. That is, the design may be checked to make sure that the system keeps errors that occur in one particular module of a subsystem from propagating into other modules of other subsystem. A checklist for a design review is given in Table 4.2.

TABLE 4.2 Checklist for a Design Review

The presentation should be rehearsed.
The time of the presentation should fall within allotted limits.
Sufficient time should be available for questions.
People who can answer technical questions should be in attendance.
All slides, transparencies, and multimedia materials must be checked for accuracy.
All slides, transparencies, and multimedia materials must be checked for spelling.
Paper copies of slides, transparencies, and multimedia materials should be available to all participants.
Someone must verify that the meeting room has all necessary equipment: microphones, overhead projector, slide projector, videotape player and monitor, computer-monitor hookup, etc.
An attendance list with names, addresses, phone numbers, and affiliations must be passed around, and all attendees must sign it.

One last step that is becoming more common in software engineering practice is putting each approved design document (preliminary, intermediate, or final) on either a local organizational network or the Internet. The design documents may be as simple as a set of attachments of files created by a word processor. They may include the project's design diagrams in some drawing tool's format. They may also be digital photographic images of other documents stored in a format such as a .gif file. In an extreme case, some of the documents may be stored in animated formats such as .jpg or .mpg files.

Why are design documents posted electronically in many organizations? The reason is simple: People lose paper documents with personnel moves, office relocation, and they generally wish to reduce the amount of clutter in offices. In addition, documents that are available on networks can be used by any personnel who have security access to the files. Using these documents can aid with revisions of the system during the maintenance phase, especially if the design documents are included in a CASE system.

On a personal note, I recently worked on a reengineering project where all the original requirements and design documents were lost and the source code had to be ported to a new machine because the existing hardware was no longer manufactured. A considerable amount of the reengineering effort was spent obtaining the system's design by analyzing the source code's organization. The project would have been much smoother if *any* written documentation for the system had been available. This loss of all paper documents associated with a project is not an unknown phenomenon. You should expect to use networks for document storage in many projects.

It is reasonable to ask what is different about design reviews in agile software development processes. Certainly, a design review should be planned, with strict rules to prevent any one person from monopolizing the review. An agile process involves constant review of code, with reviews at various intervals. Any changes to a design will almost certainly be accompanied by a much longer discussion of code already written, and any changes that favor the use of existing large-scale components or COTS products.

4.15 A MANAGER'S VIEWPOINT OF DESIGN

As indicated in the previous section, a project manager is likely to expect two things during the design process: a detailed design review and some sort of inspection. In many organizations, there will be several design reviews, often with increasingly detailed views of the design being presented. In others, particularly those using agile processes, design reviews will be a brief part of frequently scheduled meetings, where the importance of the design is largely confined to a discussion of which high-level components or COTS products can be used to create the desired system faster. Regardless of the software development life cycle used, the manager will usually consider these reviews to be extremely important. For the remainder of this section we will focus on design reviews for nonagile processes and save our additional discussion of design reviews for agile processes to the case study given in Section 4.16.

In the same way that we discussed management of the requirements process, we will continue to simplify the discussion by assuming that there is a single overall project manager for the entire software project.

The project manager will want to have every item in the requirements traceability matrix checked against the detailed design to make certain that the written system requirements will be met by every possible software system that can be implemented according to this detailed design.

Because the manager's job-performance evaluation by his or her supervisors often depends upon successful completion of a project, the prudent manager will attempt to use some quantitative methods to determine the status of his or her project. Metrics may be characterized as being used to measure a product such as a design or the process by which the product is created. We discussed several metrics that are commonly used to describe design artifacts in Section 4.13.

A manager may apply some metrics to evaluate the efficiency of the process of creating a successful design. He or she may compare the number of people working on the design team for this project with the design of other similar projects to get an informal evaluation of the efficiency of the design team. A large number of design reviews can be a negative factor if the design team does not seem to be making progress in successive design iterations.

The manager will require that the design team work with the organization's technical publications staff to ensure that the design adheres to standard formats for the company and the potential clients.

As was the case with the determination of the software's requirements, it is necessary that the design must be done using some form of configuration management in order to treat the inevitable changes that will occur.

4.16 CASE STUDY OF DESIGN IN AGILE DEVELOPMENT

We continue the discussion of our case study of agile software development, this time focusing on the issue of design. So far, we have described a typical higher-level organizational perspective on the management of agile development teams as well as provide an example of requirements gathering in an agile process.

Previously, we described two simple tools—a Comprehensive Discrepancy System and a Concurrent Version System—that were used in this case study. The Comprehensive Discrepancy System was a tool for collecting data on known instances of errors between specifications and observed results during testing and making these data available online in a simple, consistent manner. The Concurrent Version System provided configuration management support. Using the standard UNIX and Linux SCCS revision control system was not possible because of minor differences in the variants of UNIX and Linux used by different members of the agile team who often worked on multiple projects.

Both support tools had been previously developed by the overarching organization. Since many such tools have been provided as part of either advanced IDEs or CASE tools for many years, it is possible that such IDEs or CASE tools would be used in current agile projects, although that was not the case in the case study discussed here. The only equivalents were the aforementioned Comprehensive Discrepancy System and Concurrent Version System. We note that these tools would not normally be created as part of an agile process. Indeed, creation of support tools is almost never part of an agile software development process.

Note that the common experience with existing tools for Comprehensive Discrepancy System and a Concurrent Version System was a result of long-term relationships between all the technical partners involved. Expecting an agile development team with its members having access to different supporting software tools, such as might be the case for organizations with much of their software development outsourced, to be successful appears to be rather unreasonable.

As was the case described in our discussion of the case study of requirements using agile development, the system's design had not been finalized before coding began. In fact, the activity of gathering of information about existing, large-scale components in the application domain, together with appropriate COTS products, usually has not been completed by the time that the design process begins.

The term "design" is almost always used in the sense of high- or intermediate-level design in agile development. The most important steps in an agile design are to describe the large-scale components, subsystems, or COTS products, together with their data interfaces. The more detailed, lower-level levels of a design are generally omitted in an agile design process, because the intent is to reuse these existing large-scale components, subsystems, or COTS products, and avoid reiterating the details of their internal design.

Of course, this deliberate decision to not describe the internal design details of these large-scale components, subsystems, or COTS products is only appropriate if there is deep knowledge of the various assets that are available in this particular application domain. This, again, shows the critical importance of having members of the agile team who are very familiar with the software components, COTS products, and subsystems available.

4.17 ARCHITECTURE OF THE MAJOR SOFTWARE ENGINEERING PROJECT

We are now ready to describe the architecture of the major software engineering project that we will study throughout the remainder of this book. To refresh your memory, you should look at the requirements traceability matrix requirements for this system that were developed in Chapter 3 (and at the detailed requirements, if you have time). To make the architecture easier to understand, we will concentrate on the most relevant architecture types. Also, we will be content with combinations of textual and graphical representations rather than being confined to UML, because we want to make the process as general as possible.

Recall that there are several different architecture types:

- Physical architecture
- Logical architecture
- Functional architecture
- Software architecture
- Technical architecture
- System architecture
- Deployment architecture

We will begin with the simplest ideas first. The software is to run on some version of the Windows operating system. Therefore we can make use of the facilities that are available in Windows to interface to the operating system. Also, if the computer is on a network, we can make use of the network to separate the computation between a client and a server, or between multiple clients and servers. Some of the options for the physical architecture of our major software engineering project are shown in Figures 4.18 and 4.19.

One advantage of the close integration of Microsoft Internet Explorer with the operating system is that several of the user interfaces are the same for networked systems as for stand-alone ones. (This close integration was a major part of the basis for an antitrust suit filed against Microsoft by the U.S. government.) The commonality of the user interface

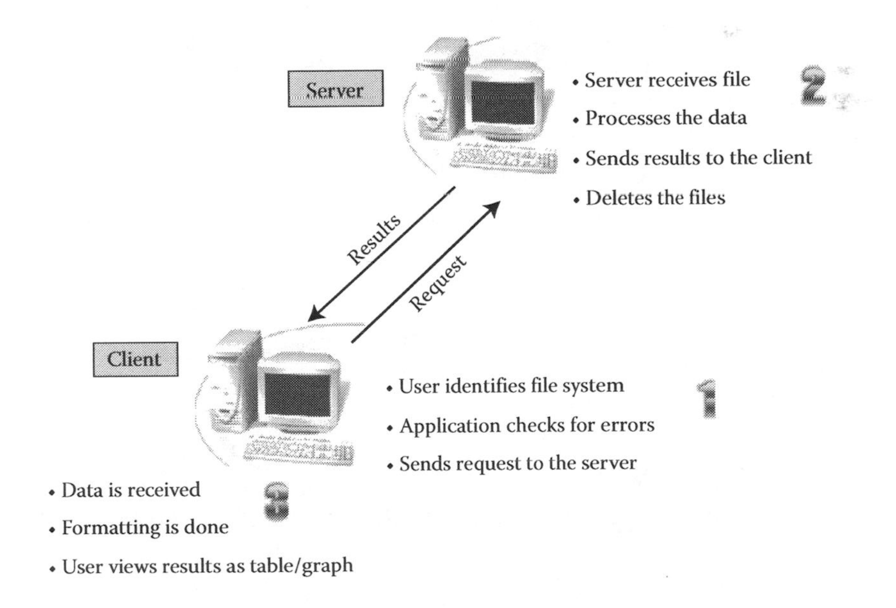

FIGURE 4.18 One possible organization for the physical architecture of the continuing major software engineering project, with heavy computation performed on the client. (Courtesy of M. Armstrong, C. Barnes, B. Fough, senior project design, Howard University.)

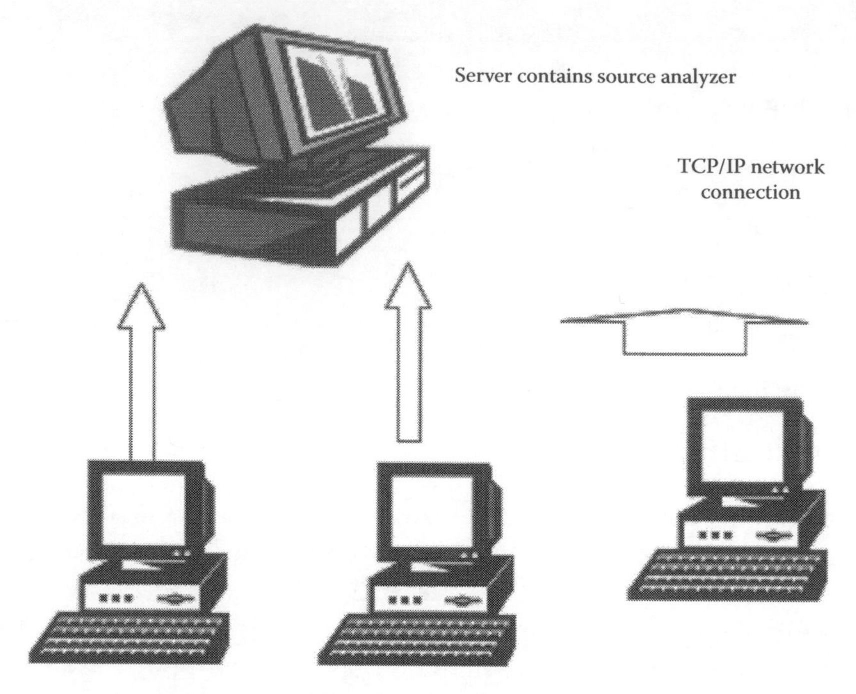
Server contains source analyzer
TCP/IP network
connection
Client machines contain the source files (C, C++ and java)

FIGURE 4.19 An alternative design for the physical architecture of the continuing major software engineering project, with computation done on the server. (Courtesy of S. Armstrong, M. Henry, A. Rogers, senior project, Howard University.)

makes both the user interface and the back end computational subsystem highly portable within the Windows environment and relatively easy to be interoperable with other (Microsoft-based) systems. Other browsers may also be used, but there may be some configuration difficulties that can arise if any of the advanced features of modern browsers (including the new Microsoft Edge) are used.

Thus, we can use existing dialog boxes for file input with very few modifications, as shown in Figures 4.20 and 4.21. The presence of the dialog box labeled "Drives" makes it clear that we can access files on local hard drives and on drives of other computers on the same network.

Recall that the technical architecture of a system involves mapping of software to the physical architecture. The decision should be made at this time. The interfaces shown in Figures 4.18 and 4.19 illustrate the feasibility of nearly any mapping to the physical architecture. In an ideal world, with considerable time, many alternative architectures would be considered. In reality, the choice might be limited to a smaller set based on the decision to have everything done on one system or on a small number of clients and servers. (Most college and universities do not devote large numbers of computers to student projects, especially if students require root administrative access in order to create their systems and configure them properly.)

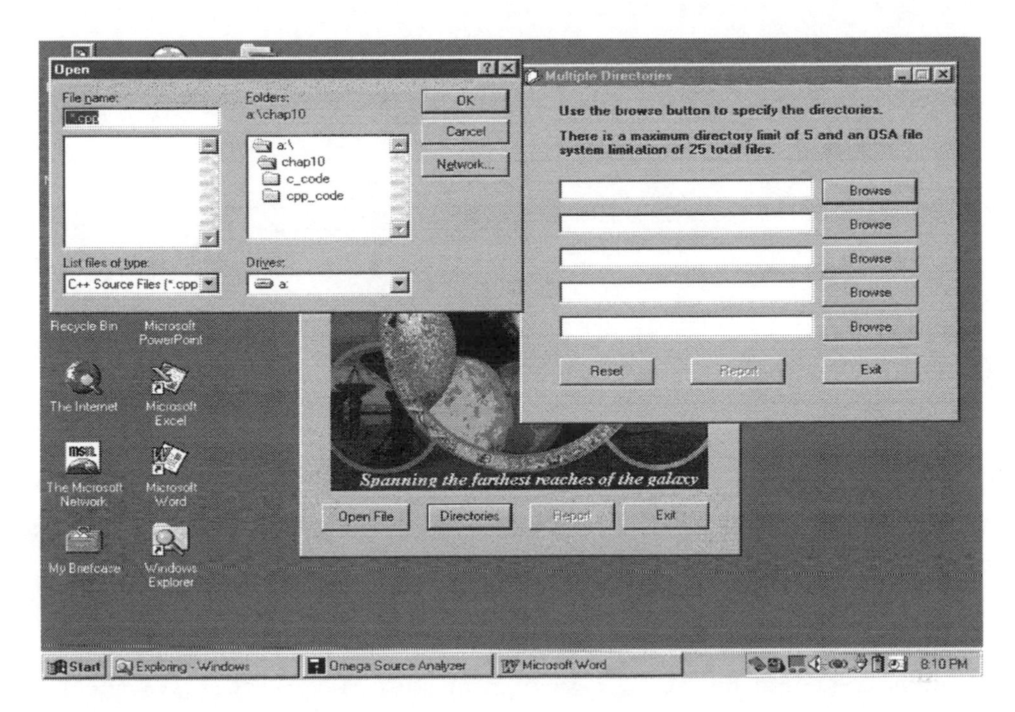

FIGURE 4.20 An illustration of how a software system can interface with an operating system.

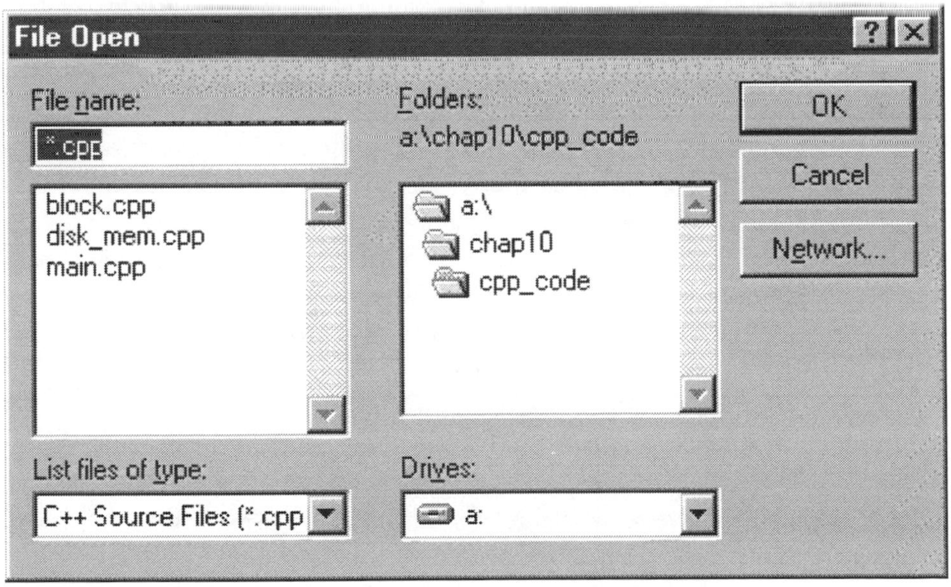

FIGURE 4.21 A more detailed illustration of how a software system can interface with an operating system.

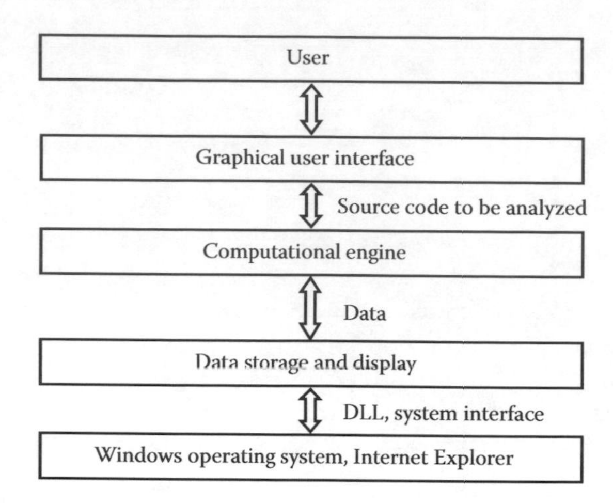

FIGURE 4.22 A high-level view of the system architecture of the continuing major software engineering project.

We will be content to indicate the system architecture in Figure 4.22. We note that the architectural design should indicate the interactions with the operating system; in this case, the decision was whether to use Java or existing Microsoft DLLs (dynamically linked libraries). (We have chosen DLLs.) The rest of the architecture should be clear from reading the next two sections.

Some of the items listed in the system architecture for the major software engineering project shown in Figure 4.22 can be expanded to create what is known as the software architecture.

4.18 PRELIMINARY DESIGN OF THE MAJOR SOFTWARE PROJECT

We now turn to an application of our design approach: our continuing software project for which we developed requirements in Chapter 3. You should review the design information and the requirements traceability matrix in Table 3.6 before continuing with the discussion. In this section, we will develop both the preliminary and detailed designs of our continuing example of a software system. There are several questions that must be answered if we are to have a proper design that will lead to an eventual software system that adheres to the software engineering goals given in Chapter 1.

We wish to develop a design that is easy to implement in an efficient manner, keeping in mind the goals of software engineering that were listed in Chapter 1. At the same time, we want our design to map to the requirements in the sense that we can determine if each requirement is met. We have seen a large amount of information on software patterns, design representations, and a large number of techniques. The sheer volume of material presented (and the clear evidence that there is much more information than could ever be put into any book) makes it difficult to know where to start.

Let us begin by trying to match our problem's requirements to one or more of the very large-scale software patterns listed in Section 4.2. We will discuss the applicability of each pattern in turn. We will consider the application of any of the more specific twenty-three design patterns in the Gang of Four book (Gamma et al., 1995) to this project in the next section.

The first software pattern describes a menu-driven user interface. It is clearly not applicable because we wish to have a batch driven system in the initial version of the software.

The second software pattern describes an event-driven system. An event-driven user interface is clearly not appropriate for our system. Modeling the system as a set of states might be possible due to the text processing necessary to compute our lines of code. However, it seems unlikely. The next two patterns also use states and seem more appropriate for our purposes.

Pattern 3 seems very appealing because a source code file can be considered to be a stream of tokens that are removed from the input stream once it is processed. There appear to be a small number of "states" and a small set of optional actions that can be taken for each state. The optional action taken depends upon both the state and the value of an input "token."

Pattern 4 seems less promising than pattern 3 in that the decision about which action to take may depend on more information than is available from just the pair consisting of the state and the input token. In this pattern, the tokens may still remain in the input stream after being processed.

Pattern 5 suggests a database system. This might be appropriate for the back end of the software. A spreadsheet also fits this pattern.

Pattern 6, a general, flexible, configurable system, seems to be inappropriate.

Patterns 7 and 8 present several reasonable possibilities. If the system has everything primarily guided by an algorithm, rather than depending primarily on data, then a procedural system is appropriate. This is the case with pattern 7.

On the other hand, a system with many relatively independent computational actions taking place might be more appropriate as an object-oriented system, which is the case with pattern 8.

There seem to be some clear conclusions and some issues left to be decided. Treating the inputs to our software system as a stream of tokens seems appropriate, as does thinking of the back end of the software as a database or at least interfacing to one.

Unfortunately, there is no obvious choice between writing a procedurally oriented or an object-oriented system, at least in our first attempt at system design.

The place to start is with the highest-level view of the system: the system's software architecture. Let us examine the basic functionality of the system. The software appears to have the basic architectural building blocks that are described in Table 4.3.

These activities seem to exhaust the basic functionality of the system. We now turn to the problem of representing these high-level activities in order to develop an initial design and to allow us to improve the initial design through iteration. A preliminary, procedurally oriented design is given in Figure 4.23.

This flowchart represents the intent of the requirements of our continuing software project. There is a common interface, a determination of the type of the source code files used as input, a separation of functionality into three different analysis subsystems, a common data collection routine, and an interface to a database or spreadsheet program.

TABLE 4.3 First List of Large Components and Software Architecture of Our Continuing Major Software Project

1. A front end that will interact with the user and manage the input files that will be analyzed by our system.
2. A user interface that will include the front end described above.
3. A set of analysis routines to evaluate the input files provided in step 1. The analysis routines will compute the lines of code for each function and the other necessary information.
4. Output routines to interface to a database or spreadsheet for later analysis.
5. A database or spreadsheet program to perform additional statistical analyses and data storage for the output obtained in step 4.

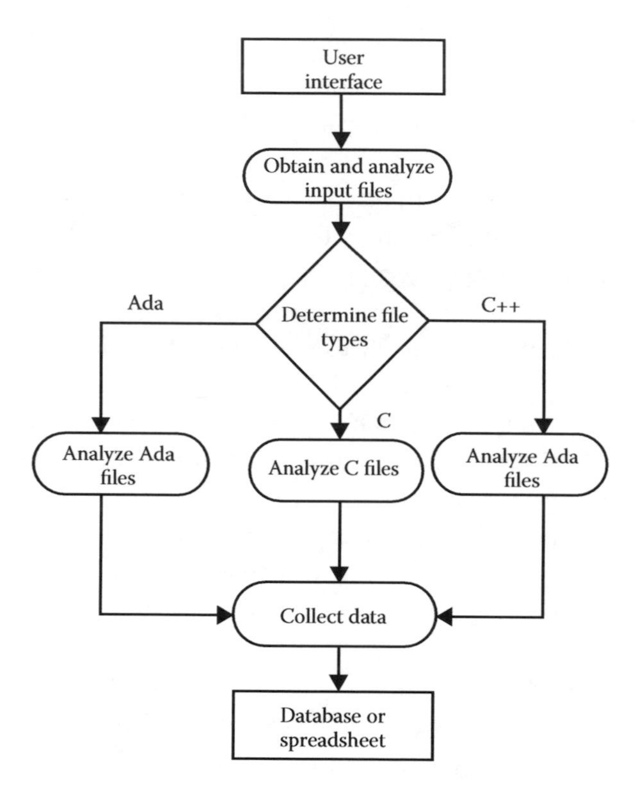

FIGURE 4.23 A preliminary, procedurally oriented design of our continuing software project.

Of course, we should not be content with this preliminary design, since we have flexibility. Unlike the situation of working on large-scale systems in government or industry, our employer is enforcing no particular design methodology. Let us consider what a data-flow-oriented design might look like for the same system. A level 0 data flow diagram is shown in Figure 4.24.

It appears as if the control-flow-based design representation provides more information than the representation using the data-flow approach. Of course, we might reach a different conclusion if we used these two design representations to develop more detailed designs. The control flow representation of Figure 4.23 also appears to lead more easily to a decomposition into subsystems.

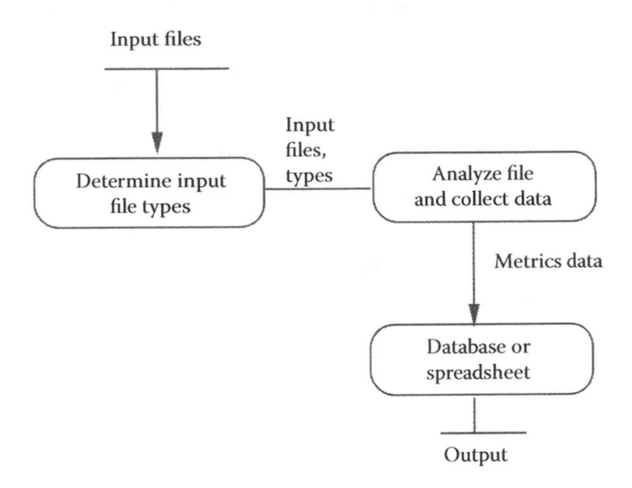

FIGURE 4.24 A preliminary data flow design (DFD level 0) of our continuing software project.

We reject an object-oriented design (not shown here) of the complete system for the same reason. Thus, we will use Figure 4.24 as the basis for our high-level system design, which will be procedurally based. There will be some object-oriented components, as we will see. Note that this design does not conflict with the requirements traceability matrix given in Table 3.12.

It is now reasonable to examine any software utilities, complete programs, integrated tools, or other resources that can be used as part of our system. If we can find such utilities, programs, or tools, then we can make our future development easier by reusing them. Recall from Chapter 1 that reusability is one of the goals of software engineering. Any satisfactory, reusable software utilities can simplify the design process by eliminating the need for detailed design functionality provided by the tool or utility. (In addition, the need for coding the functionality provided by the tool or utility is also avoided through such reuse.)

An Internet search on January 30, 1997, found seven free source code analyzers on a now-defunct website maintained by Christopher Lott when he was at Queens University in Canada. A much more recent website (found in 2014), with entries in the last few years, http://maultech.com/chrislott/resources/cmetrics/, now contains seventeen such source code analyzers. At this point in his career, Lott has become a full-time software designer rather than a developer of academic research tools or being a senior professor of computer science. Currently, he is the lead software designer for Murtha Design. He is the developer of the #1 iPad app in the entertainment category, *Drawing Pad.*

We quote Lott's aforementioned website about the portability of the source code for these metrics:

> Consequently I had to monkey with some makefiles. Packages 'ccount', 'clc', 'lc', and 'spr' are relatively simple and should not present many problems. …
> However, I have not used all of the tools extensively, so unfortunately I can't make any helpful statements about reliability or ease of use.

This quote emphasizes what should be clear to anyone who has used freeware: there are no guarantees. In the language of software reuse, the software packages from this site have not been certified except for the ability to compile them on two particular operating systems—in this case a Sun computer running SunOS 4.1.3 that was used for the first edition of the book, and Darwin, the Linux variant used on Apple computers, that was used for the second edition. (The SunOS 4.1 operating system version is also known as Solaris 1.1, which is very different from later versions of Solaris.)

Some of the same issues that we observed in our examination of publicly available analysis tools for measuring C or related language source code files also would have occurred with analysis tools for C++ source code files. The problem would occur when attempting to reuse publicly available software for larger systems other than the one that we discuss in this text.

Although we have not examined in detail the existing software tools mentioned earlier, there is sufficient information to believe that several of them provide sufficient functionality to satisfy our project. Even if the preliminary assessment of the reusable tools' quality is incorrect, we can use the existence of the tools as a guide in our design. In other words, the question of the quality of the existing reusable software tools can be deferred until the coding phase of the project. We do not need to certify the quality of the tools at this point because they do not affect the design (at this point).

We now turn to the problem of transforming the high-level design represented in Figure 4.24 into a more detailed design that is complete in the sense that the interfaces between system components are well understood and the functionality of each system component is spelled out in sufficient detail. The final design should be sufficiently detailed so that it can be implemented without excessive difficulty.

The first step is to consider the interfaces between different subsystems. There appear to be six subsystems in the complete system; they are listed in Table 4.4.

Recall that before we changed the requirements to use a graphical user interface (GUI), the requirements specified that the user interface be by means of command-line arguments. We will return to this simpler case in the exercises. Since we have illustrated the feasibility of using existing dialog boxes for file input, getting the files seems to be no problem.

However, there is one issue that may occur if the system is to be used in practice. The system requirements specifically noted that the system's input need not be checked. Thus, we do not have to worry about someone using the system with an executable file as input in this release of the project. The difficulty is that realistic software systems are usually spread

TABLE 4.4 Set of Subsystems in the Initial Design of Our Continuing Major Software Project

1. Subsystem for the user interface.
2. Subsystem for obtaining files and analyzing file types.
3. Subsystem for passing input files to data collection subsystem.
4. Subsystem for collecting data for each input file.
5. Subsystem for sending data to spreadsheet or database.
6. Spreadsheet or database program.

out over several different directories, so our software system might be given an input that is the name of a single directory. How can we incorporate the processing of directories into the design of this software?

There are several issues, depending on our knowledge of the operating system on which the software system will be used and the naming standards that we can assume have been followed for the input files. (Naming standards will be discussed in detail in Chapter 5, when we discuss coding issues.)

We will make the following assumptions about naming conventions for our input files:

- All input files written in C have names that end in either the .c or .C extensions.
- All input files written in C++ have names that end in either the .cpp or .C extensions.
- All input files written in Java have the .java extension (or .jar if they are Java class archives).
- All input files written in Ada have names that end in the .a, .A, or .ada extensions.
- All C or C++ header files have names that end in .h and are not to be processed further.
- Any files named Makefile or makefile are compilation instructions and are not to be processed further.
- Any files named readme, read.me, README, READ.ME, or ending in .doc or .txt are documentation and are not to be processed further.
- Any files with names that end in either .exe or .obj can be ignored.
- All other files are directories and may contain source code as well as other files. (This means, for example, that any "special files" such as device drivers in UNIX will be ignored.)

The assumption that the input source code was developed using a naming convention makes designing this portion of the subsystem easy. We just get the input files and pass them to the next subsystem. In the case of a directory, which can be determined by its name not matching one of the other possibilities, the contents of the directory must be searched also. The search of each directory should be recursive, so that source code files located in directories that are included in other directories will be analyzed.

A moment's look at our basic set of subsystems suggests that the first two subsystems for user interface and for obtaining files and analyzing file types should be combined. This is a natural result of our assumption that the naming conventions listed earlier were followed in the creation of the input files to our system. The subsystem for passing input files to a data collection subsystem also seems redundant. Hence, it too can be combined with the initial subsystems. Thus, the six subsystems listed in Table 4.4 can be combined into the four subsystems listed in Table 4.5.

TABLE 4.5 Final Set of Four Subsystems in the Design of Our Major Continuing Software Project

1. Subsystem for the user interface, obtaining files, and analyzing types.
2. Subsystem for collecting data for each input file.
3. Subsystem for sending data to spreadsheet or database.
4. Spreadsheet or database program.

This section will close with the following two changes to the requirements for the major project. You will be asked to modify the design to reflect the new requirements in the exercises.

The first change is that the customer now wants a web-based user interface instead of one that is PC-based. What changes need to be made to the design process? What changes need to be made to the various levels of design themselves?

The second change is that the customer now wants to use a cloud for data storage instead of one that is PC-based. What changes need to be made to the design process? What changes need to be made to the various levels of design themselves?

4.19 SUBSYSTEM DESIGN FOR THE MAJOR SOFTWARE PROJECT

A decision needs to be made before we go any further in our design of subsystems for the continuing major software engineering project. Should the subsystem designs be object-oriented, or should they be created using procedural approaches? The answer may not be as simple as you think.

An important 2006 paper by Meyer and Arnout discusses an issue that some software engineers and researchers have with patterns: that there is *less* code reuse than would occur with simply forming the code for a particular design pattern into a component that can be reused as is. At first glance this seems counterintuitive. You might recall that we previously indicated that the greatest amount of saving in both cost and time occurs when high-level components are used early in the software life cycle. Thus we should consider both possibilities: large-scale components and matching design patterns in order to make our software engineering effort more efficient. This means that we should look at the largest, most applicable previously developed component that is available to us when designing our system.

Now let us attempt to design the first subsystem. As a basis, we will use the control flow-oriented diagram given in Figure 4.23. The upper portions of the design are adequate for our purposes, except that they do not indicate the actions that will be taken in response to specific types of inputs. That is, the diagram in Figure 4.23 does not indicate any mechanism for checking the types of input files, for determining if there are directory files to be examined further, which files can be ignored, and whether there are any input files at all.

Recall that the requirements document for our software project specified that wild cards could be used as names of input source code files. This will create some problems in implementation because it will imply interaction with the operating system's command processor to parse command lines to translate wild cards into names of actual files. However, we can design this portion of the software carefully as long as we know how the operating system actually stores command-line arguments and makes them available to C or C++ programs. Thus, determining this information is an essential part of the design process at this point. The design must take into account whether performing any proposed interaction with the operating system is possible.

It is natural to ask if the detailed design for this first subsystem should be done as a pattern, one of the twenty-three we met in Section 4.2. This first subsystem has several features that suggest a use of a number of patterns: abstract factory, builder, factory method, chain of responsibility, command, and iterator. Since there are so many possibilities, it would appear that an object-oriented approach to the design of this subsystem is premature, although there would be advantages from the perspective of the treatment of exceptions.

The subsystem for collecting data for each input file will be based on the existing software tools located on the Internet that we discussed earlier in this section. Specifically, we will examine each of these tools for its functionality and determine the interface. If the tool's functionality is sufficient for our project's requirements, we will then provide an interface to the tool. If none of the tools has appropriate functionality, then at least some portion of the software to provide the necessary functionality will be written from scratch. This means that the new software tool must be designed.

We would like to put off the decision to write new software tools as long as possible, emphasizing an efficient software development process based on software reuse. However, this means that the software tools must be evaluated for three things: functionality, quality, and interface to the rest of the system. An interface control document (ICD) must be used here. Also, if the software tools are likely to change, we must institute some form of configuration management.

It would appear that the prototype design pattern might be appropriate here, because each of the types of source code files in our repository will be analyzed by a freeware tool designed for that type of source code file (C, C++, etc.). An iterator pattern might also be appropriate, because we do not know a priori just how the data repository will be organized or how our software will traverse it while gathering metric information. This is much clearer than was the case for the first subsystem.

The third subsystem to be considered will provide the connection and aggregation of the data from the various reused software tools that form the major part of subsystem 2. It is clear that the individual tools count the data differently. Thus, we need some consistency in the data definition and in the way that the data will be sent to the final subsystem, which is the database or spreadsheet. Perhaps we should view this third subsystem as suitable for a template design pattern.

There is one other point that needs to be made before we design the third subsystem. Several of the tools may not provide the desired information. Thus, there may be several pieces of information that are either not available or else have to be replaced by a default value. This seems to suggest the presence of polymorphism, which in turn suggests an object-oriented approach.

With an object-oriented approach, the most important step is determining the objects. Recall the steps suggested in Table 4.1 for determination of objects:

1. Choose a candidate to be an object.
2. Determine a set of attributes and their possible sets of values. Use the has-a relation. List all relevant transformations on the object.
3. Develop an initial set of transformations on the object to serve as member functions.
4. Determine if there is more than one example of the object.
5. Apply the is-a relation.
6. Use polymorphism and overloading of operators (and functions) to check if we have described the objects in sufficient detail.
7. Use the uses-a relation to determine all instances of client–server or agent-based relationships.
8. Review the object for completeness.
9. Repeat steps 2 through 8 for all combinations of relevant objects (triples, quadruples, and so on) until the object's role has been adequately described.

There are a few candidate objects to be considered for the fundamental object: individual functions, source code files, or collections of source code files into subsystems. As before, we use the term "function" to include "function" or "procedure." We examine each of these choices for the basic objects in turn.

If the object is "function," then the most obvious attributes and some potential values are

Attribute	Typical Value
Name	Character string
Size	Integer
Module in which located	Character string

If the object is "source code file," then the most obvious attributes and some potential values are

Attribute	Typical Value
Name	Character string
Size	Integer
Number of functions	Integer
List of functions	Linked list of character strings
Subsystem in which located	Character string

If the object is "subsystem," then the most obvious attributes and some potential values are

Attribute	Typical Value
Name	Character string
Size	Integer
Number of source code files	Integer
List of source code files	Linked list of character strings

We have to know the spreadsheet or database input format. For the Microsoft Excel spreadsheet software, input to a spreadsheet can be in several forms, including, but not limited to, the following:

- An existing spreadsheet created in Excel
- An existing spreadsheet created in another spreadsheet program for which translators are available
- A text file with commas used as delimiters
- A text file with tabs used as delimiters

We choose to use the comma-delimited form, because file names are unlikely to consist of commas. This will work for spreadsheet input. (Note: This means that the requirements must be changed to handle source code files with names that include commas.)

There are similar issues for interfacing with a database management program. Unfortunately, the use of commas as delimiters here may cause more problems, because commas are perfectly valid as punctuation marks. Most databases use a nonprintable ASCII character such as Control-R as a delimiter to separate different fields. We leave the details of the design of the interface to a database to the exercises.

The design should now be checked for internal consistency according to the principles discussed in Section 4.12. As was indicated there, any interface standards should be consistent with the software requirements of all modules or subsystems affected by the interface.

Now that we have a design, it is necessary to check the design against the requirements traceability matrix for the project. Item 9 in the requirements traceability matrix calls for an installation routine. Since we have not discussed this previously, we must now include it in our set of subsystems to be designed. The installation subsystem should be relatively independent of the other subsystems of the complete software product, since the requirements specify that only a single size of the system will be produced and, thus, little configuration is necessary. Note that the software pattern of a general, flexible, highly configurable system might be used to characterize the installation subsystem. The newest version of the requirements traceability matrix is given in Table 4.6.

TABLE 4.6 Requirements Traceability Matrix for Our Major Continuing Software Engineering Project

Requirement	Design	Code	Test
1. Intel-based	Y		
2. Windows 8	Y		
3. Windows 8 UI	Y		
4. Consistent with Excel 14.0	Y		
5. System one size only	Y		
6. 400 MB system			
7. 500 MB disk space			
8. CD			
9. Includes installation	Y		
10. No decompression utility	Y		
11. One input file at a time	Y		
12. Size of each function	Y		
13. Size of each file	Y		
14. Size of system	Y		
15. Compute totals	Y		
16. Develop std for C, C++, Ada	Y		
17. Batch-oriented system	Y		
18. Precisely define LOC	Y		
19. Measure separately	Y		
20. No error checking of input	Y		
21. Front end in C or C++	Y		
22. Batch processing	Y		
23. File names limited to 32 char.	Y		
24. Wild cards can be used	Y		
25. Dead code ignored	N		
26. No special compilers needed	Y		

You might have wondered if we need to make an explicit link to a spreadsheet program such as Microsoft Excel or if the output can be obtained by using various DLLs from the Microsoft library. Of course, the requirements say one thing, and changing them might cause difficulty.

However, if we illustrate the feasibility of using these DLL components in a design review, it is possible that a potential client may allow the requirements to be changed. Changing of system requirements is common in the market-driven software world.

You should also note that some of the issues involve setting standards and, as such, have not been considered at this point in the design process. Other requirements cannot be fully addressed in the design at this time. We leave the design of the installation subsystem to the exercises. For example, we cannot precisely determine the size of the complete system at this time. The size of the software tools that we intend to use can be determined accurately, because these software tools already exist.

We will consider the detailed design of our software system in the next section.

As we did in Section 4.18, this section will close with the following two changes to the requirements for the major project. You will be asked to modify the design to reflect the new requirements in the exercises.

The first change is that the customer now wants a web-based user interface instead of one that is PC-based. What changes need to be made to the design process? What changes need to be made to the various levels of design themselves?

The second change is that the customer now wants to use a cloud for data storage instead of one that is PC-based. What changes need to be made to the design process? What changes need to be made to the various levels of design themselves?

4.20 DETAILED DESIGN FOR THE MAJOR SOFTWARE PROJECT

The next activity for our major software engineering project is to flesh out the previous designs in order to describe the system in detail. How much detail is needed? The answer is essentially the same as the one we gave when discussing the approach of Daniel Berry to the problem of determining software requirements: the requirements are considered to be complete and specified in sufficient detail when a programmer has enough information to be able to tell if he or she can develop a system to meet these requirements. We need this level of clarity in the design.

The major distinction between this degree of detail in the requirements and a similar level of detail in the design is that the use of higher-level languages may make some of the implementation details of the system unnecessary. Allocation of resources for the detailed implementation may be unnecessary because some of the design prototypes and GUI diagrams will be sufficiently developed to make further implementation effort redundant.

Figure 4.25 illustrates the point. This figure shows the interface to the operating system and makes it clear that the output can be displayed in a simple pop-up window. Other aspects of the GUI are shown in Figures 4.26 through 4.28.

What else must be specified in a detailed design? We need to see the descriptions of the interfaces, the functionality of software we write as "bridgeware" between the components we are obtaining from the Internet and other sources. We leave this to the exercises.

It seems likely that this project is well suited to concurrent development. Thus, we need an ICD to make sure that all developers use the same common interfaces. We leave this to the exercises, too.

From a project management perspective, we should also do a status check. Are we ahead of schedule (unlikely), behind schedule (likely), or approximately on target? Have there been any unpleasant surprises, any portions of the system that were more difficult than we had expected? Does any portion of the system require extra attention, perhaps additional resources? Have technology or market pressures rendered any portion of the system obsolete? Are there any changes that could affect our ability to complete this project?

As we did in Sections 4.18 and 4.19, this section will close with the following two changes to the requirements for the major project. You will be asked to modify the design to reflect the new requirements in the exercises.

The first change is that the customer now wants a web-based user interface instead of one that is PC-based. What changes need to be made to the design process? What changes need to be made to the various levels of design themselves?

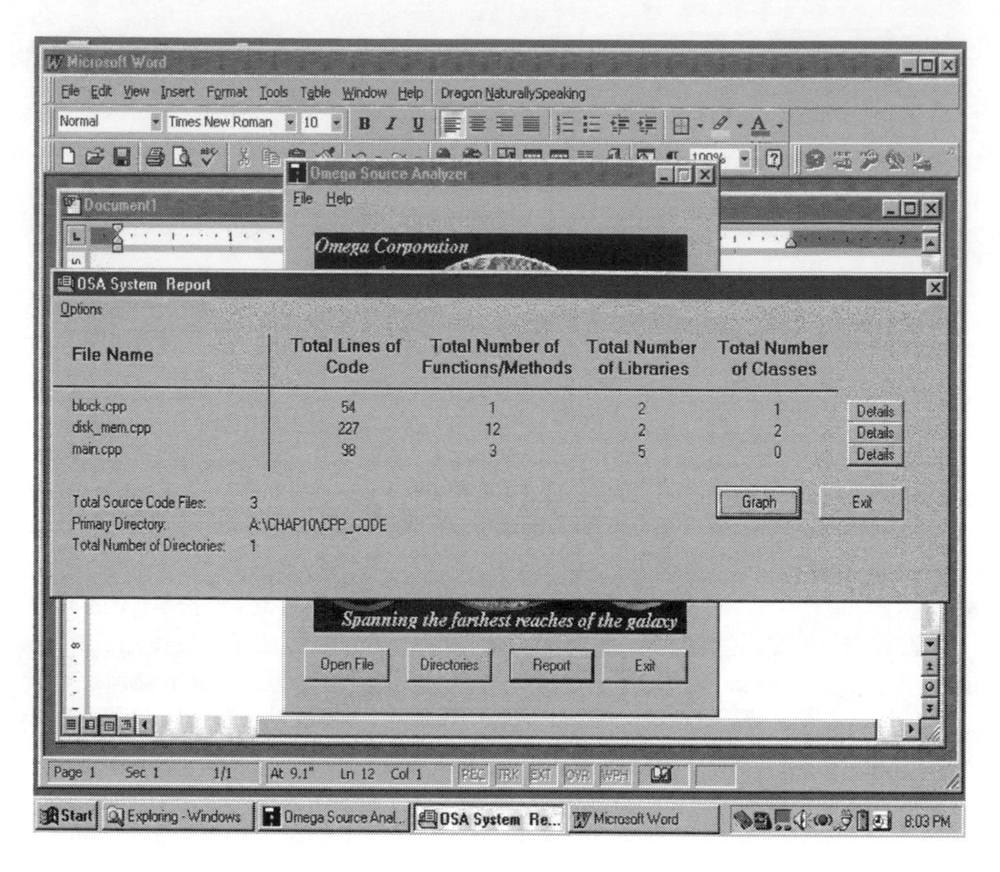

FIGURE 4.25 An illustration of system output from the continuing major software engineering project. The output is shown in a window with dialog boxes overlaid on top of another application. (Courtesy of S. Armstrong, M. Henry, and A. Rogers, proprietary software system.)

The second change is that the customer now wants to use a cloud for data storage instead of one that is PC-based. What changes need to be made to the design process? What changes need to be made to the various levels of design themselves?

SUMMARY

Design is a major part of the software engineering process. The goals of software engineering include efficiency, reliability, modularity, usability, modifiability, portability, testability, reusability, maintainability, and interoperability with other systems. Good software design techniques can support these goals.

One of the most powerful software design approaches is pattern matching. Determining that a subsystem, or even an entire system, matches a pattern can encourage reuse and make both design and follow-up implementation much more efficient and can produce better software. Common software patterns include:

1. A menu-driven system, where the user must pass through several steps in a hierarchy in order to perform his or her work.

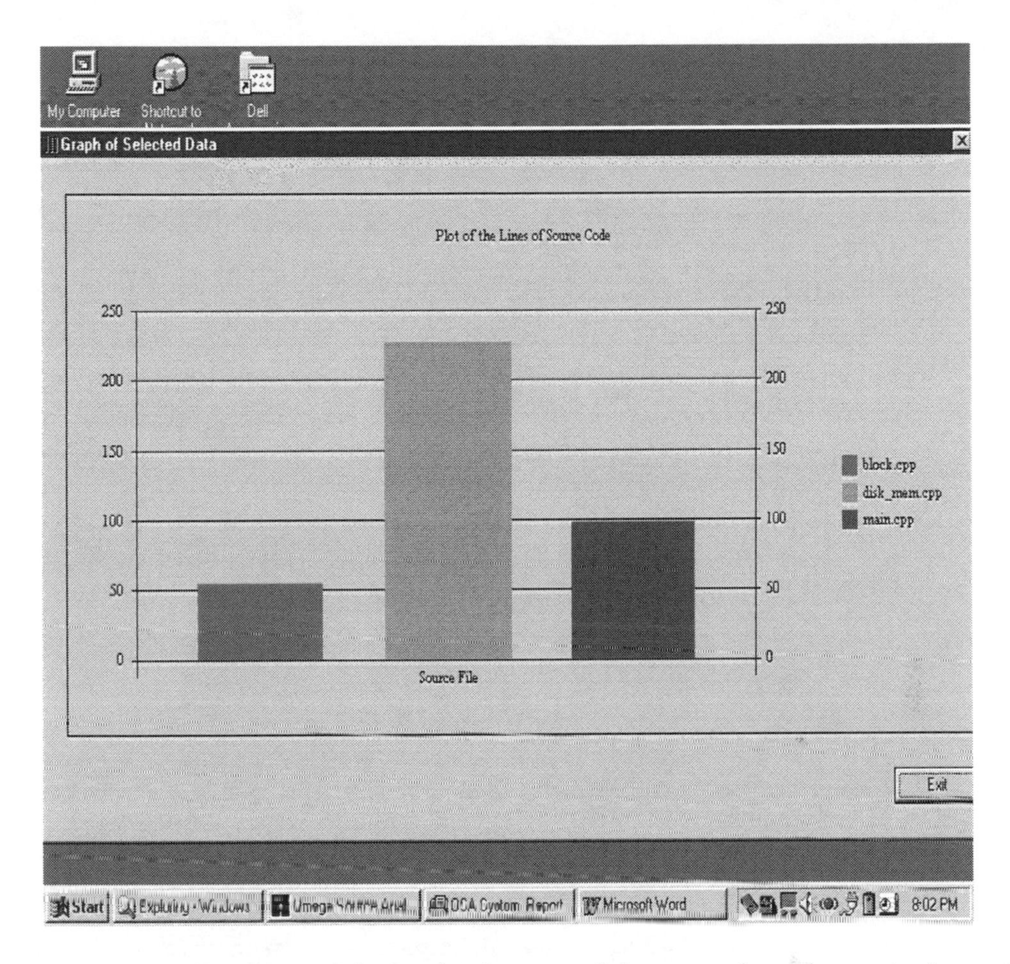

FIGURE 4.26 An illustration of the graphical output of the system from the continuing major software engineering project. The output is shown in a window with a single dialog box. (Courtesy of S. Armstrong, M. Henry, and A. Rogers, proprietary software system.)

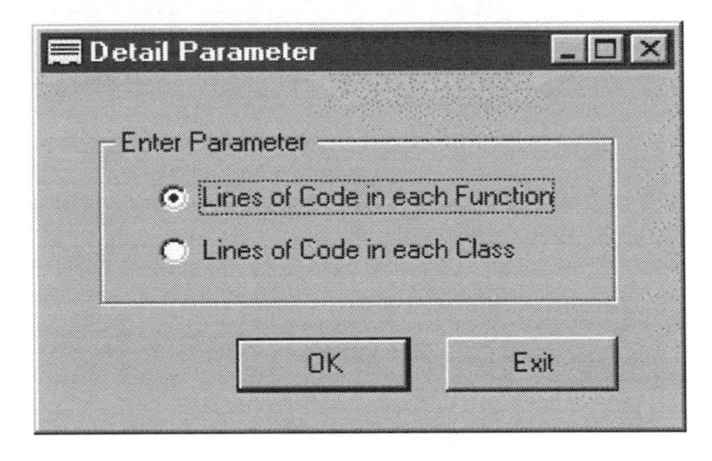

FIGURE 4.27 A dialog box to control the type of output of our system for our continuing major software engineering project. (Courtesy of S. Armstrong, M. Henry, and A. Rogers, proprietary software system.)

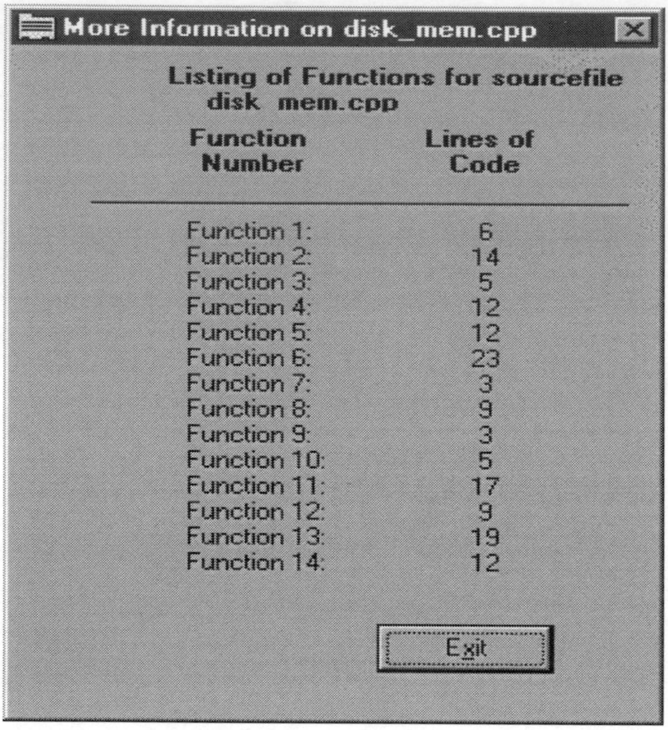

More Information on disk_mem.cpp

Listing of Functions for sourcefile disk_mem.cpp

Function Number	Lines of Code
Function 1:	6
Function 2:	14
Function 3:	5
Function 4:	12
Function 5:	12
Function 6:	23
Function 7:	3
Function 8:	9
Function 9:	3
Function 10:	5
Function 11:	17
Function 12:	9
Function 13:	19
Function 14:	12

Exit

FIGURE 4.28 Detailed output of our system for our continuing major software engineering project. (Courtesy of S. Armstrong, M. Henry, and A. Rogers, proprietary software system.)

2. An event-driven system, where the user must select steps in order to perform his or her work. The steps need not be taken in a hierarchical order.
3. A system in which the actions taken depend on one of a small number of "states" and a small set of optional actions that can be taken for each state depending on both the state and the value of an input "token."
4. A system in which a sequence of input tokens (usually in text format) is processed, one token at a time.
5. A system in which a large amount of information is searched for one or more specific pieces of information.
6. A system that can be used in a variety of applications but needs adjustments to work properly in new settings.
7. A system in which everything is primarily guided by an algorithm, rather than depending primarily on data.
8. A system that is distributed, with many relatively independent computational actions taking place.

Design representations can make a system easier to understand by using a graphical representation to provide a high-level view, hiding relatively unimportant details. Common design representations for procedurally based designs include flowcharts and data flow diagrams. Pseudocode is also frequently used.

The most important design technique for procedurally developed software systems is decomposition of existing high-level designs. For object-oriented systems, the emphasis is on developing high-quality, reusable classes that can act autonomously.

The first step in designing an object-oriented system is to determine the relevant objects. Objects have attributes, and for each instance of an object, the attributes can take on a set of possible values. Determination of the attributes of an object and the set of possible values helps in the development of the function prototypes for the class.

Objects should not be defined in a vacuum for realistic descriptions of systems. The interfaces for a particular class may make it difficult to use other classes already present in the class library. Thus, the class library must be examined, either by a catalog (listing) of the classes or by using a software tool called a library browser.

The interface of the class should be consistent with that of related classes performing similar services.

Preexisting classes in the class library should be examined for the possible availability of usable friend functions.

The development of a set of objects for a system should be an iterative process. Candidates for objects should be able to pass the "multiple example" and other tests. Attributes of objects can be found by the "has-a relationship." A derived class must be related to a base class by the "is a relationship." Other relationships between classes can be found using the "uses-a relationship."

Design representations for object-oriented systems can be either graphical or text based. Graphical ones are often based on information models or entity–relationship (E-R) graphs.

A major goal of software design is to promote reuse. Designs should be reviewed to make certain that existing reusable software components are reused whenever possible. David Weiss and C. Lai have written an important book on software product line architectures (Weiss and Lai, 1998). Much of their work is based on their experiences at Lucent Technologies and the resulting cost savings and quality improvement. An important follow-up paper by Coplien et al. appeared in 1999.

KEYWORDS AND PHRASES

Design, detailed design, design review, E-R diagram, UML, design pattern, componentization

FURTHER READING

There are many excellent books on software design. A classic one by Jackson (Jackson, 1983) describes a method that is commonly known as "Jackson Structured Design," or JSD. Yourdon (1989), Yourdon and Constantine (1979), Coad and Yourdon (1990), and Booch (1993) also have excellent books on the subject.

The classic paper by David Parnas (1972) on design decomposition issues is still worth reading. This paper provides an assessment of the effect of several alternative design decisions that can be made even in a relatively simple situation.

A new book by Joseph Bergin 2015 describes the process of developing polymorphism in objects in a manner reminiscent of the way we described requirements elicitation in Chapter 3—by a series of dialogues.

A recent paper by Peter Denning in the *Communications of the ACM* provides insight into the design philosophy of the IDEO company, which arose out of the Stanford Design Center (Denning, 2012). Particularly interesting is the recommendation for design teams to have multiple viewpoints, with ethnography and sociology especially important, given the tremendous increase in internationalization.

An accessible paper by Binder (1994) discusses the effect that design choices can have on the testing of object-oriented programs.

Jakob Nielsen (1994) and Ben Shneiderman (1980) provide excellent overviews of human–computer interaction and their roles in good software design. Their discussion applies even today to most user interfaces of software systems that run on desktop or laptop computers with substantial screen real estate. Nielsen has written a more recent book on usability of software on mobile devices with his coauthor Raluca Budiu (Nielsen and Budiu, 2012). A book by Shneiderman, with coauthor Catherine Plaisant, is in its fifth edition and is an outstanding introduction to interface evaluation (Shneidermann and Plaisant, 2010).

More detailed information on user interfaces can be found in publications such as that of the Special Interest Group in Human–Computer Interaction (SIGCHI) of the ACM and many conferences devoted to human–computer interaction, as well as some of the publications listed in the references.

The Gang of Four book provides an accessible introduction to design patterns (Gamma et al., 1995). A 2006 paper by Meyer and Arnout discusses an issue that some software engineers and researchers have with the use of patterns: that there is less code reuse than would occur with simply forming the code for a particular design pattern into a component that can be reused as is. That paper includes a detailed discussion of an explicit design pattern: the visitor pattern.

The tutorial by Ezran, Morisio, and Tully provides a classification of architecture types (Ezran et al., 1999).

For publicly available information on the Department of Defense DoDAF software development process, see the website http://dodcio.defense.gov/Library/DoDArchitecture Framework.aspx.

EXERCISES

1. Take any non-object-oriented program that you have written with approximately 50 to 300 lines of code. Write a flowchart for this program.

2. Repeat Exercise 1, writing a data flow diagram.

3. What are the objects in the system that you chose for Exercises 1 and 2? List the attributes and the set of possible attribute values for each.

4. Take any object-oriented program that you have written with approximately 50 to 300 lines of code. Write a flowchart for this program.

5. Repeat Exercise 4, writing a data flow diagram.

6. Which, if any, of the twenty-three design patterns best describe the program? Explain your answer.

7. Consider the development of an external computer system to evaluate how well a human subject is learning the use of a computer. The user is to interact with the I/O devices of monitor, keyboard, trackball, and mouse. The "user object" has an attribute called "experience level," which has the possible values "novice computer user," "frequent computer user," and "experienced computer user." List other possible attributes that might be appropriate for describing the user. (For simplicity, assume that there is a single software application running and that the user is not familiar with it.) Refer to the model of a user interface that was given in Figure 3.6.

8. Write a control flow diagram for the MacTAGS example given in Figure 4.8.

9. Examine four or five books on data structures and determine how they implement the quicksort algorithm. Classify their presentations as being control-flow oriented, data-flow oriented, procedurally oriented, or object-oriented. Compare these approaches with the algorithm in Hoare's original paper (Hoare, 1961).

10. Consider the system requirements for a simple chemical reaction process control system that were written in the form of a finite state machine with an associated state diagram. These requirements were described in Chapter 3, Section 3.8. Use this formal design representation to develop a detailed design.

11. Consider the system requirements for a simple chemical reaction process control system that were written in the form of a decision table. These requirements were described in Chapter 3, Section 3.8. Use this formal design representation to develop a detailed design.

12. Consider the system requirements for a simple chemical reaction process control system that were written in the form of a Petri net. These requirements were described in Chapter 3, Section 3.8. Use this formal design representation to develop a detailed design.

EXERCISES FOR THE MAJOR SOFTWARE PROJECT

1. Suppose that we wished to add a graphical user interface to the major software engineering project. An electronic list of such tools can be found at the URL http://www.cs.cmu.edu/afs/cs.cmu.edu/user/bam/www/toolnames.html. This website is

managed by Brad Myers, who is a well-known researcher in the area of user interface design. Indicate how the incorporation of GUI tools would affect the design.

2. Chapter 3 presented a hypothetical dialogue during our requirements elicitation. The initial requirements were later changed to allow for the web-based interface whose architecture was described in this chapter. Modify the architectures we presented here to reflect the high-level design of this (simpler) system. What changes did you make? Do you think it is easy to modify the design of a complex system to create a simpler one or to design the simpler system from scratch?

3. Expand the design of the major software engineering project to include analysis of source code files written in Java.

4. What is the effect of moving the project to a different operating system, such as one running Linux or Apple Yosemite, on the assumptions made in the design of the first subsystem for the user interface?

5. How does the design change if the data is to be collected in a database program instead of a spreadsheet?

6. Design the installation subsystem for the major software engineering project.

7. The user interface was to be written in either C or C++ to allow command-line arguments. Thus, this subsystem is very easy to design. It has two major features: interfacing with the operating system to obtain the command-line arguments and determining which arguments are to be passed on to the next subsystem. (The reader unfamiliar with the usage of command-line arguments should read Appendix B.) We note that the system's software requirements did not specify the development of a GUI for the project. Therefore, we did not consider any GUI tools. The incorporation of a GUI into the design is discussed in Section 4.17 and elsewhere in this chapter. The command-line argument interface is so simple that it does not appear to require any special design; at first glance, using the code in Appendix B as a template appears to be sufficient for our purposes.

8. In Section 4.19, we mentioned that we should consider both large-scale components and matching design patterns in order to make our software engineering effort more efficient. We chose to consider large-scale components in that discussion. Determine one or more relevant design patterns and use them to redesign this project.

9. Give the details of the design of the interface to a database for our major software project. Be sure that you specify the fields in the database.

10. This question considers the ramifications of a considerable change to the requirements on the design of the software. The first change is that the customer now wants a web-based user interface instead of one that is PC-based. Determine the changes that need to be made to the requirements to have a different type of user interface, with different issues in the possible configurations (multiple operating system versions on

the PC, multiple browsers, and multiple versions of HTML and Java). What changes, if any, need to be made to the software development life cycle model or the design process?

11. This question considers the ramifications of a considerable change to the requirements on the design of the software. The second change is that the customer now wants to use a cloud for data storage instead of one that is PC-based. Determine the changes that need to be made to the requirements to have cloud storage for the data, with different issues in scalability and the size of the cloud that is needed for the project. Also, determine the changes that need to be made to the requirements in order to have cloud storage for the source code modules being analyzed, with different issues in scalability and the size of the cloud that is needed for the project. What changes, if any, need to be made to the software development life cycle model or the design process?

CHAPTER 5

Coding

YOU ARE PROBABLY SURPRISED to see an entire chapter of a software engineering book devoted to coding. After all, you have been writing code during much of your computer science education, if not before. You may even have had some work experience or an internship in which you wrote source code. You probably think that you are a good programmer. (You probably are and will be an even better one after reading this book.) Why, then, is there a chapter on coding?

This chapter describes some typical industry practices that go far beyond what is normally required to complete typical classroom programming projects. This chapter includes a discussion of coding styles, coding standards, naming conventions, code organization, and code reviews that both promote reuse and encourage efficient program execution. Each of these topics will be discussed in a separate section. In addition, we will talk about some issues that may arise when coding the major software project.

Keep in mind that one of the major advantages of software reuse is the ability to eliminate the effort needed to create any software artifact once a larger scale artifact that was reused earlier in the life cycle (or elsewhere) has been deployed. For example, if some portion of a design is reused, then the source code that had been created along with that artifact can be reused. This is true for every type of software development life cycle, whether waterfall, spiral, rapid prototyping, agile development, or anything else. The rest of this chapter is devoted to the effort involved in creating new source code.

5.1 CHOICE OF PROGRAMMING LANGUAGE

It is important to understand that the choice of a programming language for implementation of a system is often made on nontechnical grounds. For example, market pressures, the availability of particular compilers, existence of software tools and development environments, or the experience of the software engineering staff may be more important than the perceived technical advantages of any particular language. Technical issues in programming-language selection include such things as ease of integration with existing software components, adherence to technical mandates, real-time or other performance requirements, and the need to keep staff members happy by giving them access to new technologies.

TABLE 5.1 Size of Several Computer Language Manuals

Language	Approximate Number of Pages in Language Manual
FORTRAN 77	150
Ada83	280
C (1990)	190
FORTRAN 90	400
Ada95	600
C++ (1998?)	800 (and growing)

You probably feel confident with your ability to program in at least one high-level language. You may feel that you have mastered all of one particular programming language's idiosyncrasies. You may have even mastered the idiosyncrasies of several languages. However, there are many subtle language features that even an expert may not have mastered. The primary reason for this is the increasing complexity of programming languages.

Table 5.1 emphasizes this viewpoint. The information in this table was provided by Les Hatton in his keynote lecture at the 1997 Conference on Computer Assurance (ComPASS '97) in Washington, D.C. (Hatton, 1997). Since many newer programming languages either do not have a standardized language manual or have a manual that is only available online, we will use this table only as a guide.

It is clear that programming languages are complex and that some of the more obscure features will be difficult to understand if used in programs. In addition, some of the more obscure language features may not be correctly implemented by your compiler. The following statement from the *Annotated C++ Reference Manual* by Margaret Ellis and Bjarne Stroustrup (2010) indicates the level of language complexity of C++:

> If a class base contains a virtual function vf, and a class derived from it also contains a function vf of the same type, then a call of vf for an object of class derived invokes derived::vf (even if the access is through a pointer or reference to base). The derived class function is said to *override* the base class function. If the function types are different, however, the functions are considered different and the virtual mechanism is not invoked. It is an error for a derived class function to differ from a base class's virtual function in the return type only. … The interpretation of the virtual function depends on the type of the object for which it is called, whereas the interpretation of a call of a nonvirtual member function depends only on the type of the pointer or reference denoting that object.

There is no reason to single out the C++ language on this list as being overly complex. The basic issue applies to nearly all programming languages. The potential obfuscation present in C language statements such as

```
*(--a[i++]) + = - *(--a[i++]);
```

is obvious.

Is the situation with modern programming languages any different? Probably not. Apple introduced the Swift programming language in 2014, and the difficulty in using it with somewhat limited documentation led Nick Walter to create the best-selling tutorial "Swift: Learn Apple's New Programming Language Step By Step," almost immediately after learning this language, with an enrollment of 9574 on Udemy in early 2015.

Similar complexities occur with Ada packages that have multiple combinations of *with* and *use* clauses, as in the following. (A *with* clause incorporates an Ada package into an Ada program. A *use* clause allows every function or object included within the package brought into the program by a *with* clause to be described by its simple name rather than a long name that shows the package that the function or object belongs to. This is the equivalent of describing something by its relative path name rather than its full path name.)

```
with package1; use package1;
with package2; use package2;
with package3;
package confusing is
 begin
 . . .
 package3.doit(); -- this is the one from package 3
   doit(); -- is this the one from package 1 ?
   doit(); -- is this the one from package 2 ?
end package confusing;
```

You can figure out what this software is supposed to do if you understand the naming and scoping rules of the *with* and *use* clauses in the Ada language. Still, the source code is hard to read. A similar example can be constructed in Java. C and C++ have complex scoping rules.

Java presents a special problem for software developers. The Java language itself is relatively small. Java is a purely object-oriented language. A software developer confining his or her Java classes to those provided in a standard such as the Java Development Kit (JDK), any particular version, is likely to create software that is relatively easy to port if the version is close enough to the current one.

Major factors in the success of Java are the well-defined interfaces known as application programming interfaces, or APIs. A software system that must interface to one or more existing programs may be written easily in Java if there is a smooth API. Writing the same software system may be much harder, or even impossible, to create in a reasonable time if there are no APIs available to interface to the preexisting programs. It is interesting to note that there is a currently unresolved intellectual property legal case between Google and Oracle (the purchaser of Sun Microsystems, where the original Java was created) over the ability of Google to use the same names for APIs as Oracle (Sun) but have different details for the methods hidden behind the APIs in order to use them on Android phones. Of course, the discussion about the need for adherence to APIs also applies to the use of the HTTP, SOAP, REST, or JSON protocols when interfacing objects to the cloud.

Visual BASIC (and the other languages that are usually considered to be part of the Microsoft Visual Studio) provides an easy interface to many operating system services by means of library components. For example, the user interfaces illustrated in Figures 3.14 and 3.15 (see Chapter 3) use linkages to Microsoft's library components to provide the familiar look and feel of these windows and the associated selections to be made by the mouse, which is the preferred input device for most Windows applications.

There are several popular scripting languages, including JavaScript, PHP, Perl, and Python. These languages, while powerful, have their own special nuances that can cause difficulty for the programmer not well versed in some of the more arcane features. The same could be said for almost every useful programming language.

From the perspective of providing high-quality, maintainable components, esoteric language features should be avoided as much as possible. If such features must be used, they should be avoided as much as possible, and their use should be confined to a few, well-documented and tested components.

In any event, the choice of programming language may be severely restricted by the system requirements, some of which may be in the form of an actual contract. A software system with many components, most of which are loaded or removed during system operation, is very likely to be written in Java. On the other hand, a software system that makes extensive use of UNIX or Linux system calls will probably be written in C or, less likely, in C++.

5.2 CODING STYLES

Consider the following three fragments of C source code. These three fragments are all intended to be equivalent when executed. (Recall that in the C programming language, the statement "i++" means that the variable "i" is to be incremented by 1; this is equivalent to the longer statement i = i + 1.)

```
while (i>0)
   {
   mystery (i);
   i++;
   }

while (i>0) {
   mystery (i);
   i++;
}

while (i>0)
   {mystery (i);
   i++;}
```

The three code fragments illustrate distinct approaches to formatting source code so that the body of the while-loop can be recognized easily. In the first approach, the curly braces are lined up vertically, and are indented the same number of spaces or tabs. The braces appear with no other characters on the same line.

In the second approach, the initial left curly brace appears immediately after the Boolean expression in the while-loop. The terminating right curly base is aligned with the "while" statement. The closing brace appears on a line by itself.

Finally, the third code fragment illustrates a compromise. The initial left brace appears on the line after the "while" condition, but the terminating right brace is no longer aligned with it.

The three distinct styles here should not be mixed. This means that you should pick one style and stay with it. We emphasize stylistic issues not for the difficulty in reading these trivial examples but for having to understand large amounts of source code when debugging or maintaining a realistic software system.

Which coding style should you use? The simple answer is use whatever style is specified by your organization's coding style manual. If no such manual is available to you, then emulate the style of a respected senior programmer in the organization.

There are many different coding standards that have been published. A readily available coding standard for the C++ language has been developed by the Flight Dynamics Division of NASA's Goddard Space Flight Center. This division is the parent organization of the Software Engineering Laboratory. The document can be found on the Internet by starting at the location indicated in the URL http:// www.gsfc.nasa.gov and following the appropriate links (the site is being reorganized as this book is being written). (There is another style guide at the Goddard site, prepared by the Software and Automation Systems Branch.)

Henricson and Nyquist developed a different coding standard for C++ (Henricson and Nyquist, 1996). A readily available set of a commercially created set of standards can be found at the URL http://code.google.com/hosting/. Coding standards for the C programming language can be found in similar places.

Coding standards for Ada are more formal. The most readily available coding standards for Ada can be found in the Public Ada Catalog portion of the ASSET library at the URL http://www.adaic.org/ada-resources and at the login-required site http://adapower.com.

If no coding standards exist, then look at some source code at http://SourceForge.net or follow the lead of a senior person in the organization. Above all, be consistent.

The aforementioned examples might lead you to believe that coding standards are merely a matter of formatting and indentation. There is more to style than just indentation, however. Most organizations' coding standards require a large amount of internal documentation; that is, documentation that is included within the actual source code. What follows is a list of some of the basic elements of a coding standard.

Standards for coding

1. White space
   a. Blank lines
   b. Spacing

   c. Indentation

   d. Continuation lines

   e. Braces and parentheses

2. Comments

3. Standard naming conventions

   a. Name formats

   b. General guidelines for variable names

Standards for program organization

1. Program files

2. Encapsulation and information hiding

3. Readme file

4. Header files

5. Source files

6. Makefiles

Standards for file organization

1. File prolog

2. Header file (class definition) (*.h)

   a. Order of contents

   b. Format of class declarations

   c. Method (function) documentation in header file

3. Method function implementation and documentation in source file

The amount of internal documentation required is often surprising to the student beginning a career in software engineering. The reason for this emphasis is the need to keep the documentation within the source code. This ensures that documentation will be available, regardless of any external documents that may be lost or any changes in project personnel.

Let me give a personal example. I recently worked on a project that involved moving a combination of old FORTRAN code and the assembly language used for Unisys mainframes. Most of which did not even adhere to the standards of FORTRAN 66, let alone FORTRAN 77, or the more modern FORTRAN 90, or even the most recent standard ISO/IEC 1539-1:2010 that was approved in 2010 but is commonly known as Fortran 2008. The goal of the project was to have equivalent software run on modern Linux or UNIX workstations. The project had to determine which source code modules could be moved as is, which could be moved with few changes, and which software modules had to be redesigned and rewritten for the new host computer environment. This, of course, is the classic build-versus-buy-versus-reuse-versus-reengineer problem.

The code was well documented in terms of having extensive internal documentation, with nearly two-thirds of each source code file consisting of comments. This was a lifesaver, because all external design and requirements documents had been lost during the software's twenty-five years of operation. The internal documentation made clear precisely what the purpose of each function in each file was supposed to do and how the functions fit together, in the sense that we could determine by reading each source code file. Using simple freeware tools to determine program organization were used and the results for the global description could be easily checked against the internal documentation to determine consistency.

Recall that in Chapter 2, we discussed the potential for storage of design and requirements documents on a computer network. If these documents had been available electronically, the process would have been much easier.

Documentation and code formatting obviously are important parts of coding style. There are also coding issues that affect code quality in the sense of making the code inefficient. One such factor is how many unnecessary assignments are made.

The trivial example

```
x = 1;
y = x - z;
x = y + 1;
temp = x;
x = y + temp;
```

shows the effect of an unnecessary assignment statement to an equally unnecessary variable. Compare this code fragment to the equivalent

```
x = 1;
y = x - z;
x = y + x;
```

Of course the selection of an optimizing option of a computer can remove the effect of such unnecessary assignments on the program's execution time. However, the use of optimization cannot change the increased difficulties that such unnecessary assignments will cause a software engineer trying to maintain this code.

The problems caused by poor programming style become worse if an unusual style causes a major problem in the performance because of typical virtual memory implementation. Consider the initialization of a matrix of size 1024 by 1024 in which the entries have the value 0. The double loop to perform the initialization can be written as either

```
#define MAX_SIZE 1024


for (i = 0; i < MAX_SIZE; i++)
  for (j = 0; j< MAX_SIZE; j++)
    a[i][j] = 0;
```

or

```
for (j = 0; j < MAX_SIZE; j++)
  for (i = 0; i < MAX_SIZE; i++)
    a[i][j] = 0;
```

Here, the coding examples are given in C, which stores two-dimensional arrays in what is called row-major form, with the array entries

```
a[0][0] , a[0][1], ... a[0][MAX_SIZE]
```

stored together, followed by the entries

```
a[1][0], a[1][1], ... a[1][MAX_SIZE]
```

and so on. (Technically, C does not allow two-dimensional arrays; it only allows one-dimensional arrays whose entries are also one dimensional.) The same storage pattern holds for programs written in nearly every other programming language, although the notation and the index of the starting array elements may be different from that of C. (FORTRAN is unique in that it uses a different arrangement for storage of two-dimensional arrays and so FORTRAN programs would have the loop variables reversed.)

The problem with the improper organization of the logic is that the underlying operating system must map the program's statement about the addresses of array elements in the program's logical memory space onto a set of pages that are mapped into physical memory when needed. The second nested loop has many more changes in the pages being accessed than does the first related loop. Therefore the second nested loop will probably run much slower than the first. (A good optimizing routine in a compiler may improve the run-time performance, but this improvement cannot be counted on.)

Coding styles can affect performance in other ways, as a more complicated situation shows. Here the matrix entries are either +1 or –1, with the successive entries alternating as in

```
1    -1   1    -1   .   .   .
-1   1    -1   1    .   .   .
1    -1   1    -1   .   .   .
-1   1    -1   1    .   .   .
.    .    .    .    .   .   .
.    .    .    .    .   .   .
```

The entry in the *i*th row and *j*th columns is simply –1 raised to the power of i+j. There are several options for implementation. We discuss the major ones next.

Option 1—We could use the standard mathematical library function pow() to compute the powers of –1 as in

```
a[i][j] = pow(-1, i+j);
```

in C. Unfortunately, this would involve the overhead of a function call. We might be able to use a macro on C or an inline function in C++ to perform the action performed by the function pow(). This point is discussed in the exercises.

Option 2—We could use a Boolean selection and the modulus operator % in statements such as the C code

```
if (((i+j) % 2) == 0)
   a[i][j] = 1;
else
   a[i][j] = -1;
```

or the Ada equivalent

```
if ((i+j) mod j) = 0)
    a(i,j) : = 1;
else
    a(i,j) : = -1;
end if;
```

The idea in each code fragment is to determine if the index is an even integer or an odd integer.

Option 3—We can use the single C statement using the special, but hard to read, conditional operator in the C language

```
a[i][j] = ((i+j) %2) ? 1: −1;
```

In many situations, none of these approaches is satisfactory because they introduce extra logical decisions, in addition to an addition of the two indices i and j and a division by 2. Each of these additional operations is repeated 1024 × 1024 times, which means that there are more than one million of these unnecessary operations.

It is easier to note that the entries in each row alternate in signs with the first entry in the two-dimensional array being a 1. Thus, a more efficient solution to our problem is illustrated by the nested loop

```
a[0][0] = 1;
for (i = 0; i < MAX_SIZE; i++)
 {
 if (i > 0)
   a[i][0] = a[i−1][0] * (−1);
   for (j = 1; j < MAX_SIZE; j++)
     a[i][j] = a[i][j−1] * (−1);
 }
```

Further improvements are possible, as you will see in the exercises.

Obviously there are many ways to solve this problem. Your coding style should encourage both program reliability and program efficiency. Recall that efficiency was one of the software engineering goals discussed in Chapter 1.

You should be aware that some commonly available utilities can help you improve the performance of your software. One of the most useful of these performance-improving utilities is known as a profiler. Most modern programming environments include profilers, which are typically invoked as an option when the source code is compiled.

The purpose of a profiler is to provide a software engineer with a detailed analysis of how the execution time of a running program is spent executing individual functions or program subunits. This analysis can be used to determine bottlenecks in program execution. Such bottlenecks are candidates for recoding to improve system performance.

Profilers are available on most Linux C and C++ compilers, among others. To use a profiler on a C++ program running on a Linux system when using the Free Software Foundation's g++ compiler, simply compile the program with the -p option

```
g++ -p file.cpp
```

and run the program. Each time the program is successfully completed, a file named mon.out is created. This file contains the various functions called (both user-defined and low-level system functions), as well as the time that is spent executing each function.

The file mon.out is read automatically when a profiler is used. Either the prof or gprof utilities are readily available as part of the standard software distribution on Linux systems. They are also available on most standard compilers other than the GNU compilers from the Free Software Foundation.

Details of other profiling techniques can be found in most compiler manuals.

5.3 CODING STANDARDS

In the previous section, we discussed ways that you could organize your code to make it easier to read or more efficient. Now, we will turn to the reality of modern software development, with its need for coding standards.

It is very difficult for the beginning computer science student to understand the need for such standards. Generally, he or she is able to do the assignments given by the instructor. The student is usually able to read his or her source code, even if it was written months before.

The situation is much more difficult in typical software development environments. The fundamental problem is that it is extremely hard to understand more than 10,000 lines of source code; this seems to be the limit of human comprehension. Software development organizations are doing everything they can to reduce costs and simplify their systems. At the same time, software is becoming more complex, thus forcing development organizations to produce more elaborate systems with fewer resources. The motto of many organizations seems to be "More! Better! Cheaper! Faster!"

In the 1990s the myth of the lone wolf programmer who develops large systems on a diet of caffeinated colas and junk food was exposed as just that, a myth. The myth was promulgated by the personal computer magazines of the early 1980s and was associated with the undeniable initial success of microcomputer operating systems developed by Steven Wozniak for the Apple II and Bill Gates and Paul Allen for the IBM PC.

The nature of software for personal computers has changed greatly since its early days. Almost all software has a graphical user interface and a large number of features. This is well beyond the capacity of a single programmer at this time. The book by Jim McCarthy of Microsoft describes the development environment for the successful C++ project (McCarthy, 1995). It makes clear the need for team activities.

On a personal note, I had firsthand knowledge of this point in the 1980s at NASA while working with private contractors providing government service. An employee of a well-known computer software company was notorious for producing source code that was correct, but poorly documented and inconsistently formatted. The company fired this programmer after it became clear that the lack of adherence to the company's coding standards was costing the company much more in software maintenance costs than the individual's productivity could ever justify.

Incidentally, some engineering students from Howard University who were working as summer interns were able to speed up the performance of one of his implementations of an important algorithm by a factor of more than ten. The improved efficiency made it feasible to consider using relatively low-cost personal computers for certain graphical displays instead of what, at the time, were extremely expensive graphics terminals.

It is likely that the professional programmer overlooked the potential for improving program performance simply because this code was so poorly presented and documented that even he did not understand it. Do not let this happen to you. Follow your organization's coding standards.

The same advice applies to some of the developers of software sold in app stores. You can often tell from user reviews of apps that some newer versions of an app are not as good as previous ones, which is likely due to difficulty in maintaining code because the (probably lone) developer no longer understood details of the source code he or she had written.

A coding standard is usually developed by an organization to meet its needs, and different organizations usually have different standards.

A coding standard will include the following:

- Conventions for naming variables
- Conventions for naming files
- Conventions for organizing source code files into directories
- Conventions for the size or complexity of source code
- Requirements for documentation of source code
- Elaborate style guides
- Restrictions in the use of certain functions

We now discuss each of these issues in turn.

A convention for naming variables is a natural extension of what you probably already do when choosing variable's names. A coding standard for naming variables will require meaningful names. The coding standard may require that all nonlocal variable names be in several distinct parts that are separated by symbols such as underscores, periods, or the judicious use of uppercase letters.

Different portions of a name have different meanings. The intention is to allow a person reading the code to determine at a glance that the two names `db_index` and `DbIndex` refer to an index with a database. In a similar manner, the rather long variable names `InputChannelIndex` and `Input_Channel_Index` clearly refer to an index of input channels. The name `input_channel_index` might be used for this purpose, also, while `inputchannelindex` is probably a poor choice. The naming convention for variables will include the appropriate style, generally allowing the use of (one group of) underscores or uppercase letters. We note that the pattern of starting each name within an object name with an uppercase letter, as in `InputChannelIndex`, is the de facto standard in Java and Swift.

Another issue may be addressed by the variable naming portion of a coding standard: the need to be compatible with compilers with severe limits to the length of variable names. In any event, the coding standard should address this.

File-naming standards have all the features of variable naming standards and also should reflect the contents of the file. Many standards require a single function, procedure, or object per file, others simply place a limit on the total size of a file that contains multiple functions, procedures, or objects. This is due to the need for compatibility with multiple operating systems, such as the case with Microsoft's NTFS intended for use with most Microsoft operating systems, and exFAT, which is optimized for use with flash drives. Ancient systems often even restricted file names to eight characters, with a three-character extension.

Ideally, the names of source code files should indicate the files' contents so well that it is easy to determine the appropriateness for future potential reuse of the source code file.

The issues with file naming standards also apply to organization of files into directories. Each directory represents a subsystem. Here, the most important issue is developing high-level directories whose names are related to the functionality of the subsystem whose code is included in the directory and its subdirectories.

Of course, this is precisely the way that operating systems are organized. The 3784 distinct user files that were on my PC at the time the first edition of this book was written in 1997 were organized into directories with meaningful names that represented the relevant subsystems. Most applications follow the same organizational approach.

There are several other issues related to file organization. It is easiest to have separate directories for development of the source code and for the final source code. It may even be easier to separate the object code and source code into separate directories. This means that for most realistic software systems, makefiles (or other instructions about which source code, object code, libraries, configuration files, or data files are used for installation) will use relative file names.

The use of relative, as opposed to absolute, file names, allows a directory and its contents to be used whenever it is most appropriate, as opposed to being restricted to a specific location. Appropriate use of relative file names also allows the final executable code to be placed wherever the installer wishes, instead of being limited to the current directory. We will discuss installation details in Chapter 7 when we describe the installation and delivery process.

5.4 CODING, DESIGN, REQUIREMENTS, AND CHANGE

Many software projects are so large that the entire source code is compiled relatively infrequently. For example, in many of its software systems, Microsoft creates a "build," or clean compilation and linkage of all source code only once a day. This is done because too much time would be spent compiling source code and not enough time spent thinking about it. This approach is common in organizations that develop large software systems, especially those that develop software in several geographically separate locations.

This approach requires a new way of thinking that is very different from the typical edit–compile–execute–debug cycle that is so common among computer science students. In most organizations, the software development process must be much more systematic than this hit-or-miss approach.

It is often surprising to beginning software engineers how little time is actually spent writing code and how much time is spent on other software life cycle activities. Typically, only 15 percent of the time allotted for a project's completion is devoted to coding. Of course, there are many reasons for this distribution of effort.

One problem that occurs frequently during the implementation process involves the requirements traceability matrix. The problem is that often the requirements were not sufficiently detailed to make certain that the software's implementation matched the precise requirements. In short, each requirement must be traceable to an implementation action. If not, there is no guarantee that the requirements have been met.

What about changes in design? The major deficiency of the waterfall software life cycle approach is the assumption that the requirements are so well understood that they can be set well in advance of the actual implementation or deployment of the software. The strictest interpretation of the waterfall life cycle model allows just the requirements to be changed only when major issues arise during the design process. Similarly, the strictest interpretation of the waterfall life cycle model allows the design to be modified only when major changes have to be made during the coding process. Of course, the more flexible iterative software development approaches of the spiral and prototyping models involve changing design consistently.

Changes in software design can occur during coding for several reasons:

- Changes in technology have made portions of the design obsolete or irrelevant.
- Software libraries have been changed or reorganized.
- The supposedly reusable software components do not work as promised. (This is especially true of special commercial software packages with heavy marketing.)
- The reusable software components are not of high quality. (This is especially true of public domain software or software written by companies under great pressure to maintain market share. Many such companies follow the "good enough" approach when releasing new versions of their software.)
- The compiler did not generate sufficiently fast executable code to meet the real-time requirements of the system on its target operating system environment.

Many other problems can occur during software implementation. As a member of a software team, be prepared for change; your manager should be. Each change will create pressure on a project's schedule and you may be tempted to skip some documentation and ignore coding standards when under pressure. Do not do it. Your work will not really go any faster; you will only create more work for testing and quality assurance, and your work will not be good enough to be included in a reuse library.

5.5 COUPLING CAN BE DANGEROUS

It is often necessary to consider the nature of coupling between components during the coding process. There are many coupling metrics that measure the interconnection between program units. The most commonly used coupling metrics were first described in detail by S. D. Conte, H. E. Dunsmore, and V. Y. Shen in their classic but still important book *Software Engineering Metrics and Models* (Conte et al., 1986).

Generally speaking, the less coupling between modules, the better, at least from the perspective of using modules as components. Coupling-based testing is based on the premise that system errors occur most frequently at unit boundaries and where they interface. Modules with the most coupling must require the most testing effort.

There are many types of coupling. We will present them in the order of greatest amount of coupling to smallest amount, under the assumption that we should test in the places where errors are most likely to occur. Not all the forms of coupling are present in every programming language.

- Content coupling
- Common coupling
- Control coupling
- Stamp coupling and data coupling
- External coupling
- Message coupling
- No coupling, also known as weak coupling

We now define these terms.

Content coupling is the highest level of coupling between modules. One module modifies or relies on the internal local data of another module. Changing one module's data (such as location, type, timing) will change the dependent module. We illustrate this form of coupling in Figure 5.1.

We note the following facts about content coupling.

- The flow of control is managed in a systematic way by exception handlers.
- Ideally, exception handlers bring software execution back to a "safe state," although this is not guaranteed.
- There may be problems if multiple instances of content coupling occur in a distributed system.

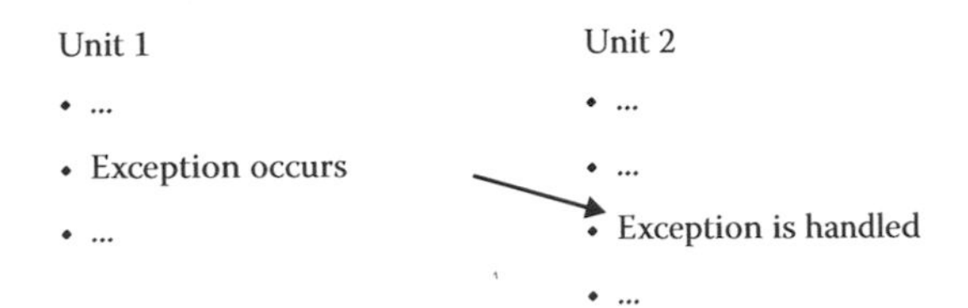

FIGURE 5.1 An example of content coupling illustrated by exceptions and exception handlers.

- Content coupling can also cause a disaster in nondistributed systems.
- The problems may be very hard to test.
- Systems with modules having content coupling can be tested using recovery testing ideas.

In languages such as C, using the commonly available low-level calls setjmp() and longjmp() to provide a lower level of exception handling than the exception handling available in most programming languages makes problems hard to test, locate, and fix. You should avoid using these nonlocal GOTO statements whenever possible. Languages such as Java and Ada, and, to a lesser extent, C++, treat such common coupling using exceptions and exception handling instead of these lower-level system-level function calls.

Common coupling occurs when two or more modules share the same global data (e.g., a global variable). In common coupling, changing the shared resource in a single module implies changing all the modules using it. We illustrate this form of coupling in Figure 5.2.

Control coupling occurs when one module is able to control the logic of another, by passing it information on what to do (e.g., by passing a what-to-do flag). Therefore, in control coupling, execution of a loop or program branch depends upon the parameter. We illustrate this form of coupling in Figure 5.3.

Data coupling occurs when modules share data through parameters. Parameters may be used in a calculation, but they do not affect logical flow (loops, branches) of the receiving module. Some researchers call this stamp coupling. (Others use the term *stamp coupling* to mean that only one field of a multifield record is actually used and use the term *data coupling* to describe the passing of parameters.) We illustrate this form of coupling in Figure 5.4.

External coupling occurs when two modules use data that is external to both. The external data may have a specified format, communications protocol, or device interface. This relatively rare form of coupling is illustrated in Figure 5.5.

Message coupling is a very low level of coupling in which modules do not communicate with each other through parameters. The communicating modules use a public interface,

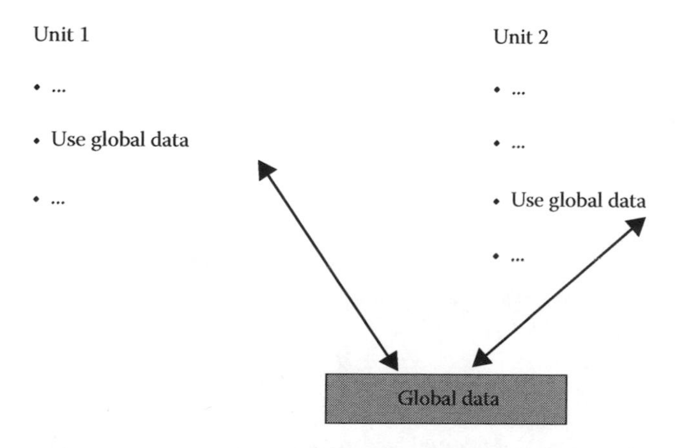

FIGURE 5.2 An example of common coupling with two modules sharing global data.

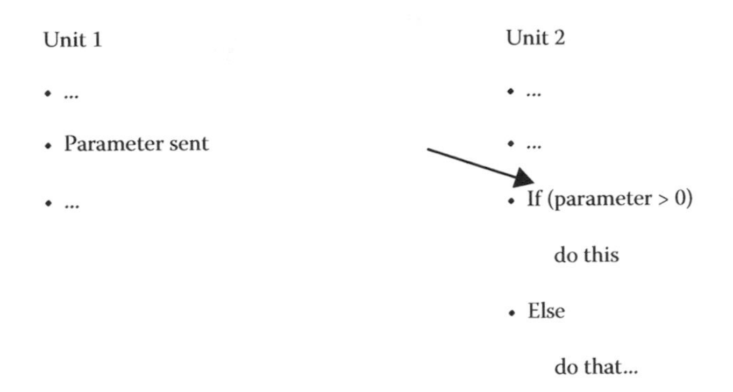

FIGURE 5.3 An example of control coupling.

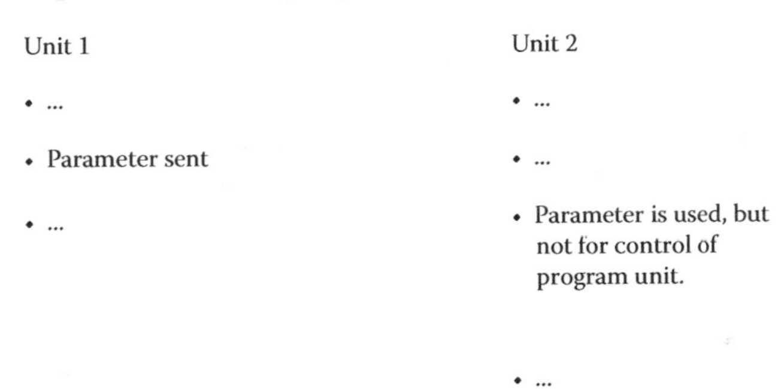

FIGURE 5.4 An example of data or stamp coupling.

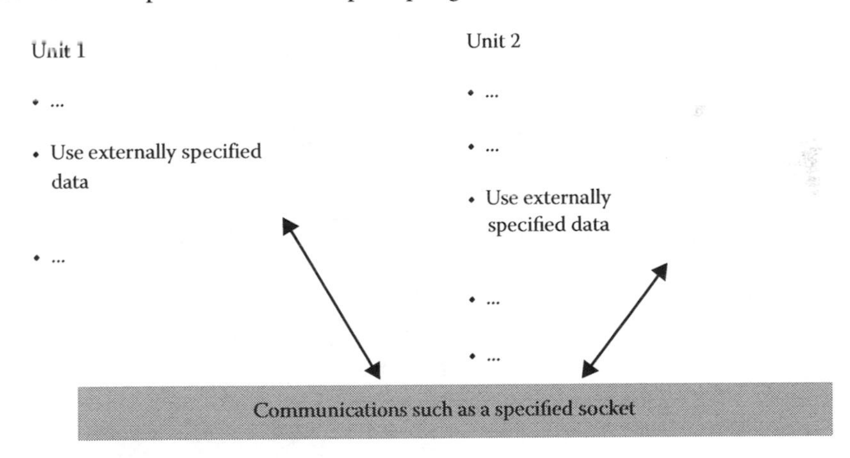

FIGURE 5.5 An example of external coupling.

such as messages or events, to exchange parameter-less messages. For obvious reasons, message coupling is more prevalent in object-oriented systems. This form of coupling is illustrated in Figure 5.6.

The lowest form of coupling is when modules do not appear to communicate with one another at all. There is a minor concern that each module might use up system resources, making it hard for the other module to execute. It can be very hard to find such errors without extreme stress testing.

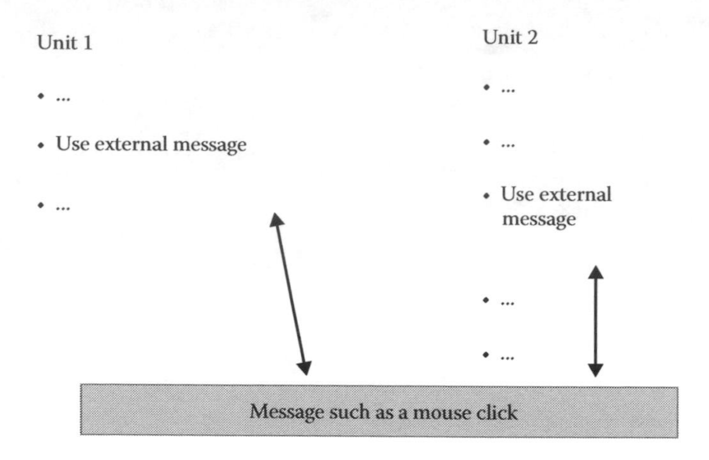

FIGURE 5.6 An example of message coupling.

Here is an example from the world of unethical hacking of weak or no coupling causing a problem via a distributed denial-of-service attack. A denial-of-service attack occurs when one or more servers, such as those of a bank or retailer, are unable to provide normal service because they are inundated by an unusually large number of requests, each of which requires some operating system resources. These resources include system data structures for control of such things as heavyweight processes, lightweight processes (threads), and the CPU time to process these requests.

Writing an effectively infinite loop for creation of processes and threads is very easy for unethical hackers to do. Such an attack from a single computer is easy for a competent system administrator to stop by simply refusing communication from an easy-to-identify IP address. One of the reasons that many unethical hackers try to take over poorly defended personal computers is that they enable the use of multiple IP addresses to try to thwart such simple defenses against denial-of-service attacks.

Some of the issues with this form of coupling are illustrated in Figure 5.7. You should try to avoid the deepest levels of coupling when you create source code. At the very least, this strategy can reduce coding effort later on.

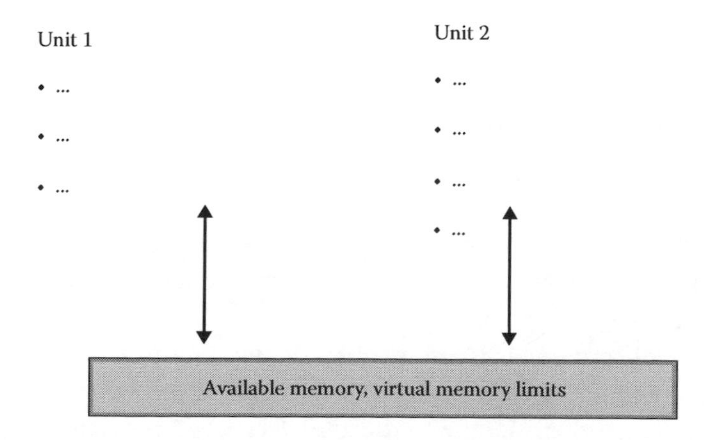

FIGURE 5.7 An example of weak (or no) coupling.

5.6 SOME CODING METRICS

Since software engineering is, after all, an engineering discipline, measurement is an important activity. In this section, we will describe several issues associated with measurement of source code.

Since our major software project develops a tool to measure the size of software source code, we will not consider size metrics in this section. Instead, we will consider two commonly used types of metrics: control flow metrics, which attempt to describe the possible execution paths of a program, and coupling metrics, which treat the interfaces between program subunits. We will describe control flow metrics first.

The McCabe cyclomatic complexity metric (McCabe, 1976) measures the complexity of the flow of control within source code. The metrics process creates a representation of the control flow graph of a program based on Euler's formula for the number of components in a graph. All statements in a program are considered vertices of a graph. Edges are drawn between vertices if there is direct connection between two statements by a loop, conditional branch, call to a subprogram, or if the statements are in sequential order. McCabe's metric is $E - V + 2P$, where E is the number of edges, V is the number of vertices, and P is the number of separate parts (the number of subprograms called), including the main program.

The cyclomatic complexity metric essentially reduces to the total number of logical predicates plus 1. As such, it is invariant under changes of names of variables and under changes of the ordering of arguments in expressions. This metric is also invariant under any changes in the format of the control structure. Thus, changing a while-loop from the form

```
while (!found)
   {...
   }
```

to one of the form

```
while (found ! = 0)
   {...
   }
```

leaves the McCabe cyclomatic complexity unchanged.

Each of these program fragments has a graph similar to the one shown in Figure 5.8. Each of these graphs has three relevant vertices (nodes), corresponding to the three identified lines in the program. There are also three edges in each graph. Therefore, both of these two loops have a McCabe cyclomatic complexity of 3 – 3 + 2, or 2. Adding nonbranching statements between the braces adds 1 to both the count of vertices and the count of edges, leaving the value of this metric unchanged. Note that changing a while-loop to an equivalent do-while-loop also leaves this metric invariant.

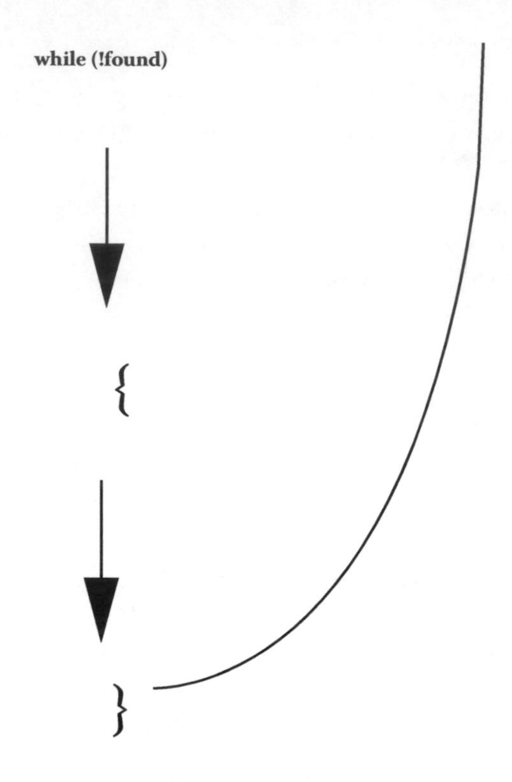

FIGURE 5.8 A portion of a program graph.

Here is an example of why this cyclomatic complexity metric is important. A 2008 Enerjy Co. study analyzed open source Java applications and found a strong correlation between high cyclomatic complexity and the probability of software faults:

- Classes with a combined cyclomatic complexity of 11 had a fault probability of 0.28.
- Classes with a combined cyclomatic complexity of 74 had a fault probability of 0.98.

The reason for this correlation was the much higher cyclomatic complexity represented more execution paths to be tested, and not everything was tested.

I encountered another example of this on a visit to HP's Medical Products Group in Waltham, Massachusetts, several years ago. (This is now part of Agilent.) I learned about a policy that anyone who created a source code module with a cyclomatic complexity greater than ten had to arrange a formal meeting with management to explain why the source code was so complex. Needless to say, everyone tried to avoid these meetings and so the source code was always simplified.

The cyclomatic complexity metric considers only control flow, ignoring the complexity of the number and occurrence of operators and operands, or the general program structure and, thus, cannot be a complete measure of program complexity. However, the cyclomatic complexity can be applied to an entire program, to a single file, or to a single module. It can also be applied to detailed designs and to the PDL (program design language) before a module is coded. It is one of the few metrics that can be applied at several places in the life cycle. Halstead's Software Science metric is based on counts of operators and operands (1976).

We now consider coupling metrics. In the simplest case considered here, coupling metrics are based on a count of the number of arguments and global variables that can be accessed by a function. A more refined analysis distinguishes between arguments and global variables that can be modified within a module, control the flow of a module, or are merely used as data sources. See Section 5.5 for a detailed discussion of coupling metrics. Note that the SANAC system, whose text-based user interface was shown in Figure 4.16 when we discussed user interfaces in Section 4.11 (see Chapter 4), described a system that developed these metrics for programs in the Ada programming language.

We now consider a metric based on the published interface between functions and subroutines. Specifically, we consider the arguments sent to each function or subroutine. The BVA metric (Leach, 1997) is based on an assessment of the number of cases required for testing of a module based on its interface and results from testing theory that indicate that logical errors in software are most likely to occur at certain boundary values in the domain of the software. It is a measurement of modularity and testability. The BVA values associated with a function's arguments are defined as follows:

- Arguments of type Boolean are given a weight of 2 (true, false).
- Arguments of type int are given a weight of 5 (MININT, −1, 0, 1, MAXINT).
- Arguments of type float are given a weight of 5 (MINFLOAT, −1.0, 0.0, 1.0, MAXFLOAT).
- Arguments of type struct are given a weight that is the product of the weights of the components.
- Arguments of type pointer are given a weight of one plus the type of the object pointed to.
- Arguments of type array are given a weight of two plus the type of the element in the array. (The difference in treatment of arrays and pointers is a reflection of common usage, not syntax, since arrays and pointers are the same idea in C.)
- Global variables that are visible to a function are treated the same way as function arguments.
- For a function with multiple arguments, the BVA value is the product of the BVA values associated with the individual arguments.
- For a file, the BVA value is the sum of the BVA values of the individual functions.

We chose to omit qualifiers such as long, short, or unsigned since the first two do not change the BVA value. The qualifier unsigned restricts the integer to be nonnegative. This is a small decrease in the BVA value; we chose to ignore it because the qualifier is rarely used and is often used incorrectly. We would use a weight of three (0, 1, MAXINT) for function arguments with the type classification NATURAL in the Ada programming language, since the proper use of this type is more likely to be enforced by an Ada compiler.

The storage class qualifiers static, register, and extern were also ignored in our BVA computations, since they specify where and how the data type is stored, not what the set of possible values is.

As an example of the calculation of this metric, consider the C code implementation of the following stack:

```
struct stack
   {
   int  ITEM[MAXSTACK];
   int top;
   };
```

There are two fields in this structured data type: an array of fixed size whose name is ITEM and whose entries are integers, and an integer variable named top. Any function that uses a parameter of this stack data type has to consider the two fields. The second field, top, has an infinite range (if we ignore the construction of a stack) and has several likely candidates to select for black-box testing. The five cases that we use are: –1, 0, 1, MAXSTACK, MAXSTACK + 1, assuming that top takes only values either inside or near the range of index values.

The array indices are tested at the upper and lower bounds plus or minus 1; the test cases are 0, 1, 2, MAXSTACK – 1, MAXSTACK, MAXSTACK + 1. Thus the total number of test cases to be added to the value of the BVA metric is 5 × 6, or 30.

The count of the number of cases for the BVA metric will be different in programming languages that support strong typing and run-time checking. For example, a definition of a stack in Ada might look like

```
record STACK is
 ITEM: array(1 .. MAXSTACK) of integer;
 TOP: integer range 0 .. MAXSTACK;
end STACK;
```

There are still two fields in this structured data type that must be considered by any function that uses a parameter of this stack data type. The second field, TOP, has a finite range and has several obvious values to select for black-box testing. The four cases that we use are 0, 1, MAXSTACK – 1, and MAXSTACK. This implementation of a stack in Ada has a BVA metric value of 4 × 4, or 16.

Note that neither count makes use of the typical way in which a stack is used (access to the stack is usually limited to the top element of the stack, which should be done only by using functions to push and pop the stack). Therefore, the BVA metric may overstate the effect of the complexity of the data, particularly in an object-oriented environment in which access to internal data of an object is restricted to specially written member functions of that object. Thus, the BVA metric is only a first approximation to a data structure-based metric. I view the lack of a consistent industry standard for a data structure-based metric that has been validated as a measure of software structure as a severe deficiency of research in this area.

5.7 CODING REVIEWS AND INSPECTIONS

There is a growing body of software engineering research literature that suggests that careful inspection of source code has a greater chance of detecting faults (at least in small programs) than does software testing. Although this hypothesis has not been proved conclusively and has not been examined in detail for larger programs, it is worthwhile to describe the review and inspection process in more detail at this point.

Fagan (1996) was the first to write extensively about inspections in the open literature and, thus, inspections are often called "Fagan inspections." Research on inspections have shown that the following types of approaches can be very helpful, whether the inspection is performed by a team or by an individual (Barbey and Strohmeier, 1994; Basili and Selby, 1985).

As with the requirements and design of software systems, source code can also be reviewed. Table 3.11 (Chapter 3) and Table 4.2 (Chapter 4) give checklists for requirements and design reviews, respectively. You should review those checklists now.

As with requirements and design reviews, we may also perform an inspection of source code. The inspection may take the following forms:

- The source code may be examined for adherence to coding standards.
- The directory structure may be examined for adherence to naming conventions.
- The inputs or outputs from each function may be checked by hand for correctness.
- The invariants of each loop may be checked to make sure that the loop exit conditions are met.
- Objects may be checked for completeness of coverage of all conditions if some of the object's methods are overloaded.

Note that there are several different ways in which the code may be examined for correctness. You should read the current literature to determine the most recent thinking on this complex subject.

5.8 CONFIGURATION MANAGEMENT

One of the most troublesome problems in coding occurs when there is more than one programmer working on the same part of the system. Even if the programmers adhere rigidly to the standards set in the interface control document, the changes made by one programmer may affect the code written by another. Often, even a single programmer wishes to go back to code written the day before. This is not always easy, due to the likelihood that many changes were made during the elapsed time period.

The situation is much worse if there are many developers working in different locations. Since this is becoming common in larger, distributed software development projects, there is a critical need for a smooth mechanism to allow changes to be made systematically and to allow the changes to be undone if necessary.

The general problem of controlling changes systematically during software development is called *configuration management*, or CM. It is pervasive in the software industry. Simply put, if you have not used configuration management, you have not worked on large software projects. You may hear the term *version control* used instead of configuration management; the terms are synonymous. We met configuration management previously in Section 2.7 (Chapter 2), Section 3.14 (Chapter 3), and Section 4.13 (Chapter 4).

There are many commercial systems and free utilities that perform CM. They all keep a master set of each document under the control of the CM system. Changes are entered with a version number and a date, and it is easy to go back to any previous version.

In most CM systems, once a file is "checked out" (being worked on by a developer) the file is logged into the CM system as being read-only for others. This blocking of edits by developers other than the one who has the file checked out helps ensure system consistency. (A few powerful CM systems relax this rule because the organizations need to have simultaneous access from locations that are widely separated, but this approach is hardly the norm.)

We illustrate the behavior of a CM system by using the powerful sccs utility developed for UNIX systems by Marc Rochkind. The acronym stands for Source Code Control System. It is one of the first CM systems and available free of charge on most Linux systems. Instead of forcing you to keep copies of your entire document every time you want to make a backup, the sccs utility only keeps copies of the changes that you have made, along with a copy of the most recent version of the document. Even though there are newer and more powerful CM systems, this one is worthy of discussion because its structure clearly illustrates the issues.

At the most elementary level, the sccs utility is used as follows: A programmer wishing to use this tool creates a directory named SCCS. The files that are to be under CM are placed in this directory. We illustrate this with the directory whose contents are shown in Example 5.1.

Example 5.1: Contents of a Directory Named SCCS That Is Used for a Project under Configuration Management by the sccs Utility

```
s.makefile
s.file1.c
s.file2.c
s.file1.h
```

You may be wondering how the files are placed into this directory and how they get such strange names. They are put there by the command

```
sccs create file1.c
```

that looks for the proper directory and places a copy of the file filel.c in this directory.

Suppose that a programmer wishes to edit the file filel.c, which is under the control of sccs. (The file is named s.filel.c in the SCCS directory.) From the parent directory, the programmer types the command

```
sccs edit file1.c
```

The sccs utility responds by opening the file for editing. The file will be visible as a collection of bytes and some indication of the version number, as is shown in Example 5.2. (The version number will change as the file goes through successive edits.)

Example 5.2: A File under sccs Control

```
/* Version 1.1 */
int main(int argc, char *argv[ ])
{int i;
for (i = 1; i <= 10; i++)
  {printf("%d\n", i);
}
```

The sccs paradigm is that you will leave your files as read-only, except when you are explicitly editing them. When one user wishes to edit a file, it is checked out of the SCCS directory by the sccs utility and no one else can edit it.

There is an obvious question to ask: How are other users able to see this file when someone else created it? There are two obvious ways that this can be done. The first way, which is fraught with potential security issues, is for the owner of the sccs directory to set permissions as rwxrwxrwx so that everyone can see the files, open them, and edit them. Permissions on UNIX and Linux systems are in triples of read, write, or execute for the owner, the owner's group, and the world. (The execute permission when applied to a directory allows someone to traverse that directory.) This setting of permissions must be done for the full path that describes the location of the sccs directory.

A better way to handle file access is to place all developers on a project into the same group. This is done by the system administrator who creates a new group shell command called newgrp, which is used to change the current group ID during a user's login session. With that choice available, the permissions need to be rwxrwxr, instead of rwxrwxrwx, reflecting that the owner and his or her new group can see and edit the files but others (the "world") cannot.

When the file is made ready for use in a system (as would be the case if the make command were typed at the appropriate directory level), the file might appear to a user in a particular version, as is shown in Example 5.3.

Example 5.3 A Final File SCCS Control (Assuming No Changes)

```
int main(int argc, char *argv[ ])
{int i;
for (i = 1; i <= 10; i++)
  {printf("%d\n", i);
}
```

When you are done editing a file, you can enter the command

```
sccs delta myfile.c
```

You will be asked to supply a comment about what changes you have made. Documenting the reason for the changes can make program maintenance much easier.

You can see how the sccs utility allows access to a file by one person at a time and allows files to revert to earlier versions if, say, a software change causes a new problem. Keep in mind that sccs was intended for use on a single computer system running the Linux precursor, UNIX, and that the dominant way of using UNIX at that time (it was first produced in 1972 and released as part of a standard UNIX distribution in 1975) was to log on using remote terminals. That is clearly not the way we do computing now!

Now let us look at some newer software for configuration management. It is clearly important to support remote access and not be tied into hard-wired terminals (or dial-in). This requires at least a client–server model, and might need something more distributed for large, geographically distributed development systems.

A recent check of a discussion on the SIGCSE mailing list listed the following essentially free tools as popular solutions to the problem of providing configuration management software within an academic environment:

- Razor (SCM Tool from Visible Systems)
- CVS
- RCS
- Git
- Github (actually, a collection of source code, which includes configuration management tools)

- Mercurial (HG Init is a tutorial for Mercurial and is available from the website http://hginit.com/.)
- Subversion
- Bitbucket
- Bugzilla

Most of these tools have graphical user interfaces, which can be a great help, since configuration management is a tool intended to aid in the software development process. If you use one of these or similar software tools for your academic software engineering projects, keep in mind the simple sccs example as a guide to what is available in your own tool and how this works.

Figures 5.9 and 5.10 illustrate some information about the Razor SCM Tool, produced by Visible Systems. This tool can be used either stand-alone or as an integrated part of a larger system. For example, Razor can be integrated with any set of files that are under a source code control system that is Microsoft compliant. In particular, Razor can work with not just source code, but also requirements and design documents.

Some, but not all, software development kits that are intended for use by a single user on a personal computer include the ability for version control that is the core of configuration management. Of course, any CASE tool intended for large-scale software development will have a strong capability for configuration management.

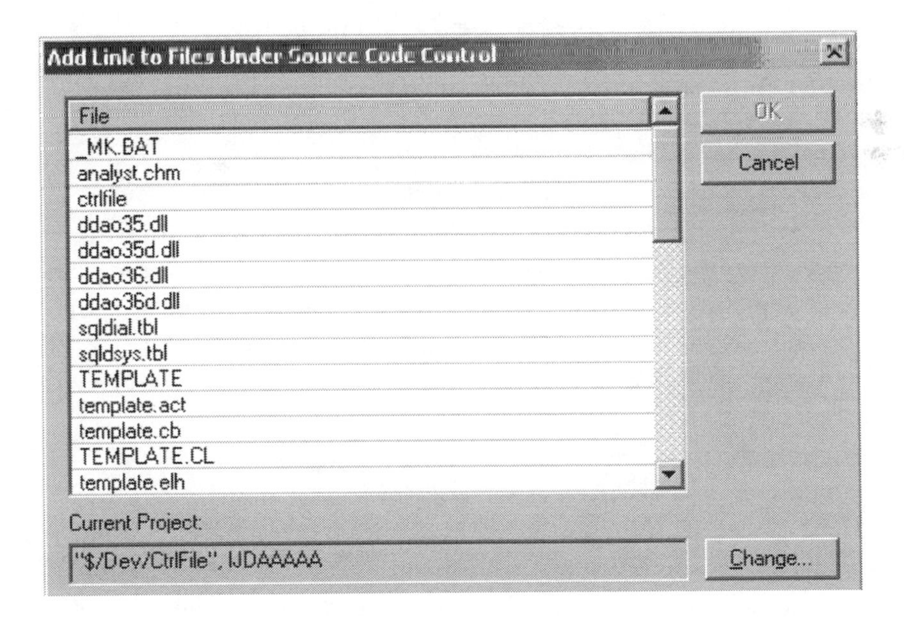

FIGURE 5.9 Razor, a commercial software configuration management tool. (Courtesy of Visible Systems.)

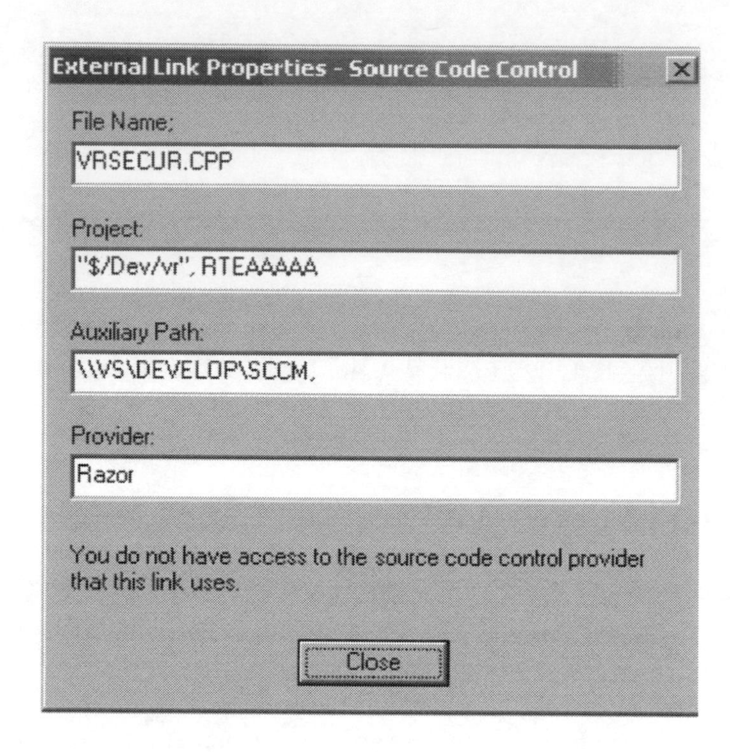

FIGURE 5.10 Another screen image from Razor, a commercial software configuration management tool. (Courtesy of Visible Systems.)

5.9 A MANAGEMENT PERSPECTIVE ON CODING

As we have discussed several times previously, once an organization is committed to a project, the project's manager wants to minimize risk and to make the consequences of any necessary risk predictable. Unfortunately, the coding phase is where any problems that had been glossed over during the requirements or design phases cannot be avoided.

In many projects, the manager is in this position because of the nature of the programming process. The difficulty is of course that the manager has to trust the promises of his or her programming team that the software will be created on time and will work. The manager often has no way to know whether these promises will be kept.

Let us expand on this statement. If the software development follows the classical waterfall approach, then the manager will not have seen any software until very close to the date scheduled for its completion. He or she cannot anticipate disaster.

The iterative software development methods may offer more comfort to a manager. Obvious disasters can be seen more quickly. Many new software development environments for object-oriented designs can show object models, with prototypes of programs readily available to the potential users very early in the software's development. Unfortunately, the hard problems rarely surface so quickly, unless the software's design is poor or incomplete.

Software implementation is always the tensest portion of the life cycle for managers. In many cases, managers depend on a carefully defined process, such as the higher levels of the Software Engineering Institute's Capability Maturity Model (CMM). This can help, because monitoring activities are built into the process model. Clearly, the manager's goal is to prevent disaster if the project is to be delivered very late or cannot be delivered at all.

Metrics are extremely important in providing a snapshot of the state of the software and the progress toward the goal of releasing a high-quality product within budget and on schedule.

The requirements traceability matrix is used to make sure that each feature that was specified in the system requirements and promised in the design has been implemented in the actual source code and tested before the final product is released.

5.10 CASE STUDY IN CODING IN AGILE DEVELOPMENT

The object of coding in an agile development process is to create a high-quality system as rapidly as possible. To this end, there will be as little effort placed on non-coding-related issues during the coding process as is humanly possible. This may mean that there may be less than the typical amount of effort placed on coding standards.

Thus, source code written by different members of the agile team may not follow coding standards about indentation and the actual physical presentation of the formatting of loops on paper printouts of the source code. Since the code is almost certain to be done using an IDE (integrated development environment), the standards, if any, will almost certainly be those used by the IDE. The use of configuration management tools for version control is almost universal.

For the same reason, the internal documentation of the code, that is, the comments within the source code modules themselves, often will be a far lower priority for the team than the actual development and testing of the source code.

The testing of the code is paramount. It is primarily focused on the interfaces between components, COTS (commercial off-the shelf) products, and subsystems. We will discuss testing in our agile development case study in Chapter 6.

It is not clear, unfortunately, whether the likely lack of adherence to coding standards in the creation of a software system during an agile development process will have any effect on the maintenance of such a system over time, especially if the newly created system has a long lifetime. For systems that are expected to have short deployment lifetimes, any maintenance problems that do occur are not likely to be major ones.

5.11 CODING OF THE MAJOR SOFTWARE PROJECT

In this section, we begin the process of implementing the design of the major project that we have been considering in the last few chapters. The most important things for us to keep in mind are that that we must adhere to the design as much as is reasonable, and that any changes must be carefully documented. After all, the goal is to provide our customer with the system that he or she wants. You should begin by reviewing the design documents and the project's requirements as necessary.

We will not present the complete source code, since it is available to you electronically. Instead, we will focus our attention on a few essential implementation details.

For example, the user interface of the system is coded in C. In Example 5.4, we show a portion of this code to illustrate the coding standard we are using for the C code.

Example 5.4: A Small Portion of the Java Code (the Front-End Interface to the Operating System) in Our Project Showing Some Coding Standards

```
import java.awt.*;
import java.net.URL;
import com.sun.java.swing.JLabel;
public class AttentionDialog extends ModalDialog
{
   public AttentionDialog(Frame parent, String title, String message,
            URL iconURL)
            {
            super(parent, title);
            if (iconURL ! = null)
                     {
                     ImageViewer img = newl mageViewer(iconURL);
                     add(img);
                     }
//
// SOME SOURCE CODE OMITTED HERE
//
okButton1_ActionPerformed_Interaction1(java.awt.event.ActionEvent
event)
{
try {
   this.dispose();
   } catch (Exception e) {
}
}
}
```

As another example, the back end of the system, which must interface to standard spreadsheets, is coded in Java. In Example 5.5, we show a portion of this code to illustrate the coding standard we would use for the Java code.

Example 5.5: A Portion of the Java in Our Project Showing the Coding Standards

```
import java.rmi.*;
import java.rmi.server.*;
import COM.stevesoft.pat.*;
import java.util.*;
import java.io.*;
import java.lang.*;
public class SoftwareEval
{
//
// SOME SOURCE CODE OMITTED HERE
//
// start class actions-------------------------------------------------------------------
// check for class definition
   lineCom.search(line);
   if(classDec.search(line) && !commentFlag &&
           !classDec.search(lineCom.right())) {
             s[i].total_classes++;
           isClass  = true;
             if((s[i].filetype).equals("C"))
//
// SOME SOURCE CODE OMITTED HERE
//
   s[i].filetype = "C++";
//
// SOME SOURCE CODE OMITTED HERE
//
}
}
```

One other example is of sufficient importance to devote space to it in this book: the interface between two high-level software components and the need for exception handling. We illustrate the interface between two such subsystems in Example 5.6. (This interface uses the Java language).

These three code samples are indicative of the coding standards used in our project. The code given in Examples 5.4, 5.5, and 5.6 also illustrates the naming conventions used in our software system. It should be pointed out, however, that the software tools used as part of our system probably used different coding and naming standards. Assuming that they work properly, it does not make sense to recode these existing software utilities just to meet our coding or naming standards. We are trying to make the software development process more efficient by emphasizing software reuse. Hence, we should treat these software tools as black boxes. The only things to worry about are the degree to which these software tools meet the project requirements and the quality of the tools themselves, including their correctness and robustness.

Example 5.6: A Portion of the Java Code (Interface between the Computational Subsystem and the Operating Subsystem) in Our Project Showing the Coding Standards

```
/* A basic implementation of the JDialog class. */
import java.awt.*;
import com.sun.java.swing.*;
public class JAboutDialog extends com.sun.java.swing.JDialog
{
//
// SOME SOURCE CODE OMITTED HERE
//
void okButton_actionPerformed_Interaction1(java.awt.event.
ActionEvent event) {
try {
   // JAboutDialog Hide the JAboutDialog
   this.setVisible(false);
} catch (Exception e) {
}
}
}
```

You should note that the approach of the previous paragraph is not satisfactory if safety-critical software is being developed. For such software applications, every component used in the system must have been certified as to both its correctness and its adherence to appropriate standards.

Let us assume that the software development has been completed, with testing and integration to follow. Of course, we must relate our design to the requirements traceability matrix for the project. The changed matrix is given in Table 5.2. Note that the requirements for a small system (limits on disk space, for example) have been deleted because of the change to a graphical, web-based user interface.

You might also note that we have changed a requirement for a user interface based on C or C++. It is clear from some of the code samples that we can implement the project in either Java or one of the "visual" family of languages such as Visual C, Visual C++, or Visual BASIC. It is probably the case that our requirements were too confining and that we should have left the functionality of the requirements to the requirements documents and kept decisions about implementation details to the design process.

On the other hand, we may have violated the requirements of our customer and made all our efforts useful. In a real-world situation, the customer would be consulted before such a change is made. Just keep in mind the possible trade-offs.

TABLE 5.2 Requirements Traceability Matrix for Our Major Continuing Software Engineering Project

Requirement	Design	Code	Test
1. Intel-based	Y	Y	
2. Windows 8.1	Y	Y	
3. Windows 8.1 UI	Y	Y	
4. Consistent with Excel 14.0	Y	Y	
5. System one size only	Y	Y	
6. Memory used	*	*	*
7. Disk space	*	*	*
8. One CD	*	*	*
9. Includes installation	Y	Y	
10. No decompression utility	Y	Y	
11. One input file at a time	Y	Y	
12. Size of each function	Y	Y	
13. Size of each file	Y	Y	
14. Size of system	Y	Y	
15. Compute totals	Y	Y	
16. Develop std for C, C++, Ada	Y	Y	
17. Batch-oriented system	Y	Y	
18. Precisely define LOC	Y	Y	
19. Measure separately	Y	Y	
20. No error checking of input	Y	Y	
21. Front end in C or C++	Y	N	
22. Batch processing	Y	Y	
23. File names limited to 32 char.	Y	Y	
24. Wild cards can be used	Y	Y	
25. Dead code ignored			
26. No special compilers needed	Y	Y	
27. Microsoft Excel 14.0 needed	Y	Y	

What do we need to focus our attention on at this point? We must know that the individual software components have been linked together, that the filters or "glueware" have been created and that the software works well enough (or at least the individual modules do) for us to show it to another person: a software tester. The testing and integration process is the next step in the development of this software system.

From a project management perspective, we should also do a status check. Are we ahead of schedule (unlikely), behind schedule (likely), or approximately on target? Have there been any unpleasant surprises, any portions of the system that were more difficult than we expected? Does any portion of the system require extra attention, perhaps additional resources? Have technology or market pressures rendered any portion of the system obsolete? (Recall that we did this status check earlier.)

In the previous two chapters, we considered the possibility of major changes to requirements in the form of either moving from a PC based system for the user interface to a web-based interface or from a PC-based data repository to one that is cloud based. Each of these changes might result in a different choice of programming language because of greater appropriateness for the particular application. We will not pursue this issue any further in this book.

SUMMARY

This chapter has considered some of the issues that can affect the maintainability, readability, and performance of source code. These issues include coding style, coding standards, and file organization.

Some commonly used software metrics for source code were also considered. The intention of these metrics is to provide some assessment of the size and complexity of the source code. The metrics discussed were

- Lines of code
- Halstead Software Science Metrics
- McCabe cyclomatic complexity
- Module interconnection metrics

In many organizations, the implementation team is responsible for providing the testing team with a system that compiles cleanly and has many obvious bugs or faults removed. In others, the implementation team is responsible for providing a software system to the testing team that the implementation team believes is free of major faults or bugs.

A software manager expects the implementation team to follow the organization's established coding process whenever possible. He or she may require the collection of several different types of metrics data for both the software source code and the process of developing it.

KEYWORDS AND PHRASES

Configuration management, coding standards, coding reviews, coding inspections, inheritance, readability, software metrics, lines of code, cyclomatic complexity, McCabe cyclomatic complexity, coupling, content coupling, common coupling, control coupling, stamp coupling, data coupling, external coupling, message coupling, weak coupling

FURTHER READING

There are few references on coding standards. The Public Ada Catalog available at the URL http://www.adaic.org/ada-resources provides some information for Ada programmers in both programming style and language philosophy. A sample C++ coding guide is described in the book by Henricson and Nyquist (1996). The book by Jim McCarthy of Microsoft illustrates the team-building activities and corporate culture that apply in many software development organizations (McCarthy, 1995).

There are several excellent sources for information on software metrics. The original paper by McCabe (1976), and a classic book by Fenton and Pfleeger (1996) and its revision (2014) are perhaps the most accessible.

Many books on particular programming languages include discussions of style issues and programming standards. The list of references includes reference manuals for the Ada programming language (Ada, 1983, 1995), the classic C programming language books by Kernighan and Ritchie (1982, 1988), several general C++ books (Leach, 1993), two classic books by Stroustrup (Stroustrup, 1991, 1994), the annotated reference manual for C++ by Ellis and Stroustrup (1990), three or four Java books, as well as nearly any related books on the shelves of your local library.

For many newer programming languages, no written language manuals are available. Therefore, the best way to see the technical details of the language implementation issues is to consult the organization originating the development and promulgation of the language. For many of these newer programming languages, see the online reference manuals at the appropriate addresses. A few are

Python—https://docs.python.org

Swift—https://developer.apple.com/library

Perl—http://perldoc.perl.org

EXERCISES

1. Examine some software you have previously written from the perspective of coding style. Are there any improvements you would make? Why?

2. Same as Exercise 1, but now consider coding standards.

3. Examine a reasonable amount of the source code available to you in a relatively large project that you did not write yourself. (You might look at software available from the Internet if you cannot find any locally.) List some of the coding standards that must have been in place during the project's software development.

4. There are several differences between the coding standards listed in this chapter. Give an explanation for these differences.

5. Determine if your software development environment has a "pretty printer" available. If so, determine if it is flexible enough to meet multiple coding standards.

6. Develop a C macro or a C++ inline function to compute the pow() function. Compare the efficiency of your iteration with that of the built-in pow() function.

7. Consider the simple C or C++ language statements:

```
a ++;
```

and

```
a = a + 1;
```

For integer variables they are completely equivalent. For pointers in C and C++, the assignment a = a + 1 indicates pointer arithmetic; the pointer a (which represents a location) is set to the location computed by the expression

```
a + sizeof(type of expression pointed to by a)
```

This equality is the basis for the equivalence of a[1] and *(a + 1) in C.

Unfortunately, there are problems with the use of these notations interchangeably in C++, due to the design of the standard C++ library. List the problems that can occur when mixing the expressions a++ and a = a + 1 in C++ programs. Do the same for a — and a = a – 1. (Recall that the standard C++ library offers five kinds of iterators: input, output, forward, bidirectional, and random access.)

8. Devise an experiment to determine which of two distinct implementations of the algorithm presented in Section 5.1 is fastest. Recall that the purpose of these algorithms was to initialize a square matrix to alternating ones and negative ones. Implement the algorithm so that the code runs as fast as you think is possible. Then analyze the code to determine if there are any additional improvements that you might have missed.

9. Use the profiler option on your compiler to examine the code fragment at the end of Section 5.1. (As stated in Section 5.2, the purpose of a profiler is to create executable code that includes timing information on the functions and operating system-level operations that the program spends most of its running time executing.) The profiler output should include sufficient information to determine the parts of the program that take the longest. Rewrite these parts and run the new program through the profiler again. Compare the two outputs.

10. Write a program that will solve a system of linear equations of the form AX = B, where A is an *n*-by-*n* array of floating point numbers that has all entries other than those on the main diagonal equal to 0, X is an *n*-by-1 array of floating point numbers that represents the array of unknowns to be solved for, and B is an *n*-by-1 array of floating point numbers.

11. The next program is an example of poor programming practice. It is based on an actual program that was used to control the display of a moving object. The main consideration at that time was speeding up the program as much as possible. That is your objective here. Some of the code was the actual code used in the first attempt to perform the desired action. I added a few nasty features to slow the program. Try to find as many ways as possible to speed up the code. You should concentrate on minimizing the number of floating point operations. There are at least nine separate improvements that can be made.

    Note: The functions move_to() and draw_to() were actual graphics functions; use the ones given here to simulate the time that such functions take.

```
#include <stdio.h>
#include <math.h>
#define PI 3.14159
static double old[3][3] =
   {
   {0.0, 0.0, 0.0},
   {0.0, 0.0, 0.0},
   {0.0, 0.0, 0.0}
   };
static double new[3][3] =
   {
   {1.0, 0.0, 0.0},
   {0.0, 1.0, 0.0},
   {0.0, 0.0, 1.0}
   };
static double trans[3][3] =
   {
   {0.0, 0.0, 0.0},
   {0.0, 0.0, 0.0},
   {0.0, 0.0, 0.0}
   };
static double x, y, z, theta = PI, phi = PI, psi = 2* PI ;

/* function prototypes */
void get_angles(void)
void move_to(double x, double y, double z);
void draw_to(double x, double y, double z);


int main(void)
{
int i, j, k, count, p, q, r, s, t, u, v, w, a, b, c, d, e, f, g, h, l, m, n, o;
 for (count = 1; count < = 500; count ++)
   {
   get_angles();
                get_transformation_matrix(theta,phi,psi);
   for(i = 0; i < = 3–1; i ++)
      {
      for (j = 0; j < = 3-1;  j++)
         {
      new[i][j] = 0.0;
      for (k = 0; k < = 3-1; k++)
```

```
            new[i][j] = new[i][j] + new[i][k]* trans[k][j];
      new[i][j] = (new[i][0]*0.9 + new[i][1]*0.9 +new[i][2]*1.2) / 4.60;

                x = new[0][0];
                y = new[0][1];
                z = new[0][2];
                }
        }
        if (count %2 == 0)
          move_to(x,y,z);
        else
          draw_to(x,y,z);
}             /* end of count loop */
}                    /* end of main */


/* This is a poor simulation of a random number generator - note the
   range of values of theta, psi, and phi.  */
void get_angles(void)
{
static int i;
float result = PI;


if ( i = =–1)
   i = 1;
theta = result /(i+6);
phi = (theta)/( i +2);
psi = (((psi))/(i + 4));
i = i + 1.000;
}



/*----------------------------------------------*/
get_transformation_matrix(double
     theta,double phi,double psi)
{
int i,j,k ;
/* a lot of matrix multiplication */
trans[0][0] = cos(theta);
trans[0][1] = sin(theta);
trans[1][0] = – sin(theta);
trans[1][1] = cos(theta);
```

```
trans[2][2] = 1.0;
trans[0][1] = trans[0][1] * cos_phi;
trans[0][2] = trans[0][1] * sin_phi +
   trans[0][2] * cos_phi;
trans[1][1] = trans[1][1] * –sin_phi;
trans[1][2] = trans[1][1] * sin_phi;
trans[2][1] = trans[2][2] * –sin_phi;
trans[2][1] = trans[2][2] * cos_phi;
trans[0][0] = trans[0][0] * cos_psi +
   trans[0][2] * sin_psi ;
trans[1][0] = trans[1][0] * cos_psi +
   trans[1][2] * sin_psi ;
trans[2][0] = trans[2][0] * cos_psi +
   trans[2][2] * sin_psi ;
trans[0][2] = trans[0][0] * –sin_psi +
   trans[0][2] * cos_psi;
trans[1][2] = trans[1][0] * –sin_psi +
   trans[1][2] *cos_psi;
trans[2][2] = trans[2][0] * –sin_psi +
   trans[2][2] *cos_psi;
}


/* Don't change this function – it does nothing but simulate the cursor
   moving time. */
void move_to(double x,double y,double z)
{
int i;
for (i = 1;i < = 1000000;i++)
      ;
}

/* Don't change this function – it only simulates the cursor moving. */
draw_to(double x,double y,double z)
{
int i ;
for (i = 1;i < = 1000000;i++)
      ;
}
```

CHAPTER 6

Testing and Integration

AT THIS POINT, YOU have developed your software in the sense that the individual modules compile and seem to be correct. The system itself might compile successfully, link with all necessary libraries to produce an executable file, and might even have produced some correct output. Even after these successes, there are some questions that should be asked about the software's condition at this point in its development:

- Would you be willing to release the software now, knowing that the software performing its functions correctly will determine your or your organization's economic future?
- What is the justification for your answer to the previous question?
- If you are not willing to release the software as it currently stands, what procedures would you take to improve the software's correctness?
- How much would the procedures cost to improve the software's quality?
- How long should the additional testing take?
- When should testing stop? In particular, is there a point beyond which further testing will not improve the software's quality enough to make it economically feasible?
- What can be done to ensure that changes made to parts of the software during its testing do not cause problems in other portions that have already been tested?
- How would your answers to any of the preceding questions change if you knew that the software was to be used in some safety-critical application where human life was at stake?
- How should sequences of input commands and expected interactions be combined into a test script in order to aid in showing the software's response? How many test scripts are needed?

- How should the outputs obtained by running the software on these test scripts be stored and analyzed?
- In addition to test scripts, which parts of the testing process should be automated, using modern, general-purpose software testing tools; which parts should be automated using software testing tools specifically designed for your application or environment; and which parts should be done by hand? The automation of the inputs to these modern testing tools is often called a test harness.
- Can the software system be released to a select team of potential customers or to an independent internal testing team prior to its release? This type of testing is called alpha testing and is typically used only for very large systems with known customers.
- If the software has passed alpha testing, can an improved version of it be released to a (probably larger) select team of potential customers prior to its general release? This type of testing is commonly called beta testing.

The purpose of these questions is to illustrate the importance of one of the two major subjects of this chapter: software testing. It is clear from even this brief introduction that a systematic approach to software testing is required.

The goal of software testing is to discover defects in software, not to show that none are present. That is, software testing cannot prove that software is correct (meets its specifications) for any realistic system. This is due to the large number of possible execution paths, possible combinations of function arguments, and the general complexity of various program statements. All that software testing can be expected to do is detect existing defects. There is a corollary to this: If software testing does not find many errors, then the testing process is likely to be at fault, at least for many software systems.

The statements in the previous paragraph suggest that the best we can hope for is some disciplined procedure that can detect the most likely sources of errors in software in as efficient manner as possible. It is becoming clear that the best defense against residual software errors (those that remain after completion of a phase of the software life cycle or release of a system) is proper design and coding practice.

Occasionally, the relationship between software testers and developers is tense. There are good reasons for this tension. The goal of the testing process is to uncover faults left by the developers. The goal of the development process is to enable the developers to produce a satisfactory version of the software within budget and on schedule. Thus, the two goals are somewhat in opposition. Balancing them is one of the goals of the project management.

Software testing occurs at several stages in the life of a system; these stages are frequently called the module level, unit level, and system level, depending on the size of the item being tested. As we have seen, the different activities of the software's life cycle may occur at different times, depending on the life cycle approach used (waterfall, prototyping, spiral, agile, open source, market-driven, etc.).

There is an interesting discussion about the issues in automation of the software testing process in an article in the July 2014 issue of *Communications of the ACM* in the context of testing an online team play soccer match at the game development company Electronic Arts. The article is in the form of a discussion between Michael Donat, Jafar Husein, and Terry Coatta (2014). Because there are eleven players on each side on the soccer field (known as a "pitch"), Electronic Arts did not wish to require twenty-two testers at all points of the game. The article describes decisions about when to automate the testing and when to do it by hand.

In general, the range of tools to automate all or part of software testing is rather large. Software testing tools can be as small as simple shell commands to execute test scripts, test harnesses that may have been incorporated into an IDE (integrated development environment), or as large as CASE (computer-aided software engineering) tools. Both open source and commercial software testing tools are available. For simplicity, we will not use specific tools in this chapter except for purposes of simple illustration.

The second major topic covered in this chapter is software integration. Software integration is the process of combining individual software modules and subsystems into a single working product. Software integration is not a trivial task, especially if the individual software units were developed in different locations or even by different companies and organizations that are separated geographically. Integration into fully distributed systems, software that lives primarily in the cloud, and Software as a Service (SaaS) can cause special difficulties. This coordinated situation in software integration is typical in much of modern software development. As was true of software testing, there are issues in which parts of the integration process can be automated and others that must be done manually.

6.1 TYPES OF SOFTWARE TESTING

There are several ways of viewing software from the perspective of testing. For software that is written in procedural languages such as C, Ada, Modula2, or FORTRAN, the fundamental unit is the function. Some languages such as Ada distinguish between a function, which may perform a computation and return a value, and a procedure, which performs a computation without returning a value. (FORTRAN uses the term *subroutine* instead of *procedure*.) Languages such as these are often called procedural or imperative languages. We will use the term procedural in this book. We will also use the term *module* to refer to a function, procedure, or subroutine as necessary.

For procedural languages, there is a natural hierarchy of items to be tested:

- Module
- Unit
- Subsystem
- System

The preferred testing approach is to develop a systematic way of testing an individual source code module. The tested modules are then combined into larger testable groupings that are then called units, subsystems, or systems, depending on how large they are or how much of the eventual complete software system is being combined. The process of combining tested modules into larger ones, especially systems, is called integration. The most common terms used to describe testing in this hierarchy are *unit testing* and *system testing*.

Once a software system has been through the gamut of tests, it may be subjected to additional testing to determine how the software functions with heavy load on the computer (in the case of a multiprocessing system). Another test of the software may be to examine system behavior if the essential tables and other data structures of the software are close to being filled up. This additional testing is called *stress testing* or *performance testing*.

The situation is quite different for object-oriented software, such as those typically developed in the C++, C#, Java, Smalltalk, or Eiffel programming languages. This is even truer for the Swift programming language recently released by Apple for iOS development. The object-oriented programming paradigm is that a program consists of a set of independent objects that communicate and cooperate by sending messages to one another.

For object-oriented languages, there is a natural hierarchy of items to be tested:

- Class
- Class interaction with other classes that are either directly accessed by, or directly accessing, the class
- Subsystem
- System

Note that the object hierarchy obviously needs to be modified when the objects may be distributed themselves or if they may include, or be included in, complete systems and subsystems.

Even now, the theoretical testing basis for object-oriented programming is in its infancy relative to the theory of testing procedural programs and, thus, we will be content to discuss only the most general integration issues in this circumstance.

Of course, a major factor in the popularity of several modern programming languages is the collection of powerful libraries available in some environments. The use of Apple Xcode for iOS development, Android APIs, Microsoft Foundation classes, Microsoft.NET framework or the various X-Windows toolkits, for example, provides a rich collection of utilities. Often, unless the application is safety-critical, such readily available facilities may be assumed to be correct during the testing process. Testing is hard enough without trying to test everything.

Unfortunately, there are even fewer theoretical discussions or published practical guidelines for testing functional programs written in languages such as Lisp or for testing logic programs written in languages such as Prolog. We will not discuss testing issues for programs written in functional or logical languages any further in this book.

Most testing techniques are often described as using either a black-box or a white-box approach. *Black-box testing* refers to the concept that a module is to be tested as to how well it meets its specifications, without any attention being paid to the structure of the actual source code within the module. In Chapter 5, Section 5.6 we presented the BVA metric measuring the complexity of, say, a single unit of software by how many test cases are necessary for entire coverage of the unit's input. Since this metric requires only knowledge of the external interfaces, this is an example of measuring the amount of black-box testing needed, the so-called boundary testing.

In *white-box testing*, the structure of the module's source code is used together with the module's specifications to guide the test cases used to test the module. A simple example of this approach would be a program that takes as its inputs a student's grades during a semester and a weighting scheme of how much each grade counts, and produces the student's semester letter grade as an output. Here, the test cases might be chosen to exercise the grade calculation and also to check for errors in data entry. Thus, it might be appropriate to separate testing for bad data (negative grades!) and output ranges for all of the possible output grades. This is often called equivalence partitioning, since many possible inputs would lead to the same output grade, and not all of them need to be tested.

We will illustrate both white-box and black-box testing methods in this chapter.

6.2 BLACK-BOX MODULE TESTING

We first consider the topic of black-box module testing. This testing method is based on a comparison of the software's performance on a test. Black-box testing requires several things, which are listed in Table 6.1. The first two items are generally available during the testing process. However, if they are not available, then there is no point in continuing to test, since we cannot tell if our results mean anything. We, thus, turn to the essential issue of selection of a set of test cases: a test suite.

The first thing to note about black box testing is that, in general, exhaustive testing is impossible due to the complexity of arguments to software modules. It follows that most nontrivial software cannot be tested completely. Therefore, a compromise must be reached between the impossible goal of exhaustive testing and the desire to release an acceptable version of the software product within a reasonable time. This compromise is why software testing is often considered to be an art, not a science.

We now illustrate the issue that was raised in the previous paragraph. Consider for a moment the problem of testing a function that has a single Boolean argument. Clearly, there are only two possible test cases and, thus, exhaustive testing is possible for this function.

On the other hand, consider a function with a single integer argument. Because of the nature of computer arithmetic, there is a finite number of potential values for the

TABLE 6.1 Inputs to a Black-Box Test Plan of a Module

1. A module to be tested.
2. A set of specifications for the module's output on particular inputs.
3. A set of test cases to indicate that the module is correct. This set of test cases is usually created by the tester.
4. A method of comparing actual results with results expected from the module's specifications.
5. A method of storing test case results for further analysis.
6. A set of software tools to aid in automation of the testing process.

integer argument. However, it is not reasonable to test the function for all possible arguments.

The situation is more complicated for functions with arguments whose type is one of the following: floating point, arrays, pointers, or some structured data type. The correctness of a software module may depend upon the architecture of the underlying machine, especially the hardware representation for floating point numbers and the arithmetic algorithms encoded in the computer's microcode.

It is clear that exhaustive testing is generally not appropriate for most software. In general, exhaustive black-box testing of software modules is impossible because of the large, even infinite set of possible test cases. Thus, test cases must be reduced in realistic software projects.

Since we have shown that exhaustive testing is not realistic, it is reasonable to ask what sort of test cases might be used. There is no absolute answer that will apply to every testing situation in every possible environment or application domain. Perhaps the best we can do is to apply a reasonable set of guidelines. A set of guidelines for a minimum amount of testing is given in Table 6.2.

The motivation for our choices for testing numerical types and array indices is the commonly held belief that most software errors occur at the boundaries and are of the off-by-one type.

Note that this guideline yields a small number of test cases only if the number of arguments to a function is small. A function with three floating-point arguments will give rise to 5 × 5 × 5, or 125, possible test cases. This is a very large number and, in reality, a much smaller number will be used in most software organizations. Reducing the guidelines for a floating-point variable to 4 or even 3 reduces the number to 64 or 27, respectively. This still may be too many for a number of environments.

If the function has multiple arguments that are of different types, the total number of test cases is the product of the number of test cases for each argument, which is exactly the same approach we used for multiple arguments of the same type.

TABLE 6.2 Some Guidelines for Test-Case Selection in Black-Box Module Testing

Argument Type	Test Cases
Boolean	True, False
Integer	MIN_INT, -1,0, 1, MAX_INT
Positive integer	1, MAX_INT
Nonnegative integer	0,1, MAX_INT
Float	MIN_FLOAT, -1.0, 0.0, 1.0, MAX_FLOAT
Double precision	MIN_DOUBLE, -1.0, 0.0, 1.0, MAX_DOUBLE
Character	Printable, "escape sequence," null-byte
Array index	Min index, max index
Array	Test arbitrary array index
Pointer	NULL, normal
Structured type	One or more test cases for each field of the structure, depending on the type of the field
String	NULL, length 2, MAX_STRING_ LENGTH

Note that production quality software development environments usually include a test driver utility to compute the results of running these test cases. Thus, the number of test cases is not completely unreasonable. However, if the number of test cases is too large for the test utility, it may be necessary to substitute some paper-and-pencil tests.

Here is a simple example of how black-box testing works. The code in Example 6.1 is intended to compute the factorial of a nonnegative integer whose value is given as its only input argument. The problem with this source code is a common one in C: the test for equality in the if-else statement (line 8) has as its argument the assignment statement n = 0 instead of the expected Boolean comparison n = = 0. (This unfortunate notation has been the source of many unexpected errors in C programs. Many, but not all, C compilers produce a warning when they encounter this pattern.)

We can see this logical error easily, particularly if we have had it pointed out to us. However, our problem is different with black-box testing. The real question to be considered is: How do we detect this logical software fault without being able to see any of the source code for this module? That is, how do we determine the appropriate test cases using only information in the interfaces of the function and the function's specifications? We note that this function will fail nearly every type of testing. In particular, it will fail the simple test case n = 2.

Example 6.1: An Incorrect C Function Intended to Compute Factorials

```
int fact(int n)
{
if (n < 0)
   {
   printf("Error-negative int\n");
   return (-1);
   }
else if (n = 0)
   return 1;  /* 0! is defined as 1 */
else
   return (n * fact (n-1))
}
```

If we wanted to make sure that our simple factorial function was robust, we could, of course, partition the arguments that serve as test data for this function for the case of negative numbers (to see if the function halts with an error), any legitimate input, and anything larger than MAX_INT (to see if overflow is detected when it occurs).

Another, more complex, example is due to Pleszkoch, Linger, and Hevner (1992). They examined several different implementations of the specifications given in Example 6.2 as part of their research on the use of a formal program transformation scheme during software reengineering. What test cases should we develop for this function, knowing only the function's specifications? Some cases come to mind:

- All entries in the matrix A have positive values.
- All entries in the matrix A have negative values.
- Each row in the matrix A has entries that are positive and entries that are negative.
- Each row in the matrix A has entries that are nonzero and entries that have the value zero.
- Problems can occur because of special positions in the first and last rows.
- Problems can occur because of special positions in the first and last columns.
- There may be system limitations on the number N.

You may be able to think of some others.

We will defer presentation of the source code for one implementation of this function until we discuss white-box testing in the next section. (The code will be given in Example 6.3.)

Example 6.2: Specifications for a Sample Function

For each i in 1..N, search A(i,1), A(i,2), A(i,3),..., for the first non-zero entry. Place the position of the first non-zero entry in W(i), and the type (+1 for positive, −1 for negative) in T(i). Assume that there will always be a non-zero entry. All entries in the matrix A have the value zero.

6.3 WHITE-BOX MODULE TESTING

White-box testing differs from black-box testing in that the suite of test cases is chosen using additional information obtained from the internal structure of the source code. This type of testing requires the items listed in the set of testing guidelines provided in Table 6.1, especially a set of specifications for output of the module based on certain inputs. Additional guidelines for white-box testing are almost always based on the information in Table 6.3.

A source code branch point occurs when a decision is made. This can be in the form of a simple "if," or an "if-then-else" structure. A branch point also occurs for each option in

TABLE 6.3 Typical Inputs to a White-Box Test Plan of a Module

1. A module to be tested.
2. A set of specifications for the module's output on particular inputs.
3. The number of branching decisions in the source code.
4. The number and types of loops in the source code.
5. The number of function calls in the source code.
6. The number of nonlocal gotos in the source code.
7. Recursion requires extra attention to make sure the module terminates.

a multiple decision statement such as a "switch" statement in C and C++ or a "case" statement in some other languages.

Nearly all white-box testing guidelines require that each branch of a decision statement be tested. The guidelines (Computer Science Corporation, 1991) are typical. In fact, these guidelines only require testing each branch of a decision.

We note also that the code of Example 6.1 contains recursion. The recursion must itself be tested for a proper connection between recursive function calls and to make sure that the recursion terminates properly. In our example, but not in general, these cases would also be tested if the module had used iteration instead of recursion.

We now turn to the discussion of white-box testing of the code for which the specifications were given in Example 6.2. The source code to be considered is given in Example 6.3 and is based on source code originally designed by Pleszkoch, Linger, and Hevner (1992) to illustrate some issues in code restructuring. Note that the code was not intended as an illustration of production quality code.

We can draw a graph that describes the logical flow of control by having each statement of the program correspond to a node on the graph with arrows linking each program statement to all program statements that can follow it logically. Most program statements will be linked to the program statements that follow them directly. Others will also be linked to statements that begin or terminate loops, or to branches of selection statements. The program graph for the code of Example 6.3 is given in Figure 6.1.

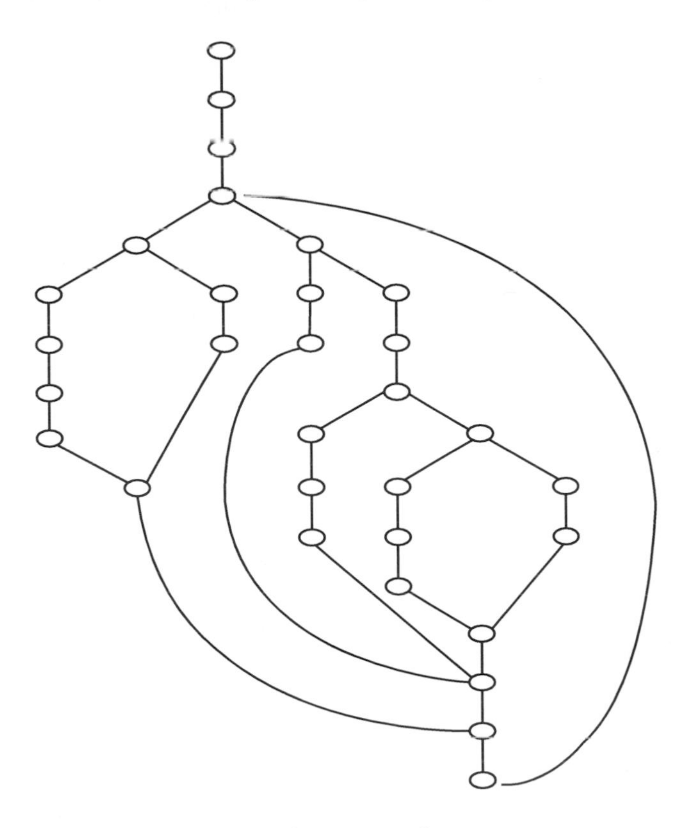

FIGURE 6.1 The program graph for the code in Example 6.3.

Example 6.3: Source Code for the Sample Function That Was Specified in Example 6.2

```
i = 1;                                  // f1
done = false;
new_one = true;
while (!done)
  {
    if (new_one)
      {
      if (i < = N)                      // p1
        {
        j = 1;                          // f2
        new_one = false;
        rest = false;
      else
        done = true;
      }  //end if
    else
      {
      if (rest)
        {
        j = j + 1;                      // f3
        rest : = false;
        }
      else
        {
        W(i) = j;                       // f4
        if A(i,j) > 0                   // p2
          {
          T(i) = 1;                     // f5
          new_one = true;
          }
        else
          {
          if (A(i.j) < 0 )              // p3
            T(i) = -1;                  // f6
            new_one = true;
          else
            rest = true;
          } // end if
      }  // end if
  } // end if
```

```
        if new_one
        i = i + 1;                          // f7
        }       // end if
    }           //end if
}               // end main loop
```

Note that straight lines without branches do not contribute to branching. We can simplify the branching using the reduced program graph, which is shown in Figure 6.2. A reduced program graph is obtained from a program graph by collapsing all nodes that have only one arrow emanating from them.

We can see from the reduced program graph in Figure 6.2 that there are four logical branches in the code and that each of them must be tested. Even the simple error state due to the input of a negative integer must be tested. A very common programming error is to simply print the error message and not set the function's return value to communicate the error to a calling function or routine.

It is important to distinguish between branches of decision statements and the number of execution paths in a program. A program will always have a finite number of branches, but the number of potential execution paths can be much larger. The number can be infinite, as illustrated by the simple example

```
i = 0;
while (i > 0)
  {
  i = i +1
  do_something();
  }
```

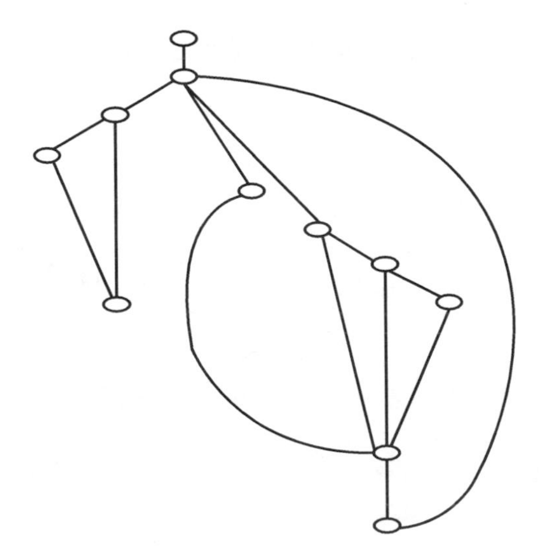

FIGURE 6.2 The reduced program graph for the code in Example 6.1.

TABLE 6.4 Suggestions for the Number of Test Cases to Be Used to Test Programs with Loops

1. Test the program with 0, 1, or some arbitrary number of executions of the body of the loop.
2. Test that the loop always terminates (assuming that termination is expected).
3. For nested loops, test both the outermost and innermost loops. If three or more loops are nested, test the individual inside loops at least once.

Clearly, we are not going to test an infinite number of program execution paths. Examination of the program graph in Figure 6.1 makes it clear that testing every program statement at least once (known as *every-statement coverage*) or at least testing each program branch (known as *every-branch coverage*) are highly desirable. Ideally, these levels of coverage can be created by selecting a compiler option and using a debugger to check the values of key variables at specific places in the program's execution, or by using features of the IDE or CASE tool used to create the software. Without such automation capability, we have to select our test cases primarily by hand.

We need some way to determine a set of test cases for programs with loops and branches. One common approach to white-box testing of programs with loops is given in Table 6.4. You are probably thinking that this process will generate too many test cases in many practical situations. For nested loops, we have suggested at least nine test cases. Even more test cases have been suggested for loops with more nesting levels. If the bodies of the loops contain branching statements, the number of test cases goes up astronomically. Clearly, we must place some sort of limit on the number of test cases for white-box testing, just as we did for black-box testing. Certainly, the number of test cases can be reduced by careful application of precondition and postcondition reasoning and including this with the source code. We will discuss this matter in Section 6.5.

6.4 REDUCING THE NUMBER OF TEST CASES BY EFFECTIVE TEST STRATEGIES

Both testing methods described so far in this chapter, black-box and white-box testing for procedures and functions, as well as most techniques for testing objects, have suffered from the problem of yielding far too many potential test cases. Let us see just how much of a problem this is for realistic software.

We will consider a software system of approximately 100,000 lines of code. We will assume that there are 1000 functions or procedures, each containing 100 lines of code. Let us suppose that the source code for each function contains a doubly nested loop in which the body of the innermost loop contains an if-else statement, a nested loop with a three-way branch in the loop body, and a stand-alone, three-way branch statement. Suppose that each of these functions has three arguments: a boolean, a float, and a pointer to a structured type with two fields, each of which is a floating point number. Pseudocode for one such function is given in Example 6.4.

Example 6.4: A Sample Function That Is Difficult to Test Completely

```
float f(int bool, float x, struct f_type *p)
{
   int i,j count;
   float temp;

   count = 0;
   for (i = 0; i < 10; i = i + 1)
      {
      if (bool = = 0)  / * false */
               do_something (bool);
      else
               count = do_something_else (bool, x);
      }
   for(i = 0; i < 10; i = i +1)
               for(j = 0; j < 20; j = j+ 1)
                        {
                        switch (i +j)
                                 {
                                 case 0 :
                                          do_something (bool);
                                          break;
                                 case 1:
                                          do_something_else(bool, x);
                                          break;
                                 default:
                                          do_something_different (p);
                                 } /* end switch */
                        } /* end for loop */
----------------------------other code here---------------------------------
switch (count)
   {
   case-1:
      printf ("error\n");
      exit (1);
   case 0:
      printf ("Count = 0\n");
      return (0);
   default:
      temp = x*x;
      return temp;
```

```
    } /* end switch */
} /* end function */
```

How many test cases do these functions have? Exhaustive testing using all possible function arguments is clearly not feasible because of the presence of the floating point and structured arguments. Even using the suggestions given in Table 6.4, we would have two test cases required for the boolean argument, five test cases for the floating point argument, and eleven for the pointer to the structured type. The number eleven is obtained by using the value of five for each of the two fields of structured type, adding these two numbers, and then adding one because of the indirect access by pointers. Thus, black-box testing would require a total of 2 × 5 × 11, or 110, test cases if we use the reduced number given in Table 6.4.

Computing the number of white-box test cases is only slightly more difficult, using a count of the number of branches and the guidelines given in Table 6.4 for programs with loops. We have two test cases for the body of the initial for-loop because of the if-else statement. The loop itself is to be tested at least three times, giving six cases for the first loop.

The body of the innermost of the nested loops contains a branching statement with three possible choices. The innermost loop is to be tested at least three times, with a similar number for the outer loop. This yields 3 × 3 × 3, or 27, test cases suggested for this doubly nested loop that contains a branching statement.

The final branching statement requires three additional test cases, making a total of 6 + 27 + 3, or 36, test cases for white-box testing. Thus, white-box testing yields a smaller number of test cases for this sample function than does black-box testing for our hypothetical example.

How long will testing take? There are 1000 functions, with 36 white-box test cases per function. Each function contains 100 lines of code. Let us estimate that each line of code corresponds to 10 assembly-language instructions. This estimate allows for the assignment statements and the need to increment the loop control variables. It greatly undercounts the assembly language equivalent of the function calls and, thus, gives a conservative estimate of the time needed to test each function:

1000 calls of functions × 36 test cases/function × 100 lines of code/test case × 10 instructions/line of code × 2 (number of comparison of results with actuals) = 72,000,000 instructions

(For black-box testing, the conservative total estimate is 220,000,000 instructions to be executed.) For a computer rated at a speed of 100,000,000 million instructions per second, or 100 MIPS, this is, at most, a few seconds.

However, we have ignored the time to develop and perform the test comparisons. If we add the time to assign values to each of the test arguments, the time increases considerably. Since the test input data and the data results would certainly have to be written to one or more files, and file operations are an order of magnitude slower than computer memory instructions, the time can get quite large. The determination of test cases itself can be quite time consuming, as we have seen. To make matters worse, this is just one function. Clearly, this testing process is too inefficient.

The solution of course is to have a rational test strategy, often known as a *test plan*. Such a strategy should reflect the amount of effort the software producing organization wishes to spend on software testing.

We will now briefly describe some simple software testing strategies used by various companies. The purpose of the discussion is to show how the formal approach to software testing is combined with a realistic assessment of the impossibility of complete testing and the desire to make software testing more efficient by uncovering the largest number of software defects in the shortest amount of time.

For one company in the aerospace field, the primary method used is white-box testing. For module testing, each decision branch of each module must be tested in a separate test case and a document describing the results of the test must be approved by a manager. No attempt is made to test loops by a separate test team. It is assumed that the developers will have tested loops within their own source code before releasing it to the test team.

A major telecommunications company had a similar process in which it demanded that any subcontracting organization produce a testing plan and test cases before starting the testing process. The plan had to consider the following as a minimum:

- Computer hardware
- Related hardware
- Testing tools
- Test drivers and stubs
- Test database to contain results of testing in a searchable form
- Other special software developed to aid in the testing process
- Testing reviews and code inspections

Other organizations have similar testing processes, with processes for testing of financial and safety-critical applications being the most complex.

6.5 TESTING OBJECTS FOR ENCAPSULATION AND COMPLETENESS

Software developed using the object-oriented paradigm should be tested differently from software developed in procedural languages. Object-oriented programs generally consist of separate objects that communicate by sending messages to one another. The first emphasis of object-oriented testing is to ensure the completeness and correctness of objects. Because of the potential for reuse of objects, testing for completeness is an essential part of testing object-oriented software. Objects can be used in situations other than the ones for which they were originally created. We illustrate this point by an example.

Because C++ is descended from C, even simple input and output causes some problems for testing of C++ programs that use both C and C++ I/O functions. The primary difficulty at this point occurs in operator overloading.

We present two simple checklists of items to be considered when testing C++ programs. For simplicity, we will concentrate on those object-oriented features specific to C++, which allows overloading and polymorphism.

The first checklist describes the steps necessary for testing programs that use an overloaded operator. This checklist is given in Table 6.5.

For example, an operator such as + is already overloaded for the int, float, and double data types. If we have some new type for which the operator is redefined, then we must test the + operator for the following combinations of arguments:

int and new_type

new_type and int

float and new_type

new_type and float

new_type and new_type

double and new_type

new_type and double

We need not test the cases that can be assumed to have been previously tested, such as

int and int

float and float

int and float

float and int

float and double

double and float

TABLE 6.5 Suggested Checklist for Testing Overloaded Operators

1. Make a list of all overloaded operators. (The operators we discuss in this section will already be defined for some of the standard types.)
2. For each overloaded operator, make a list of each type of operand that can be used with that operator. Be sure to include all predefined types that the operator can be applied to.
3. For a unary operator, test the operator on the new object type.
4. For a binary operator, list all possible combinations of types of the operands.
5. Test each possible combination of allowable operands.

TABLE 6.6 Testing of I/O Functions in C++ Programs

1. Replace all printf(), putc(), putchar(), and puts() function calls with the operator << and either cout or cerr, as appropriate.
2. Replace all scanf(), getc(), getchar(), and gets() function calls by the operator >> and cin.
3. Include the header file iostream.h.
4. Rerun the program and compare the results on some standard inputs with those obtained previously.
5. If there are differences, apply manipulators to the I/O stream in order to obtain correct results. Be sure to include the file iomanip.h if you use any I/O manipulators.

The second checklist describes the steps necessary for testing C++ programs that use the standard I/O functions of the C language instead of those available in C++. This checklist is given in Table 6.6.

These checklists can help avoid some of the more common errors that beginning C++ programmers make. Consider the object-oriented program shown in Example 6.5. Note that we have three overloaded operators in this program: +, >>, and =. The first two have been explicitly declared as functions that can access all of the data in the class Complex. (Nothing in the class Complex has been declared as being private.) We have used the default overloading of the assignment operator =.

Example 6.5: An Example of Testing Object-Oriented Programs

```
#include <iostream.h>
class Complex
{
public:
   double real;
   double imag;
   Complex(){ double real = 0.0; double imag = 0.0;}

   double realpart (Complex z) { return z.real; }
   double imagpart (Complex z) { return z.imag; }
   void print(Complex z) {cout << z.real<< " + " << z.imag <<"i" << endl;}
};

Complex operator + (Complex a, Complex b)
{
   Complex C;
   C.real = a.real + b.real;
   C.imag = a.imag + b.imag;
   return C;
}

ostream &operator << (ostream & stream, Complex a)
```

```
{
   cout << a.real << " + " << a.imag << "i" << endl;
   return stream;
}

main()
{
   Complex z1, z2;
   float x1, x2,y1,y2;
   cout << "Enter the real and imaginary parts of "
      << "the first complex number" << endl;
   cin >> x1 >> y1;
   cout << "Enter the real and imaginary parts of "
      << "the second complex number" << endl;
   cin >> x2 >> y2;
   z1.real = x1;
   z1.imag = y1;
   z2.real = x2;
   z2.imag = y2;
   z1.print();
   z2.print();
   cout << z1 << endl;
   cout << z2 << endl;
   z2 = z1 + z2;
   cout << z2;
}
```

Three functions are presented for testing purposes. The member function print() is included to make sure that different functions work correctly. It is much more difficult to find errors if the I/O functions do not work properly unless a debugger is available. The output function call corresponding to this function's definition would have been

```
z2.print(z2);
```

The functions realpart() and imagpart() were also used for testing the program. Some testing was done using the default conversions of predefined arithmetic arguments in C++.

6.6 TESTING OBJECTS WITH INHERITANCE

The availability of inheritance causes special testing problems for programs in C++ or other languages that support inheritance. One issue is what needs to be tested in a derived class (assuming that the base class was sufficiently tested). Clearly, testing a derived class requires complete knowledge of the base class from which it was derived.

We will assume that the logic of the member functions has been tested (at least by hand) and that the major concerns are the interfaces between objects.

In order to test a derived class, we should first test all possible member functions not derived from the base class to make sure that there is no interface problem. This is the easiest situation to test, since there is no possible ambiguity between member functions of the same name.

After we have tested the new member functions of the derived class, it is time to test the member functions that have been inherited. These should be tested by making sure that the correct information (base class or derived class) is being used.

Additional problems arise when testing multiple inheritance in C++ programs. Certainly, every test that was applied with single inheritance should also be applied here. In addition, each case of possible ambiguity should be tested carefully. This complexity has encouraged most C++ programmers to avoid multiple inheritance entirely. Multiple inheritance implies that a portion of a class may be inherited along different inheritance paths. Many compilation systems handle this potential ambiguity by using a precedence scheme to select proper values. The testing difficulty is due to the potential that a slight change in one of the inheritance paths can lead to an entirely new derived class. Many newer object-oriented languages do not support multiple inheritance, so there is probably even less of a reason for coding using multiple inheritance.

The following passage from the *Annotated C++ Reference Manual* by Margaret Ellis and Bjarne Stroustrup (1990) illustrates some of the difficulty due to inheritance, at least in C++. The passage addresses some questions concerning the scope of operators and variables:

> When virtual base classes are used, a single function, object, type, or enumerator may be reached through more than one path through the directed acyclic graph of base classes. This is not an ambiguity. The identical use with nonvirtual base classes is an ambiguity; in that case more than one sub-object is involved. …
> When virtual base classes are used, more than one function, object, or enumerator may be reached through paths through the directed acyclic graph of base classes. This is an ambiguity unless one of the base classes dominates the others. The identical use with nonvirtual base classes is an ambiguity; in that case more than one sub-object is involved. (p. 204)

The problem is with the complexity of the C++ programming language, not the Ellis and Stroustrup exposition. The complexity indicated in this passage suggests that certain constructions should be avoided for C++ classes in order to make resulting C++ programs easier to maintain.

Recall that C++ allows the use of "friend functions." In C++, friend functions are not members of a class, although they have access to the private and protected member of the class in which they are declared. The use of friend functions can cause a breakdown in the modularity of an object-oriented program. Any testing of friend functions should include code reading to make sure that no private or protected data is used as an lvalue (on the left-hand side of an assignment statement). Other testing should make sure that the program using the object and the friend function avoids hiding one function by another. We note that Java does not allow either friend functions or any important non-object-oriented constructs.

Note that there is a potential "combinatorial explosion" in the number of possible test cases for classes with many member functions or many levels of inheritance. If the testing seems excessive, begin with reading the code and determining if good coding practices were used. Follow this by careful testing of those objects that you believe are the most complex (because of their internal structure or their most likely use within C++ programs).

6.7 GENERAL TESTING ISSUES FOR OBJECT-ORIENTED SOFTWARE

As we have seen, testing of non-object-oriented programs can be done using either white-box or black-box methods. The distinction is made between these two approaches depending on whether the selection of appropriate test cases is based on the internal structure of the code. Of course, hybrid testing methods are possible.

Software testing occurs at several stages in the life of a system; these stages are frequently called the module level, unit level, and system level. The corresponding stages for testing object-oriented programs are method-level testing (which considers individual transformation on a class), class-level testing, module-level testing, and system-level testing. (The term *module* is loosely defined as a collection of related classes and a few related functions.)

The class is the most important unit in object-oriented programming. Unfortunately, the member functions of a class (the so-called methods that are used for a class) can be combined in arbitrary ways. As such, the possibilities for testing a class are enormous, especially if the class is relatively complete in the sense that it contains many member functions, perhaps with polymorphism and operator overloading.

One view of the test process for a class is that it is a search process for the order of methods with various collections of arguments that give errors (Lorenz, 1993). A common strategy for testing of objects includes (Smith and Robson, 1992)

- Encapsulation
- Minimalization
- Exhaustion of a depth
- Inheritance
- Interactive

The encapsulation strategy for testing objects makes use of the abstraction used in defining an object. Any set of abstract methods that can be legitimately applied to the abstract object should correspond to a set of legal combinations of transformations on the class.

The minimalization strategy develops the smallest number of test cases for the class being tested where the errors that occur can be overridden by any child class of the tested class.

The exhaustion of a depth strategy considers all legal combinations of methods allowed for an object, with the number of methods in any chain of methods being less than or equal to the previously determined value of the depth.

The inheritance strategy is used for subclasses that are inherited from a parent class. A list of methods to be tested is kept and tests are performed on the class depending on the test results for the parent class. As indicated earlier, this is much easier if multiple inheritance had been avoided in coding.

The interactive strategy requires the tester to determine which methods to test based on an assessment of the relative complexity of the internal data structures in the class. The method can be summed up as "use your intelligence guided by your experience."

Note that the exhaustion-of-a-depth strategy can be implemented mechanically with relatively little difficulty. Note also that the inheritance test strategy has potential for automation, but that the other test strategies do not. In general, automatic testing tools for object-oriented programs are either nonexistent or are very primitive (at least compared to the automated testing and analysis tools available for procedural programs in good software development environments).

Several features of object-oriented programming that encourage efficient program development cause some difficulties in testing. We will describe some of these in the next few paragraphs.

As stated in the previous section, multiple inheritance implies that a portion of a class may be inherited along different inheritance paths. In some cases, if the code is newly written, it may be better for the test team to ask the coders to recode to avoid this. If the code has been used without incident for many years, the best choice may be to leave the code alone. We will discuss this more in Chapter 8 when we discuss perfective maintenance.

Polymorphism also causes problems. We must determine that when an object receives a message, the object is in the correct form to receive it and will react in the intended manner.

There are also potential problems in concurrent systems, since the creation or destruction of a particular form of an object may not occur instantaneously and, hence, there may be timing problems. These types of timing problems are notoriously hard to detect by testing. Note that these issues are not caused by the classes themselves but in the way that they may be combined.

The lack of an implied order in which the methods used for a class are applied appears to make it difficult to use "test data." It is also difficult to use the control flow or data flow techniques commonly used in procedural programming.

For systems of reasonable complexity, there is a trade-off between using many previously tested classes either directly or by inheritance, and using more complex objects that require extensive testing. Because testing complex objects is difficult, it appears that the better approach is to use more but simpler objects, including those in a class library of data structures. There is little hard data from multiple similar software projects available as yet to support this view, but it is the growing consensus of a large group of project managers and software engineering researchers.

6.8 TEST SCRIPTS, TEST HARNESSES, AND TEST PLANS

This chapter considers three essentials of systematic testing: test scripts, test harnesses, and test plans. These three essentials are presented in increasing order of complexity.

The first we consider is a *test script*. A test script is nothing more than a simple, elegant way of providing a smooth way of getting data into an existing program unit. The major operating systems (Windows, Linux, Apple Darwin, etc.) all have the capability of executing commands using what is often called a terminal window. In Chapter 4, we briefly discussed a program named "sanac." If we have written this program to receive the name of an input file as its input data, then a simple I/O redirection, which works on all terminal windows of the major operating systems, such as

```
sanac < inputFile1 > outputFile1
```

saves the output of the sanac program on an input file named "inputFile1" to an output file named "outputFile1," which is saved for examination.

Using a simple loop allows the execution of the sanac program on, say, 100 files, with the ">" replaced by ">>" allowing the (big) output data file to be analyzed either by hand or, preferably, by other software.

(We could have replaced the line sanac < inputFile1 with sanac inputFile1 if we had chosen to use of command-line arguments, which we discuss in some detail in Appendix B.)

A *test harness* is another way of automating the testing process. Typically, a test harness provides a set of input values for the arguments to a function and produces a systematic way of saving and examining the function's output on these inputs. This is a typical feature of many modern IDEs and nearly all CASE tools.

A *test plan* is more than a collection of test cases and a database to contain the results of these test cases. It is a strategy for systematically examining the software to detect faults, then analyzing these faults and determining an appropriate set of actions. There are several elements that affect a test plan:

- There should be a precise list of the requirements that must be tested. Without this list, there is no way to tell if the requirements have been met.
- The test plan must be consistent with the goals of the organization. If the most important goal of the organization is to have a minimal test of the software's essential features as determined by the marketing department, the test plan should test only those features as a first priority. (These essential features may be tested within twenty-four hours in many organizations.)
- After all the requirements in the most essential set of requirements have been tested, the next set of test cases should consider the cases that most users are likely to want.
- The remaining set of test cases should be the ones that would exercise those features of the software that are very unlikely to be encountered in practice. The basic idea is to only test these other features if time permits.

- The test plan should specify the operational environment. Given the huge range of hardware in the personal computer world, this probably means testing the performance of the software in a minimal configuration, the maximum configuration of the hardware, and the most common configuration. (Here in the case of a single desktop computer, the term *configuration* refers to a combination of memory, hard disk space available, monitor, video memory card, and sound card. The term obviously has different meanings for laptops, network computers such as the Google Chromebook, network arrangements, or cloud computing; these meanings will not be described here.)
- The plan must allocate a sufficient amount of time to be used for testing (at each phase).
- The set of personnel available for testing must be known and part of the plan.
- The plan must ensure that there are adequate hardware resources available for testing.
- The plan must make sure that there are adequate software resources available for testing, usually in the form of test drivers and testing tools.
- There must be a sufficient amount of available free computing cycles resources that are to be used for testing.
- The plan must address the available tools that can be used for testing data analysis.
- The plan must take into account any existing standards and practices manual.
- The plan must consider the type of software (object oriented, procedural, mixed)

By now, you should be convinced that exhaustive testing is impossible for any realistic software. Thus, there are some trade-offs that must always be made. These factors must be part of a test plan.

What does a test plan look like? It might be as simple as: test each branch at each decision point, or test that each function is called with the correct arguments. It might be to not test certain software components believed to be error-free during their long lifetimes.

One thing that every test plan must include is regression testing. This term refers to the need to make sure that, when a module, unit, object or similar item is changed in response to the result of a particular test, then everything in (at least) the next largest program unit must be subjected to that test and also to every previous test that has already been passed. This is to make sure that the act of fixing one particular module, unit, object or similar item does not cause a new fault elsewhere in the encompassing unit.

The important thing is that a realistic plan must be developed and followed during the software's testing. The plan may be influenced by the organization's typical software practices.

Note that the Bugzilla software, available as a free download from http://www.bugzilla.org, provides a simple way to track software bugs.

6.9 SOFTWARE INTEGRATION

Suppose that we have tested each of the software modules individually for a program written in a procedural language such as C or Ada. Alternatively, suppose that we have tested each of the classes that are to be used in an object-oriented program written in C++, Java, Swift, or similar. The same issue occurs if the program contains multiple classes. We have to have some assurance that the entire program works correctly as a unit.

In addition, we have to make sure that our program functions correctly as a system. That is, we have to make certain that our program works properly with any software package that we intend to use with our program. Recall that this property, which is known as interoperability, was one of the goals of software engineering discussed in Chapter 1.

The process of ensuring that programs that are created from individual program modules that work together as a unit is called *software integration*.

Of course, many software systems, including the one we are developing as part of our major software engineering example, consist of both object-oriented and procedural portions. Hence, it may be necessary to employ both procedurally and object-based software integration techniques. The beginning software engineer may be surprised at the formality in which many organizations approach the issue of software integration. Software integration is much more than simply getting the individual modules to compile and link together into a complete system. The final software system must meet the requirements that were set for it initially. This testing of the final software is called *system testing* to distinguish it from the topic of unit testing, which is what was discussed in Sections 6.1 through 6.8.

Software integration is not always a smooth process, particularly in modern development environments in which efficiency requires reuse of existing software modules or even existing COTS packages. A major problem can occur if one or more modules are extremely difficult to integrate with others. If the project's management has determined that the reason is the lack of adherence to standards or the low quality of the module in the sense of a large number of software defects, then the original coding team, design team, or requirements team may be called on to help with the problem.

There are three common approaches to software integration:

- Big bang
- Bottom-up
- Top-down

Each of these approaches has considerable advantages and disadvantages and, therefore, each has its proponents and detractors. We describe each of these approaches in turn.

The big bang approach to software integration is so named because it attempts to integrate the previously tested software modules and existing COTS products into a complete system without any preliminary software integration activities. The basic idea is to see if the system works as an entirety without any changes to individual modules or configuration. Ideally, the newly created system will work perfectly. The big bang approach is illustrated in Figure 6.3. In the figure the boxes represent individual software modules, subsystems,

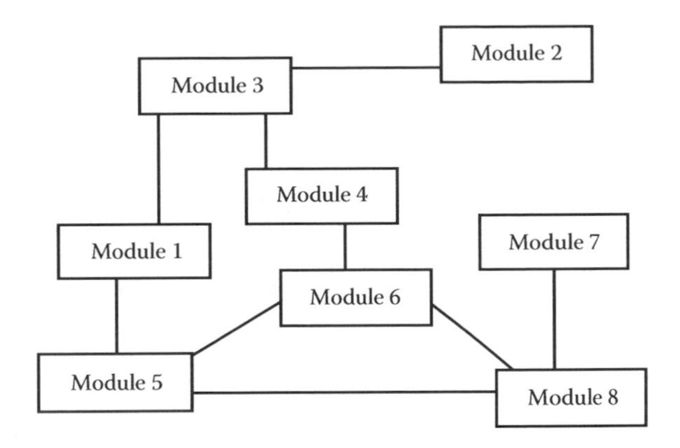

FIGURE 6.3 The big bang approach to software integration.

or COTS products. The lines connecting the boxes indicate a linkage between the modules, subsystems, or COTS products, but do not represent any type of control flow or data interface. The diagram indicates no special order, which is appropriate because the big bang approach does not distinguish between modules in terms of relative importance or their relation to the control flow or data flow of the intended system when integrating them.

More commonly, many changes are necessary for the complete system to work properly. The system integration team is then tasked with the problem of determining which modules are causing the integration difficulty and fixing the problems.

Note that in most actual software systems, as opposed to the type of software projects generally done by beginning students, getting the compiler to link the component source code modules is the easiest part of the problem. This is true, regardless of the integration method used. We thus consider only the difficulties caused by logical errors in software modules, incorrect module interface standards, or errors that are due to faulty design or requirements.

The major advantage of the big bang approach is the potential for reduction of system integration costs if the system design was sufficiently modular and the individual software components adhere to appropriate interface and performance standards.

The disadvantage of the big bang approach for many organizations is the length of time and the amount of resources needed to determine which modules actually caused the problems in integration. The techniques used include the use of debuggers and other software tools to determine the exact place where difficulties occur. This can be a very time-consuming task.

Since one of the goals of proper software project management is to minimize risk, the big bang approach to software integration is probably not appropriate unless the organization has experience with the amount of effort needed to interface individual modules. This would be the case if the application domain was well understood and relatively static, and the organization had considerable positive experience with the team that was responsible for creating the modules to be integrated. (Note that many organizations contract out all or part of their software development. In addition, many organizations produce software components at geographically diverse locations, and the quality requirements of a particular location that produces safety-critical software may be much higher than a location that does not.)

We now consider two software integration approaches that can reduce risk but will be more costly than the optimal case of the big bang approach, because they require additional software to be written in order to perform the integration.

The bottom-up approach is based on the assumption that a software system is made up of a collection of simple, lower-level software modules and that these simpler modules should be put together to form a system.

The bottom-up integration process is iterative and goes something like the following. Note the continued reliance on the system architecture.

1. The lowest-level modules in the system are determined, preferably from an examination of the system architecture. If the system architecture is not sufficiently detailed, then these modules can be located by examination of the system's call graph either by using a tool such as the standard UNIX cflow utility or examining the software by hand.
2. Drivers are written for each of the lowest-level modules. A driver is a function that passes arguments to other functions or procedures and determines the return values (in the case of functions). We will give an example of a driver function in Section 6.18, when we discuss integration of our major software engineering example system.
3. The lower-level modules are now tested, together with the new driver functions. After the lower-level modules are tested in this way, they are deemed to be correct and the next step of the software integration process starts.
4. The driver functions are replaced one by one by the next higher-level modules in the software architecture. After the driver modules are replaced by the actual modules at the next higher level, new driver functions are written in order to control the actions of all modules below the driver in the system architecture.
5. Steps 3 and 4 are iterated until the top-level module replaces the last artificially created driver.

Note that at each iteration the system will consist of a set of tested, integrated modules, with small driver functions at the top. The lower-level routines will be correct and, thus, the intermediate steps in the software integration process will always result in a system with correct outputs from lower-level routines such as device drivers. The ability to have actual outputs is a major advantage of the bottom-up method of software integration.

Unfortunately, the user interface of the eventual system will not be tested or even fully functional until the entire integration process is nearly complete. Thus, a manager or customer will not be able to see the "look and feel" of the system until it is complete. This, together with the cost of writing multiple drivers, many of which can be quite complex, are the primary disadvantages of the bottom-up process of software integration.

The bottom-up approach is illustrated in a series of diagrams illustrated in Figures 6.4 through 6.7. The four diagrams are intended to show what was not obvious from Figure 6.3: that there are lower-level modules in the system and these will be the first ones integrated into a larger system. Modules 2 and 7 are the lowest levels of this system.

FIGURE 6.4 Illustration of the first step in the bottom-up approach to software integration: determination of bottom-level modules and linking each of them to an artificial driver module.

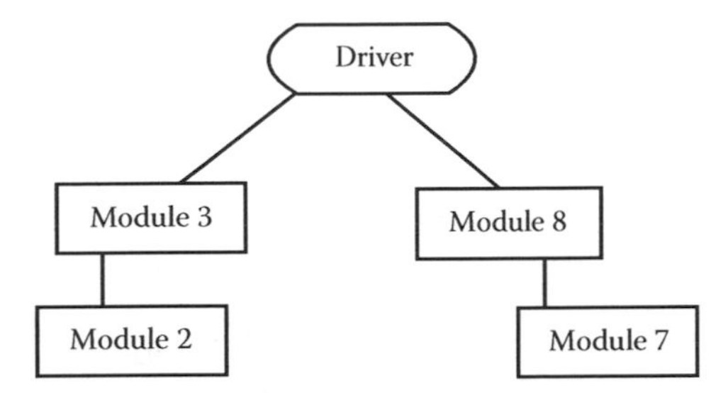

FIGURE 6.5 Illustration of the second step in the bottom-up approach to software integration: integration of modules that are just above the bottom level with artificial driver modules.

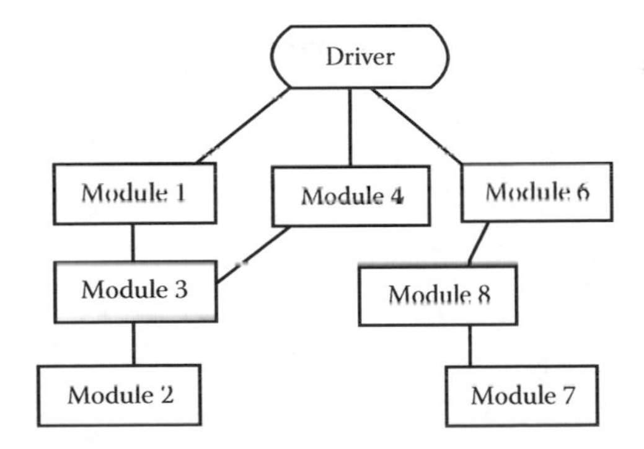

FIGURE 6.6 Illustration of the third step in the bottom-up approach to software integration: integration of the next level of modules with artificial driver modules.

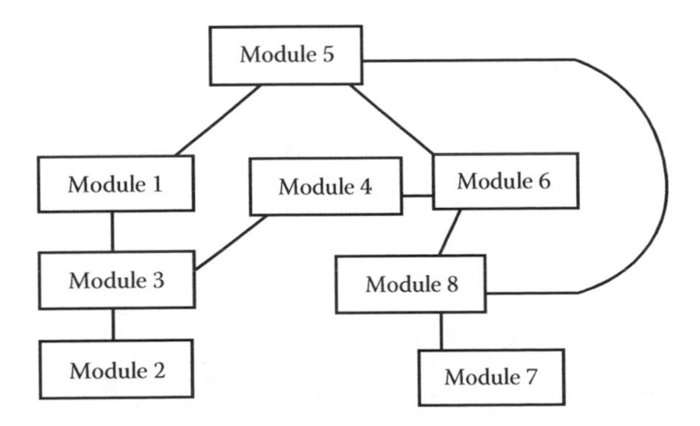

FIGURE 6.7 Illustration of the final step in the bottom-up approach to software integration: integration of the top-level module into the system with all artificial driver modules replaced by actual modules.

The top-down integration process is also iterative and goes something like the following:

1. The highest-level module in the system is determined, preferably from an examination of the system architecture. If the system architecture is not sufficiently detailed, then this module can be located by examination of the system's call graph either by using a tool such as the standard Linux and UNIX cflow utility or examining the software by hand.
2. Function stubs are written for each of the second-level modules connected to the top-level module. A function stub is a function that receives arguments passed to it from other functions or procedures, and produces the return values (in the case of functions). An example of a function stub is given in Section 6.18, when we discuss integration of our major software engineering example system.
3. If the system is primarily object-oriented, the notion of a function stub is replaced by a much simpler mock object that models (nearly all) the behavior of the eventual object that will be placed within the system without the overhead of the full object's effect on system performance.
4. The top-level module is now tested, together with the new function stubs. After the top-level module is tested in this way, the tested version of the system is deemed to be correct, then the next step of the software integration process starts.
5. The function subs are replaced one by one by the next lowest-level modules in the software architecture. After the function stubs are replaced by the actual modules at the next lower level, new function stubs are written in order to control the actions of all modules below the currently tested portion of the software in the system architecture.
6. Steps 3 and 4 are iterated until each bottom-level module replaces the bottom-level, artificially created function stub.

Note the continued reliance on the system architecture. You should also compare the steps to those of the bottom-up approach.

Note that at each iteration the system will consist of a set of tested, integrated modules, with the actual user interface at the top. Unlike the case with a bottom-up approach to software integration, a manager or customer will be able to see the "look and feel" of the system until it is complete. This can be very reassuring to management.

Unfortunately, the interface to lower-level routines will never be a completely accurate description of the program's computation until the integration process is complete. Because lower-level routines are only stubbed in, none of the intermediate steps in the software integration process can possibly result in a system with correct outputs from lower-level routines such as device drivers. The inability to have actual outputs is a major disadvantage of the top-down method of software integration. This, together with the cost of writing multiple function stubs, are the primary disadvantages of the bottom-up process of software integration.

The top-down approach is illustrated in Figures 6.8 through 6.10. Recall that for this example, module 5 was the highest-level module.

You should note that Figure 6.10 does not use a stub for module 7. Since module 7 was at the lowest level in the sense that it called no other modules, its integration into the system is complete at this stage. The other low-level modules will be integrated later in the process.

There is no need to represent the final step of the top-down integration process; it is the same as the diagram given in Figure 6.7 because it represents the end result of the software integration process.

You should note that the bottom-up and top-down approaches can be employed in the same hybrid integration process. The goal is to reduce the amount of extra drivers and stubs needed when using one of the other integration methods alone.

It is worthwhile to consider each of these integration approaches in the context of software reuse. The big bang approach can lead to serious problems if large-scale software components have not been certified as to having high-quality and standard interfaces.

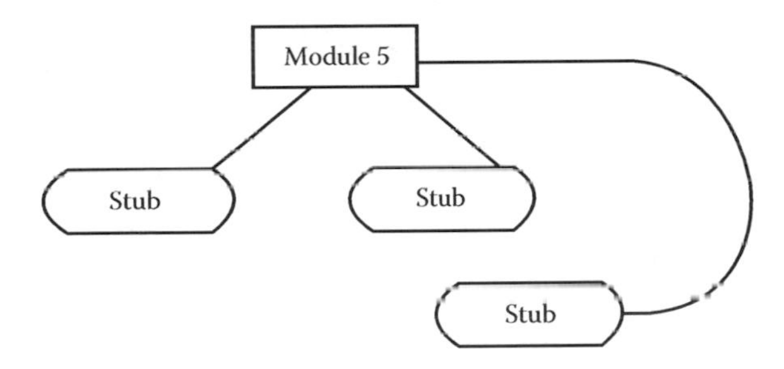

FIGURE 6.8 Illustration of the first step in the top-down approach to software integration: integration of all software modules below the top-level module with artificial stub modules.

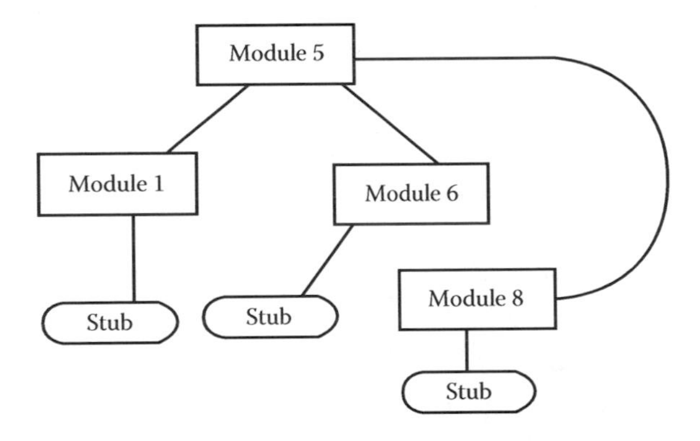

FIGURE 6.9 Illustration of the next step in the top-down approach to software integration: integration of all software modules at the next level below the top-level module with artificial stub modules.

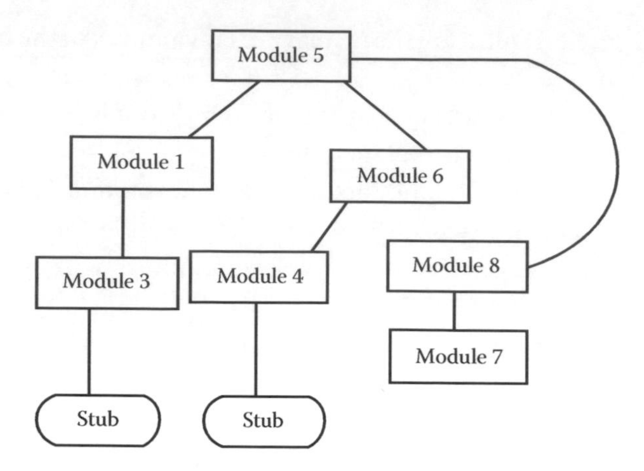

FIGURE 6.10 Illustration of the next step in the top-down approach to software integration: integration of all software modules at the next level down with artificial stub modules.

The bottom-up approach is best suited to reuse of software components that are fairly fine grained in the sense that they have a clearly defined, but narrowly limited, functionality. The likelihood of having a large number of such functions to characterize, catalog, maintain, and access is small if there is not a large reuse library of such components.

The top-down approach is best suited to a situation in which there are a few, higher-level components. This can avoid the overhead of maintenance of a reuse library. Reusing large, high-level components has the potential to reduce system costs. However, finding an exact match of an existing high-level software component to an actual set of requirements is relatively rare.

We now turn to a discussion of integration issues for object-oriented programs. It is clear that complete testing of all possible interactions of objects is essentially impossible.

The source code we used for some simple arithmetic with complex numbers that was given in Example 6.5 illustrates this effective impossibility. Consider the possibilities: each complex, real, double, or integer operand must be paired with a second operand of the same set of types. This is a minimum of sixteen possible cases, each of which must be tested to have even one example of each type tested. There are sixty-four cases if we allow multiple types of the results of the arithmetic operations, with one test case for each pair of possible argument types and four basic arithmetic operations. This is entirely too much.

Of course, integration issues are even more complex for software that has both procedural and object-oriented portions. We advocate striving for even a higher degree of modularity than normal in this situation. Considering the portions separately as much as possible seems to be the best approach, but there is little research to guide the practitioner.

In some circumstances, another integration technique, known as plug and play, may be used. It is easiest to explain the technique by an example. Some software systems, such as those used for aircraft control, can never be taken off-line. For such systems, new modules that add functionality or modules that are replacements for faulty ones must be integrated during regular operation. This requires a very careful integration process that we will not discuss here. Of course, new web pages are constantly added to the World Wide Web without affecting essential service, but this is essentially updating the data because of the HTML standard.

As a final note on the technique of software integration, the coding standards for file organization and naming discussed in Chapter 5 can be very helpful in the integration process.

6.10 CLOUD COMPUTING AND SOFTWARE INTEGRATION: SOFTWARE AS A SERVICE

One of the major advantages of cloud computing is its support for Software as a Service (SaaS). Let us review some of the major attributes of SaaS. SaaS is based on the belief that the interaction between software and hardware in modern network-centric systems should be viewed as a set of somewhat independent services, and makes use of cloud computing to allow users to access the service provided on their web browsers. This notion helps with abstraction and separation of concerns, but places serious loads on both system performance and the treatment of unexpected software interactions that have not been described in APIs. These issues show up in software system integration, both initially and during the maintenance phase. We will discuss software maintenance in the next chapter.

For convenience, we repeat the list given in Chapter 4, Section 4.6 of the most typical APIs used in SaaS, together with some information on the status of the standardization of the protocols and where technical details can be accessed:

HTTP (Hypertext Transfer Protocol)—The Hypertext Transfer Protocol format was standardized by the W3 consortium and can be accessed at http://www.w3.org/protocols.

REST (Representational State Transfer)—The Representational State Transfer format was standardized by the W3 consortium (Erl et al. 2013) and can be accessed at http://www.w3.org/protocols.

SOAP (Simple Object Access Protocol)—SOAP was formerly known as the Simple Object Access Protocol, but is now simply known as SOAP. The SOAP format was standardized by the W3 consortium and can be accessed at http://www.w3.org/TR/soap.

JSON (JavaScript Object Notation)—The JSON Data Interchange Format was formalized in 2013 by ECMA and can be accessed at http://www.ecma-international.org.

Note that the standards have evolved over time, which presents a configuration management problem for integration with cloud servers and SaaS. See reports of the Internet Engineering Task Force at http://www.ietf.org for the most up-to-date information on protocol standardization status. Note also that there are many requests for comments (RFCs) on some of the websites mentioned in this section.

A fundamental assumption in SaaS is that the services are to be available anywhere, anytime. The cloud server (or, more properly, a collection of cloud servers) is presumed to always be operating. New services are to be integrated into a SaaS system without disruption. This is analogous to a problem solved long ago for computer hardware in which failed disks and processors can be swapped out and replaced with properly operating units, without requiring the overall hardware to stop operating. Large servers can be thought of as consisting of multiple "blade servers," which are essentially integrated disk-processor combinations that can be removed or inserted without requiring shutdown and reboot. In the hardware industry, this is called a hot swap.

Here are some reasons why there are potential problems during the integration process for SaaS systems.

1. The SaaS system may have been implemented as a single entity, which has a specific combination of hardware, software, and underlying operating system. There may be many replicated versions of this single entity that are hidden within the cloud and, therefore, are not easily visible to individual users on their web browsers. This can lead to inconsistencies if there are so many replicated versions that changing all of their services takes too long a time. This is only a problem for SaaS systems of very large scale.

2. If one or more changed components of the entities making up the SaaS system is very large, such as a back-end database, the time for integration may take too long from the perspective of many users.

3. The SaaS system may reside largely inside of a company's firewall. If the software that uses the service uses a combination of software operating inside the cloud and specialized software interacting with a user's browser, there may be issues due to browser security policies or configuration settings.

4. There may be security problems if multiple applications from different organizations are running on the same cloud server. Many large-scale government and healthcare data applications require the use of dedicated hosting clouds such as those offered by Terremark (a subsidiary of Verizon) to avoid this problem.

6.11 MANAGING CHANGE IN THE INTEGRATION PROCESS

We will now consider some issues that occur in software integration, regardless of the approach taken. When a particular software module is changed because of a problem that occurs during system integration, it is necessary to make sure that no new defects are introduced into the system at the same time that another module is fixed. The interactions between software modules are often quite complex and changes frequently occur. A basic problem that we must address is how to incorporate change into the software integration process.

The change problem can be classified as falling into one of two categories. If the problem occurred because a particular software module was changed, we need to only consider those modules that may be affected by the change. The bottom-up and top-down approaches to software system integration limit the effect of changes to modules that have already been integrated. Thus, only modules at a lower level than the faulty module must be changed in a bottom-up approach. On the other hand, only modules at a higher level than the faulty module must be changed in a top-down approach. Of course, there is no limitation on changes that occur when the big bang approach is used, which is another objection to the big bang approach to software integration.

You should note that a configuration management system, such as we discussed in Chapter 5, is essential to control change during integration.

It is clear that we must perform exactly the same tests as before in order to ensure that no new errors are introduced. The test log and test results must be used to compare the results before and after the change. This step is called *regression testing.* It should be noted that this use of the term *regression* has only a tenuous connection to a term of the same name that is common in statistics.

There is another issue in software integration that might be considered as "software change in the large." In this situation, major subsystems or even entire COTS products are changed during the integration process. Perhaps a new version of an operating system or a database software package has been released, and the software under integration has to interface with them. This is a relatively common situation in modern competitive software development environments. In this situation, some sort of configuration management is necessary.

The situation is even more complex if the components integrated were written in different programming languages. You are almost certain to run across this problem in your career. Why? Because of the huge amount of code already written. A check of the website Ohloh.net in May 2014 said that there were 21,500,060,755 lines of open source code that it tracked in public repositories. This included about 4 billion lines of C code, 1.5 billion lines of C++ code, and 1 billion lines of Java code.

Even with such tools and standards as CORBA, the Common Object Model, Interface Definition Language (IDL), and the various "virtual machines" such as the Java virtual machine, there will always be problems due to language differences. Think about the ways that different languages treat exceptions, for example. Do you throw an exception, try an exception, catch an exception, accept an exception, declare an exception, raise an exception, handle an exception, or perhaps use nonlocal C system calls such as longjmp and setjmp? Not all of these may be familiar, but all will cause you difficulty if you try to integrate programs that are written with a mixture of programming languages. See the article by David Chisnall for more information on the problems that can occur when systems interface components written in multiple languages (Chisnall, 2013).

At most colleges and universities, computer science students typically compile their programs frequently, making changes to code in response to logical errors that were detected. There are even more compilations during the code entry process, when many students use the compiler to detect typographical or syntax errors.

Most professional software engineers work differently. They may spend much less of their time recompiling and fixing source code, and much more time reading it and trying to get the logic correct the first time. The time spent waiting for programs to compile is better spent in more productive activities, at least in the view of most practicing software professionals.

Indeed, many software systems are so large that compilation occurs infrequently. The compilation and linking process may take so long that it only takes place once a day. It would be useless to have each of one hundred programmers spending precious CPU cycles on compilation, when each one might change a single source code file and therefore none of the programmers knows the correct state of the software at any time. In many organizations, the term *build* is used to describe a compiled version of an entire system.

A complex software system may have many builds before it is released. For example, the source code for Microsoft Windows NT, version 3.51, indicated that this release was build number 1381. (This was the first version of the source code for a Windows operating system provided to me as a university professor and researcher.) Since Microsoft builds its software daily, this indicates the total time that elapsed between the first clean compile of the first version (which was probably a version intended for Microsoft internal use only) and this particular release of the operating system. Restricting compilation of the entire system to once a day is due to the size of the software system. The complete source code and documentation for this release of Windows NT required four compact disks. Most operating systems for desktops and typical laptops have grown considerably over time, so the time needed to create a build will also increase.

Of course, the assumptions of the agile development process preclude the creation of software systems with a build time that may take, say, a complete day.

6.12 PERFORMANCE AND STRESS TESTING

Let us suppose that all relevant software modules and objects have passed their unit tests and been integrated into a complete system. Let us also suppose that the software has passed all integration tests. Before the software is released, it often must go through what is known as *performance testing*.

A performance test is just that—it is a means of checking that the software will perform at appropriate levels. There are often requirements specifications that state how fast certain operations will be performed. A typical performance specification might be something such as "the system will process 50,000 transactions per second" or "the system will produce an output of 50 MB per second." Clearly, these kinds of requirements are common in real-time systems. They are tested by using typical inputs that may be simulations of actual ones expected in operation.

If the software fails a performance test, the performance must be improved. The typical way to do this is to use a profiler. A profiler is a software tool that allows a software engineer to examine where a software system is spending most of its time during program execution.

Profilers typically "instrument" object code by inserting statements at each function entry point or invocation of an object. These inserted statements control timing information, so that the number of times each function is called and the amount of time spent executing each function can be computed and displayed as necessary. Many profilers also compute the amount of time spent executing specified loops within functions.

Profilers often can be run by setting options on compilers, such as gcc -p file.c on the Free Software Foundation's gcc compiler. This inserts the timing mechanisms in the object code.

On a Linux system, running the executable followed by the gprof command to produce the profiler output might yield something like the following:

Function	No. Calls	Time	Cumulative
main()	1	1:00:00	100.0%
slow()	5,000	30:00	50.0%
slower()	10,000	27:30	45.8%
slowest()	100,000	2:30	4.2%

This output indicates that the software is spending most of its time in the functions slow() and slower(), but that only a tiny portion of the execution time is spent in the function slowest(). A person wishing to improve the performance would only look at the modules slow() and slower() rather than the remaining one that accounts for only 4.2 percent of the execution time, with only a tiny amount of time for each of the 100,000 calls to it.

A related type of testing is *stress testing.* In this type of testing, the intention is to see what happens when the software system is pushed beyond its limits.

Here is a typical scenario. Suppose that the software meets the requirement of processing 50,000 transactions per second. What happens if the load doubles to 100,000 transactions arriving per second? We expect delays. However, we do not want transactions to never get processed by the system because they never get written to a buffer. We do not want the software to crash under excess load, instead it should respond gracefully.

You should note that heavy loads overwhelm many systems. Even the most casual user of the Internet has experienced long delays due to servers being overloaded and unable to process requests efficiently under heavy load.

Many techniques for improving software fault tolerance are used to improve operation under stress. Techniques such as exceptions in Java, C++, and Ada; rollback techniques such as those of Randell (1975); and denial of service in networks are common approaches to preserve system integrity under stress.

6.13 QUALITY ASSURANCE

Ask this basic question about the software you are developing: How can I (or my organization) be sure that the software is of sufficient quality to be released? Of course, there must be a systematic process for making sure that everything works as it should. This systematic process of improving quality is called *quality assurance*, or QA.

Quality assurance can be performed by the development and testing teams. Often, it is done by a separate team given just the responsibility of guaranteeing quality. There may be a need for, say, ISO9000 quality certification, depending on the project's initial requirements as set by the customer.

The QA team may examine the data on the development process to determine if there are many errors that are showing up during the testing and integration phases. The number of software faults detected before release, often referred to as *internal faults*, should be decreasing to indicate that many of the faults have been already found and removed. If there is much volatility in the number of faults found in the software, there is reason to suspect the quality of the software. We will discuss this point more systematically in the next section, when we describe software reliability.

The QA team will apply a considerable number of sophisticated statistical tests and techniques with the goal of measuring the quality of the system and comparing it to acceptable standards.

6.14 SOFTWARE RELIABILITY

Electrical and mechanical equipment is known to follow a bathtub curve during its useful life. That is, the equipment is most likely to fail at the beginning of its placement into service because of poor installation, faulty design, or one bad component. After the equipment has been "burned in," it typically works well for some interval of time and then begins to have more need for repair, because different components wear out. Eventually, the cost to repair the equipment becomes too high and it is taken out of service. The situation is shown in Figure 6.11.

The term *software reliability* appears confusing at first glance. Software does not wear out (although the medium on which the software resides may). Thus, software reliability does not seem to be measurable. However, software errors and the lengths of intervals of correct operation are measurable quantities.

Software reliability models are based on statistical estimates of the failures remaining in software after each checkpoint in the development life cycle. In this context, the term *checkpoint* can refer to the correction of an error in the software at some phase of the testing process, in some release of the software, or at the time of some external event.

Software reliability models such as those of Musa et al. (1987) or Musa (1993) are the basis for the estimates, which are frequently measured as the mean time between failures, or MTBF.

The basic premise of software reliability is that faults that can cause failures will remain in the software regardless of the efforts used to detect and correct them. Reliability theory estimates the remaining failures, the so-called residual faults, by using existing fault data collected at the checkpoints to develop a probability model of the distribution of the software's errors.

It is important to note that the data used in the reliability model is collected before the software is released. The data collected is placed into an appropriate probability distribution

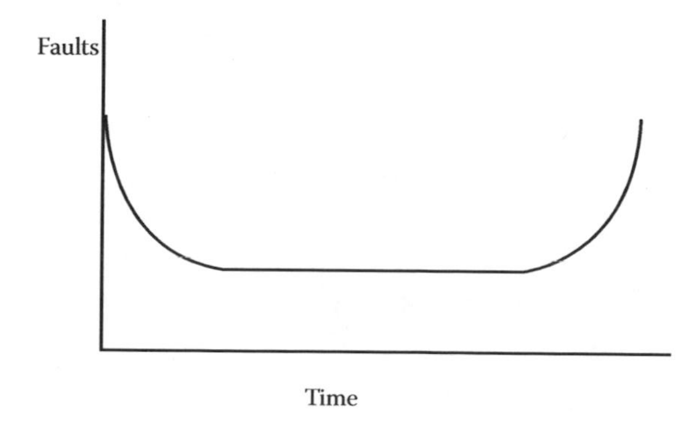

FIGURE 6.11 The typical bathtub curve for malfunctions of electrical and mechanical equipment.

model that is then used to predict the behavior of the software after delivery. The organization used for the reliability model in this type of situation is described in Figure 6.12.

The reason for application of a software reliability model to a software system is to determine the probability distribution of software faults over time. The key variable is the time between distinct software faults, or the interfailure time. This means that the number and timing of faults occurring at each checkpoint in the development process is obtained and the data is then analyzed. The checkpoints can be continually examined during the testing process at regular time intervals or at milestones in the development of the source code. A commonly used milestone is the completion of an internal release for the software. Ideally, the information is collected on a per-system or, better yet, on a per-module basis. For simplicity of discussion, we will restrict our attention to a single reliability test of an entire system.

The method of data collection is interesting. The objective is to count the number of failures at each checkpoint in a way that is consistent with the way that the software will actually behave when it is placed into service. In practice, this means that an operational profile must be obtained.

An operational profile is a set of inputs to the software, based on an assessment of how the software will be used in practice. This involves setting likely usage percentages, with sequences of appropriate user inputs also selected. The set of selected inputs is written to a file, which is used as a script to drive the operation of the software that is to be tested for its reliability. Any necessary external data files would also be included. The performance of the software on all the scripts is evaluated for software faults. The scripts should be considered as a stream of inputs to the software.

Clearly, the selection of items in the input stream used for the operational profile is guided by how the software is actually used. Often a profiler is used to determine precisely which functions in the software are exercised in "typical" uses. Many of the choices of inputs are randomly generated. For more details on this process, see books by Musa, Iannino, and Okumoto (1987) and Fenton and Pfleeger (1996).

Once the data is collected, it must be analyzed. The idea is to fit a probability distribution to the data, so that errors that will occur in the future can be predicted. The most commonly used probability distributions are the Poisson and exponential distributions. See

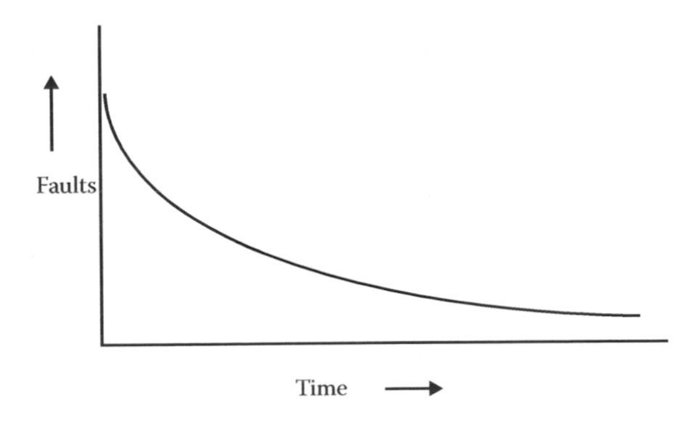

FIGURE 6.12 A typical probability distribution for software reliability.

any good book on applied probability and statistics for more information about estimating probability distributions (Box et al., 1978; Mont, 1991).

Organizations that collect reliability data should do so using any internal data about errors, as well as modification requests to fix errors after release of the system. The data should be entered into a database and the error prone modules should be checked. This data will help in flagging modules that are likely to cause problems in a reuse situation. It will often help in flagging modules and systems that are so complex and difficult to modify and test that they should not be modified unless the change provides extremely large benefits for the organization producing or using the software.

There are several issues to be addressed in reliability modeling. For example, we must be certain that the data is accurate and pertinent. Using only fault data forms in which all information is filled out is not accurate because it may give a biased sample. Test data is not relevant if it attempts to predict failure under operational loads that are different from testing loads.

The recommended practice for reliability is (IEEE, 1988)

1. Estimate size of source code.
2. Estimate fault density (faults per KLOC). This is best done by reusing the fault density from projects that are similar in their requirements, their development methodology, and their programming environment. If no estimate is available, then a number in the range of 10 (for routine programs) to 1 (in a disciplined environment, with programmers experienced in the application domain) should be assumed.
3. These two numbers are multiplied to give the expected number of faults at the beginning of formal testing.
4. Select a model for the reliability data's probability distribution, generally the Musa basic model or the Jelinski-Moranda model. (Several other models, including the Keiller-Littlewood model, could also be used.)
5. Determine the key values:
   - Initial number of faults.
   - Probability of executing a specific fault during a single execution (the fault exposure ratio). A good default value is 5×10^{-7}. This allows the expected time interval between successive software faults to be modeled.
   - Time for which prediction is to be valid.
   - Failure probability per fault and unit time.
   - Initial failure rate.

The most important uses of reliability measures are a general assessment about the system's quality and an indication of when to stop testing (when the expected number of

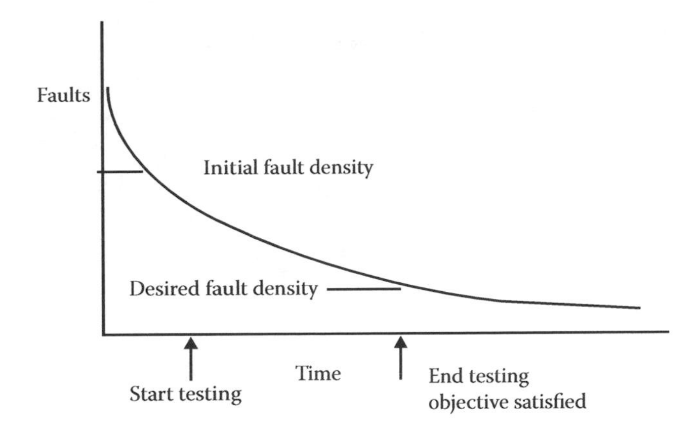

FIGURE 6.13 The use of reliability models.

errors remaining meets the objective error rate). This information can indicate where the testing process should be stopped, depending on the ultimate reliability goal of the number of defects remaining per 1000 lines of code.

A typical situation is shown in Figure 6.13. The curved line in this figure represents the probability that a particular number of faults (as measured on the vertical axis) remains in the code at the time indicated by the horizontal axis.

Software fault information is almost always available for software developed in-house. The fault ratio (number of software faults per KLOC) is probably well known and readily available. If not, it can be easily deduced from known size information and testing data. Clearly, any potentially reusable software artifact should be subjected to a severe test of its fault ratio. (Note that detailed reliability information is not likely to be readily available for software that is not developed in-house. It may not even be available for locally developed software if the organization's software development standards do not require the keeping of reliability data.)

In the absence of any information about reliability of the software, it is probably appropriate to assume that it is of the same quality as most other software used for a project. That is, you should assume that its fault ratio is the average for your other systems for similar applications. Be careful of comparing fault ratios between different application domains. There is considerable variation.

Of course, this reliance on statistical information before software is released may raise some ethical issues. We will discuss this briefly at the end of the next section.

6.15 A MANAGER'S VIEWPOINT ON TESTING AND INTEGRATION

Even more than in the early stages of the life cycle, a software manager does not want any unpleasant surprises in the testing and integration phase. The testing process should proceed in an organized, systematic way. A manager will want to review the test plans and some of the test data before testing starts. He or she will also want to make sure that the requirements traceability matrix is checked, perhaps leaving the responsibility for such testing to the QA team. The manager will also insist on regression testing; this is especially critical when an iterative software development process is followed.

Of course, if high-quality software components have been reused without change while building the software system, they probably do not need to be tested, which can cause a major cost saving during testing, making testing managers very happy.

Metrics data should be kept and used during testing; the manager will usually arrange to have troublesome modules identified at this stage, with reallocation of resources if necessary.

The integration process must also proceed in a systematic way. Since modules are likely to be changed during integration, the prudent manager will also insist on regression testing during integration. As was true during testing, the manager will arrange for metrics data to be collected and used during integration. This will also help to identify modules that are troublesome at this stage.

Of course, a manager will also want the process to be efficient, with testing proceeding on schedule. The testing arena is one of the few areas of software engineering where there is agreement about what is desired from a minimal set of tools:

- Test drivers
- Stub generators
- Test harnesses (to execute functions and objects without creating wrapper code)
- Test data management tool
- Regression tester
- Path analyzer (to determine all decision points)
- Static analyzer (to determine anomalies in the source code's organization)

For many software systems, especially those in safety-critical environments, a fault inspection tool will also be used. Such software tools deliberately place faults within code to determine if faults in subsystems can be isolated or if the effects of the faults propagate throughout the entire software.

Can a program be released with known faults? Yes. In fact, it happens all the time. The faults uncovered by testing may be considered insignificant to program function and hence may not be worth the additional testing efforts, given competitive pressures and market imperatives. Of course, this does not, or at least should not, apply to safety-critical software, which must adhere to a higher standard.

Some features may not be implemented in order to gain market share. This was the case with the first release of Microsoft Visual C++, version 1.0, which was released without an implementation of the important C++ construct known as templates. It was felt that being first to market with a C++ compiler with integrated toolkits was a better strategy than waiting for the full incorporation of C++ templates into the next release. This strategy was obviously a success.

There was no ethical issue, because product specification and documentation made it clear at the time that C++ templates were not included in the first release. However, there would have been serious ethical concerns if it had been claimed that templates were available in that release!

6.16 CASE STUDY IN TESTING AND INTEGRATION IN AGILE DEVELOPMENT

As we saw in our discussion of our agile development case study in Chapter 5, Section 5.10, testing in the agile development process is intertwined with the coding process. Recall that a considerable portion of source code in an agile process may be used to interface between large-scale software components, COTS products, or even entire subsystems.

Untested code is of little use. Source code is tested within test harnesses and much of the testing process is automated in order to make the process as efficient as possible.

Based on what we have described as the "learning" of the agile development team on successive projects, we would expect a decrease in testing time as the team gained more experience. (We reject the possibility that testing time is reduced because the team was tired of testing!)

Figure 6.14 shows the pattern of improvement of the testing process using agile methods. The later projects, largely developed after the team became more adept at employing agile development processes, are shown at the right-hand side of the graph. Note that the graph shown in Figure 6.14 includes both internal and external testing, with the percentage of total testing that is internal also decreasing over time. As before, the term *internal* refers to the time period before any version of the software had been released, and *external* refers to the period after at least one version of the software has been released.

It is appropriate at this point in the discussion of our agile development case study to summarize the major lessons learned. Of course, as is always the situation with case studies, the experiences of this particular team in this particular collection of projects in this particular application domain may not be universal, but they do provide a guideline.

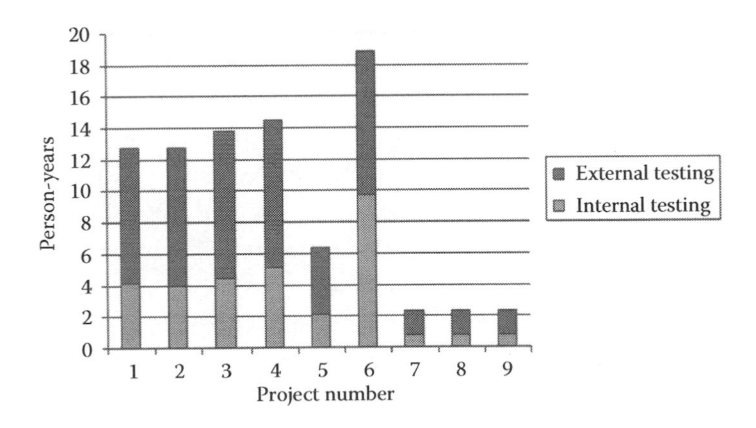

FIGURE 6.14 Decreasing time to test the system as a function of the team's increased experience in the use of agile methods.

- The agile team consisted of people who had considerable deep knowledge of the application domain that had been gained through multiple projects and they all had worked together before.
- Senior management was supportive of both the initial decision to use an agile approach and to allow the effort to continue even if there was a major initial concern about who is in charge.
- The teams were self-organizing, relatively flat in structure, which previously was not the norm in this environment.
- The teams "learned," as evidenced by decreased time to create requirements and decreased testing time.

There is one final note on the testing approaches used in the agile project described in this case study. One approach to agile programming is to create a collection of tests for which successful passage guides the software's requirements and coding. This method is called *test-driven development*. It is clearly related to the idea that if a test case cannot be described for a requirement, the requirement is somewhat vague and underspecified. We discussed testable requirements in Chapter 3, Section 3.13. I am not aware that this particular agile method was used explicitly in the case study described here. For more information, consult the references.

6.17 TESTING THE MAJOR SOFTWARE PROJECT

We will now apply the techniques of software testing to the large running example that we have been considering throughout this book. Since our system consists of both object-oriented and procedural portions, we will need to employ both types of testing techniques. In addition, we should consider both white-box and black-box methods.

Before we start the testing process, it is a good idea to review the requirements traceability matrix for this project. The goal is to make sure that at least the specific system requirements are tested before system integration begins. We will show the completed requirements traceability matrix in the next section after the system is integrated.

It makes sense to consider the system as being comprised of the subsystems designed in Chapter 5. Thus, the subsystems can naturally be grouped into those that require procedurally based testing using black-box or white-box methods and the interface to the spreadsheet that will use an object-oriented approach to testing. We note also that some requirements are easy to check, such as the ones that require a specific hardware platform and operating system environment in which the software will run.

Let us recall our basic design strategy. We wanted to view the entire software system as being comprised of a set of modules, many of which were obtained from other sources. Clearly our testing strategy should be influenced by the reuse-based strategy that we used to design and implement the system.

What are the natural consequences of this approach? The first thing that comes to mind is that the software utilities we used should be considered as black boxes, with only the interfaces needing to be checked. This is also the approach that we should take when testing the "glueware" used to combine these utilities. It is clear that black-box testing is a natural choice for most of our system.

What about white-box testing? This method appears to be especially appropriate when there are many execution paths and complex internal logic. This seems to characterize the software intended to provide an interface to the operating system for the linking of the names of input files to later routines in the program. Recall that this was to be done by means of command-line arguments. The requirement to allow wild cards in the input file names and to treat multilevel directories properly suggests a complexity that makes checking each path especially important.

Of course some object-oriented methods must be applied to the back end of the system. It seems as if the main questions to be answered about this subsystem's correctness are the completeness of the member functions of the appropriate classes and the proper assignment of default values to member functions.

This section will close with the following two changes to the requirements for the major project. You will be asked to modify the testing and integration processes in order to reflect the new requirements in the exercises. The first change is that the customer now wants a web-based user interface instead of one that is PC based. What changes need to be made to the testing and integration processes? The second change is that the customer now wants to use a cloud for data storage instead of one that is PC based. What changes need to be made to the testing and integration process?

6.18 INTEGRATING THE MAJOR SOFTWARE PROJECT

We will now apply the techniques of software integration to the large running software example that we have been considering throughout this book. The assumption is that all the modules have been tested individually and that all have passed their tests.

In order to avoid writing a large number of artificial drivers or stubs, we will follow an integration approach that is a hybrid of top-down and bottom-up techniques. We will start with a top-down approach. The interface will be checked to see if the file names are properly handled. Each input source code file will be checked to see if the first subsystem recognizes the files and the appropriate directories. This is little removed from the testing techniques that we would apply to this subsystem if we tested it by itself.

Once we have tested the initial subsystem properly and linked to the operating system by means of command-line arguments, we will then begin linking the analysis subsystem modules in groups. Naturally, we will consider each of the analysis tools as a single entity for the purpose of software integration.

There is one point to be discussed before we begin. MS/DOS has a facility to connect the output of one executable file to the input of another using pipes, in much the same way that UNIX allows the pipe method of interprocess communication. Commands such as

```
type file.txt |more
```

allow multiple actions to take place with simple commands. It is conceivable that we could link file analysis tools by using a command such as

```
dir | analysis1 | analysis2 > outfile.xls
```

where outfile.xls is a spreadsheet input file. We choose not to do this because the system is very fragile in the sense that the entire process will terminate if there is an unexpected error in one of the component processes. We will develop the code incrementally using more program-oriented linking to make an entirety.

The integrated system grows incrementally. The first modules we link in to our desired system are those in the subsystem to analyze C source code files. We develop the glueware to bridge any gaps between what is produced by the subsystem we developed to interface to the operating system and the analysis tool we use for C programs. This is tested in the most rudimentary way, using a set of test files that was used for unit testing and checking that the software runs to completion and produces the same output. Our primary concern is the appearance of two different functions named main(), one in the initial subsystem and the other in the software tool for analyzing C programs. The easiest thing to do is to rename the main() function in the downloaded utility software system. A name such as C_main() might be appropriate as the new name.

Precisely the same steps must be followed in order to integrate the other software utilities into an entire package. This again is a top-down approach to software integration.

The integration of the object-oriented portion of the software into the entire system will be done from the other direction. The bottom-up approach will be followed, linking the output to a spreadsheet.

Finally, the two collections of software modules connect, then mesh together, and we have a complete system.

The completed requirements traceability matrix is given in Table 6.7. Note that we can now measure the size of our system to see if it fits on a floppy disk and the executable file satisfies the one-megabyte limitation. (Of course, it did not. We eliminated this requirement when our customer asked for a graphical, web-based user interface.)

From a project management perspective, we should also do a status check. Are we ahead of schedule (unlikely), behind schedule (likely), or approximately on target? Have there been any unpleasant surprises, any portions of the system that were more difficult than we expected? Does any portion of the system require extra attention, perhaps additional resources? Have technology or market pressures rendered any portion of the system obsolete? (Recall that we did this status check earlier.)

As was the case with Section 6.17, this section will close with the following two changes to the requirements for the major project. You will be asked to modify the testing and integration processes in order to reflect the new requirements in the exercises. The first change is that the customer now wants a web-based user interface instead of one that is PC based. What changes need to be made to the testing and integration processes? Are any new tools needed? The second change is that the customer now wants to use a cloud for data storage instead of one that is PC based. What changes need to be made to the testing and integration process? Are any new tools needed?

TABLE 6.7 Completed Requirements Traceability Matrix for Our Major Continuing Software Engineering Project

Requirement	Design	Code	Test
1. Intel-based	Y	Y	Y
2. Windows 8.1	Y	Y	Y
3. Windows 8. UI	Y	Y	Y
4. Consistent with Excel14.0	Y	Y	Y
5. System one size only	Y	Y	Y
6. One MB system	*	*	*
7. One MB disk space	*	*	*
8. One 1.44 MB floppy disk	*	*	*
9. Includes installation	Y	Y	Y
10. No decompression utility	Y	Y	Y
11. One input file at a time	Y	Y	Y
12. Size of each function	Y	Y	Y
13. Size of each file	Y	Y	Y
14. Size of system	Y	Y	Y
15. Compute totals	Y	Y	Y
16. Develop std for C, C++, Ada	Y	Y	Y
17. Batch-oriented system	Y	Y	Y
18. Precisely define LOC	Y	Y	Y
19. Measure separately	Y	Y	Y
20. No error checking of input	Y	Y	Y
21. Front end in C or C++	Y	Y	Y
22. Batch processing	Y	Y	Y
23. File names limited to 32 char.	Y	Y	Y
24. Wild cards can be used	Y	Y	Y
25. Dead code ignored			
26. No special compilers needed	Y	Y	Y

SUMMARY

Testing strategies for classes are different from those for testing ordinary procedurally developed programs. Some common techniques for testing object-oriented programs and classes are encapsulation, minimalization, exhaustion of a depth, inheritance, and interactive methods.

Software testing can be white box or black box, depending on whether the details of the source code are used to determine the test cases or just the module interfaces.

Software integration is the process of combining modules, objects, subsystems, and/or COTS products to make complete systems. Integration can also proceed in a top-down manner, in which case lower-level modules must be stubbed in.

On the other hand, bottom-up software integration requires creation of higher-level drivers. Hybrid approaches to software integration are also possible.

Regression testing is the process of ensuring that changes made to fix one software fault do not incur other faults. It is essential during software integration.

Software reliability is a statistically based technique used to determine when testing is sufficient to release the software.

Management of software that is changed during integration can benefit immensely by being controlled by configuration management.

KEYWORDS AND PHRASES

White-box testing, black-box testing, every-statement coverage, every-branch coverage, unit testing, module testing, test script, test harness, test plan, alpha testing, beta testing, data partitioning, regression testing, system testing, stress testing, performance testing, usability testing, top-down integration, bottom-up integration, big bang integration, test-driven development.

FURTHER READING

There are many excellent general books on testing software, including two editions by Beizer (1983, 1990), Howden (1987), Myers (1979), DeMillo et al. (1987), and Marick (1997). Miller and Howden have a tutorial book on software testing that includes many of the earlier research papers in the field (Miller and Howden, 1978). Most of these classic references concentrate on procedural programs and include little information about testing object-oriented programs. Two especially informative recent books on this subject are by Whittaker, Arbon, and Carollo (2012) and Kaner and Fiedler (2013).

A discussion between Michael Donat, Jafar Husein, and Terry Coatta illustrates some issues in the practice of online game testing at the company Electronic Arts (Donat et al., 2014).

There are several other URLs for information on software testing. The STORM software testing home page at Middle Tennessee State University contains links to other websites for software testing and provides a directory of researchers in the field of software testing. The URL is http://capone.mtsu.edu/storm. The link http://www.gimpel.com points to a testing tool, PC-lint, for C and C++. This is a great aid in providing the information so helpful to users of the UNIX utility lint in terms of lack of consistency between program modules.

An interesting perspective on software integration in combination with product-line architectures in which new components are developed both for their immediate utility and for future use within a growing collection of software artifacts and systems for which these artifacts are intended to be used can be found at the following URL at the Software Engineering Institute: https://www.sei.cmu.edu/productlines/frame_report/softwaresi.htm.

Numerous examples of the test-driven development paradigm can be fund in the 2003 book by Beck (2003). See also the book by Freeman and Pryce (2009).

Using "crowds" for software testing is a relatively new concept. The book by Mukesh Sharma and Rajini Padmanaban describes some issues that may arise when using this crowd testing technique for testing commercial software (Sharma et al., 2014). (Some early adapters of buggy software products that they purchased might have felt that they also were crowd testers.)

A 2005 book by Beth Gold-Bernstein and William Ruh describes software integration at an enterprise level, with a heavy stress on database integration (Gold-Bernstein and Ruh, 2005). A more recent book by Pawel Czarnul, *Integration of Services into Workflow Applications*, addresses some issues in Software as a Service (SaaS) (Czarnul, 2015). Unfortunately, there are few other readily available sources of information on software integration.

EXERCISES

1. List all the black-box test cases for the source code given in Example 6.3. Which test cases do you think are the most important? Why?
2. List all white-box test cases for the source code given in Example 6.3. Which test cases do you think are the most important? Why?
3. What is the minimum number of test cases to ensure that every path in Example 6.3 is tested?
4. What is the minimum number of test cases to ensure that every statement in Example 6.3 is tested?
5. Write a test harness to exercise the testing of the code fragment in Example 6.3.
6. Examine a common software application. Determine a set of black-box test cases that you would use if you were in charge of testing this software.
7. Devise a new class to describe floating point numbers. The precision of the floating point numbers will be part of the class. The three precisions to be considered are float, double, and long double.
8. Assume that you have created the class described in Exercise 4. Assume also that functions have been written to perform all arithmetic operations on this class and that these functions overload the +, –, *, and / operators. What combinations of arguments must be tested for the overloaded + operator for this class?
9. List the special cases that should be tested for correct treatment of overloading of the + operator in Example 6.5. Also, list the special cases needed for testing of the << and assignment operators.
10. List all test cases that you believe are necessary to test the objects defined in Example 6.5.

EXERCISES FOR THE MAJOR SOFTWARE PROJECT

1. Apply the techniques of software testing described in this chapter to the large running software example that we have been considering in this book.
2. Apply the techniques of software integration described in this chapter to the large running software example that we have been considering in this book.

3. This exercise considers the ramifications of a considerable change to the requirements on the design of the software. The first change is that the customer now wants a web-based user interface instead of one that is PC based. Determine the changes that need to be made to the requirements to have a different type of user interface, with different issues in the possible configurations (multiple operating system versions on the PC, multiple browsers, and multiple versions of HTML and Java). What changes, if any, need to be made to the software development life cycle model or the testing process as part of these changes? Are any new tools needed?

4. This exercise considers the ramifications of a considerable change to the requirements on the design of the software. The second change is that the customer now wants to use a cloud for data storage instead of one that is PC based. Determine the changes that need to be made to the requirements to have cloud storage for the data, with different issues in scalability and the size of the cloud that is needed for the project. Also, determine the changes that need to be made to the requirements in order to have cloud storage for the source code modules being analyzed, with different issues in scalability and the size of the cloud that is needed for the project. What changes, if any, need to be made to the software development life cycle model or the integration process as part of these changes? Are any new tools needed?

CHAPTER 7

Delivery, Installation, and Documentation

7.1 DELIVERY

Even the best software is of no use until it is made available to potential users. Let us suppose that the software system has been designed, developed, and tested according to the system requirements. It is now time to make the software available to its users. In this chapter we consider software as an artifact to be delivered, installed, and documented.

Of course, software may be designed for the international market, and the need for localization (changed character sets, making sure that cultural norms are not violated by any images or icons within the software, etc.) is obvious. A discussion of these issues is beyond the scope of this book.

The software is usually delivered according to one of the following three scenarios, depending on the nature of the relationship between the customer and the software development organization:

1. If there is a single user or a very limited number of users, then the software is usually delivered on CD or DVD, with the delivery format determined by the project's requirements. The software delivery team is responsible for making certain that the software can be given to the single user and that the software can be installed correctly on the user's machine.

2. If there are many potential users of the software and the software is to be sold in stores or by other retail mechanisms, the software is usually delivered on CD or DVD, with the format determined by the marketplace. The software delivery team is responsible for making certain that the software can be duplicated properly so that it can be distributed. An installation script or installer is provided by the delivery team. The responsibility for installation rests with the user.

3. If there are many potential users of the software and the software is to be delivered electronically, using FTP or similar file transfer programs, there is no need for multiple copies of the software to be created. In this case or even an app store, the responsibility of the software's delivery team is to make certain that the software can be obtained properly. An installation script or installer is provided by the delivery team. The responsibility for installation rests with the user.

We will now discuss each of these delivery scenarios in turn.

The first delivery scenario describes a software system that is intended for a single customer or, at most, a few customers. Any customer is likely to have personnel trained in the use of this software. If the software is not delivered to the customer according to the specifications in the contract's requirements, the customer will not make the final payments. Thus, the customer will inform the software development organization that delivery was not successful.

The typical problems with installation in this scenario are

- Failure of the physical medium on which the software resides
- Incorrect placement of the software into directories, resulting in incorrect directory names
- Incorrect use of names of subdirectories, making subsystems inaccessible
- Incorrect access permissions, making files inaccessible
- Incorrect formatting of the software, such as using incorrect conversion tools for compressing or uncompressing files
- Incorrect documentation of the delivery instructions

Each potential problem should be considered in a predelivery run-through during which the installation steps are carefully tested.

The second delivery scenario of a CD or DVD is especially appropriate for shrink-wrapped software that is sold in stores or by mail. Each of the points raised earlier in this section clearly also applies in this scenario. In addition, the software-producing organization must take into account the lack of a direct developer-customer relationship in this situation. Therefore, the developer cannot have detailed knowledge of each potential customer's hardware environment or the software that has already been installed on the customer's computer.

The lack of control of the user's environment can create serious problems. Users may have great difficulty installing the software because their environment is different from the one in which the software was developed.

Users may also vary greatly in their computer expertise. (Most of the readers of this book have heard the [probably apocryphal] story of the person who called a personal computer vendor's hotline because the "cup holder" was broken. Of course, the cup holder was the CD drive!)

In the second delivery scenario, the most important additional step for the organization developing the software is to do multiple dry runs during the predelivery process, using a variety of environments and users.

The third delivery scenario best describes software that is to be made available to potential users electronically over a medium such as the Internet or an app store. All the problems described in the second scenario apply for this delivery scenario as well. There are similar problems also due to the variation in operating system and application environments.

There are two new problems that can occur in the third delivery scenario: collection of payment for the software and protection of the host computer used as a remote server for the software. A discussion of electronic payment for the software is best left to experts on Internet commerce. In any event, anything you read in a book is likely to be out of date soon, given the explosive growth of the Internet and changes in credit cards.

The server that contains the software will have some form of remote access, using anonymous FTP or some other method of file transfer. As such, the server is at least slightly vulnerable to intrusions. Thus, a decision must be made as to the relationship between the risk of having intruders trash an entire system and providing no access. The use of such security devices as firewalls is highly recommended. See the book on computer security by Charles Pfleeger, Shari Lawrence Pfleeger, and Jonathan Margulies (2015) for more information on this topic.

Determining which delivery scenario applies is usually easy. After we have addressed the points raised earlier, we must decide upon proper file formats and compression/decompression utilities. At one time, it was often necessary to reduce the size of files to be able to put them on fewer floppy disks, special drives (such as Zip), magnetic tapes, or CDs. Now, the issue is trying to reduce the software in size by compression in order to fit it on a single CD or DVD.

Reducing the amount of traffic over the Internet is also a good idea, so we might wish to limit the size of downloads. Users of smartphones may have noticed that upgrading a smartphone's operating system often requires that the smartphone be connected to a power source, rather than depending on batteries, and that Wi-Fi is preferred over even the fastest current 4G LTE networks for major operating system software updates.

We will be content to list a few utilities that might be used when selecting a format in which to deliver software. A brief list of formats for source code and other text files is given in Table 7.1. We have not considered special formats and standards such as JPEG, MPEG, TIF, or GIF for storing graphical or other nontextual material.

We will now give an example of how such compression, delivery, and decompression might be used and created. My colleague Will Craven at Howard University developed a set of UNIX shell scripts for file transfer that provide a smooth interface to some standard UNIX utilities that have complex interfaces and are not "user friendly." The shell scripts work on most Linux environments.

The scripts use the Bourne shell. To understand how the scripts work, recall that in the Bourne shell, lines beginning with the # symbol are comments and not the start of a Twitter address. A command with the two symbols # and ! at the start specifies which of many command shells are used in the syntax of the shell script.

TABLE 7.1 Some Utilities Used for Compressing, Decompressing, or Storing Software Files

Utility	Operating System
tar	UNIX, Linux
bar	UNIX, Linux
rar	UNIX, Linux, Windows
jar	Many
cpio	UNIX, Linux
gzip	Many
gunzip	Many
Pkzip	Microsoft Windows, MS-DOS
Binhex	Macintosh

Note: Most Apple computers run a form of Linux on which operating systems such as Yosemite are built.

The first script is given the name "stow." The script will allow a user to place all files in a directory in a single archive file that is in the UNIX cpio format. This format is compressed and thus reduces the amount of disk space needed for storage and the amount of traffic on a network.

Before using the stow utility, the user must create a directory with all necessary files. The user then changes the working directory to the parent directory. The stow utility is used with the syntax

```
stow  directory_name  archive_name
```

The source code for stow is

```
#! /bin/sh
# 2-7-88 (wdc)Put all files under specified
# directory into archive
#
case $# in
 2) if [ -d $1 ]; then
   if ['du -s $1|cut -f1' -lt 'ulimit'];then
      find $1 -print -depth | cpio -oac > $2
   else
      echo "$1 is too large to stow"
   fi
   else
      echo "$1 is not a directory"
   fi;;
 *) echo 'Usage:stow source destination';;
esac
# 11-16-88 (wdc) -- insert 'ulimit'
```

```
# 10-19-89 (wdc) -- -a option on cpio
# 9-26-94(wdc)--depth option on find
# no -v on cpio
# 10-4-96 (wdc) -- force /bin/sh
```

Let us examine the shell script that makes up the stow utility. The comments (lines beginning with the # sign) help explain the software's operation. The script was created in 1988 by Will Craven. The exclamation point immediately following the # symbol on the first line indicates that the Bourne shell is to be used, rather than the C, bash, Korn, or POSIX shells. This change was added in October 1996. Note that this statement has the effect of constraining the environment of a user to one that is controlled by the software designer.

The case statement suggests that we are looking to match the shell variable $#, which represents the number of arguments to the shell script other than its name, with the value 2. If there are precisely two arguments to stow, the first argument, $1, is tested to see if it is the name of a directory. If the second argument is the name of a directory, then control passes to the nested if statement.

The nested if statement is complex, testing for disk usage with du and cutting off a recursive directory search if the number of disk blocks for the directory specified in the first argument, $1, exceeds the system limit on file sizes, which is given by the standard UNIX and Linux utility, ulimit. The find command as given here prints the names of all files and passes them to the cpio archive, which is stored in the file whose name was given in the argument $2.

Note that the stow utility was modified several times, and that each modification was documented carefully. This is an excellent example of proper software maintenance.

Although the shell programming in the stow utility is complicated, it is simpler to remember and more intuitive than a call to the single shell command

```
cat directory_name|cpio -oac > archive_name
```

The meaning of the o, a, and c options to the cpio command are difficult to remember for the nonexpert UNIX or Linux user.

The second script described in this section is given the name “huput.” The script will allow a user to move an archive file that is in UNIX cpio format to the special rje directory of someone else (usually an instructor). The purpose of an rje directory is to allow individuals to place files in a different user's directory. The name rje is a vestige of the early time-sharing multiprocessing period in operating systems design; the name stands for remote job entry.

The huput utility will change permissions of each of the component files in the archive to those of the person to whom the archive is submitted. This utility is used with the syntax

```
huput archive_name source_name
```

The source code for huput is

```
#! /bin/sh
# 10-6-88(wdc)send file(s) to rje directory
#
case $# in
 0|1) echo "usage: $0 file[file ...] user";;
 *) list=$*
   shift 'expr $# - 1'
   login=$1
   if pw='grep "^${login}:" /etc/passwd'
      then rje='echo $pw | cut -d: -f6'/rje
   if [ -d $rje -a -w $rje ]
      then set $list
       while [ $# -gt 1 ]
         do file=${rje}/'basename $1'
         if cp $1 $rje
           then chown $login $file
                echo $file
         fi
       shift
      done
   else
   echo "Can't write to rje of login\07"
   fi
   else echo 'No such login!\07'
   fi;;
esac
#
# 10-4-96 (wdc) -- force /bin/sh
```

The huput utility is also a UNIX shell script. It has a series of nested if statements within the major case statement. The first shift statement uses the value of an arithmetic expression (included in backquotes) as an argument. This is a tricky construction. See Chapter 2 of my book *Advanced Topics in UNIX* for more information (Leach, 1994) or the second, AfterMath edition (Leach, 2012).

The value of the login argument is checked against entries in the password file/etc/passwd using the UNIX pattern matching utility named grep. If there is a match, then the files are sent to the rje directory of the person whose directory is the target. The chown utility is used to change ownership of the archive file to the recipient.

The final script described in this section is given the name "unstow." The script will expand a single archive that is in the UNIX cpio format into a directory that contains the files in the archive. The unstow utility is used with the syntax

```
unstow archive_name
```

The source code for unstow is

```
#! /bin/sh
# 2-7-88 (wdc) get all files out of archive
#
case $# in
 1)   cpio -icdmv <$1;;
 *)   echo 'Usage:  unstow source';;
esac
#
# 10-4-96 (wdc) -- force /bin/sh
```

Fortunately, this script is simpler than the previous two. As before, the first statement in this script forces the use of the Bourne shell. It also constrains the user's environment, at least temporarily. The case statement is used for control of execution flow and the script is well documented.

Obviously different delivery scripts are needed for different environments. The scripts presented here are highly UNIX-oriented and may not apply directly to any other environment. However, the principles are the same for many delivery situations:

- It is always necessary to give the user or installer access to the files so that any permissions can be changed.
- The user's environment is temporarily constrained for ease of programming.
- It is necessary to provide wrappers to hide users from arcane operating system utilities, especially those with many options.
- It is easier to create an archive and transfer it, rather than ask the user to transfer multiple files.

As a final note on delivery, the use of archives and compressed formats can reduce network traffic and is, therefore, considerate of other users.

7.2 INSTALLATION

Installation of software can often be quite time-consuming. For example, installation of an operating system is often a sequential process of "bootstrapping." That is, a portion of the operating system (a so-called mini-kernel) software frequently must be loaded first. After this, the first portion is used to install a second portion (the entire kernel and the root file system). The second portion is then used to install additional portions (such as the complete/var file system), and so on.

Because of the issues previously raised in this chapter concerning potential failures during delivery, successful software installation often depends on having precise knowledge of the software that has already been installed on the system. Inconsistencies must be noted, particularly with major upgrades of software. The best installation software utilities often include an installation log to aid in troubleshooting.

We note explicitly that the same ideas, if not the same specific details, apply to such utilities as the "Setup" and Windows Installer for Microsoft Windows (multiple versions), Stuffit Expander, and the VISE installer for the Mac, and others. These utilities are often invoked with formatted text files.

As an example of the use of the Setup utility, a small portion of the input file for installation of the OSA analyzer developed by Aaron Rogers, Sean Armstrong, and Melvin Henry that we discussed earlier is given next. Note that the file indicates the name of the executable file, path names where library files can be found, links to dynamically linked libraries (DLLs), and a check of the Windows Registry to determine if the most recent versions of library files are present or need to be updated.

```
[Bootstrap]
SetupTitle=Install
SetupText=Copying Files, please stand by.
CabFile=Omega Source Analyzer.CAB
Spawn=Setup1.exe
Uninstal=st6unst.exe
TmpDir=msftqws.pdw
[IconGroups]
Group0=Omega Source Analyzer v1.0
PrivateGroup0=True
Parent0=$(Programs)
[Omega Source Analyzer v1.0]
Icon1=""Omega Source Analyzer.exe""
Title1=Omega Source Analyzer v1.0
StartIn1=$(AppPath)
[Setup]
Title=Omega Source Analyzer v1.0
DefaultDir=$(ProgramFiles)\osaproj
AppExe=Omega Source Analyzer.exe
AppToUninstall=Omega Source Analyzer.exe
```

In the remainder of this section we will describe the dmalloc shareware solution to some issues with most standard C library memory allocation functions. This software is available from the site dmalloc.com using FTP. This software is a CASE tool that improves on the C malloc() function. We note that shareware is a simple solution to the problem of obtaining payment. The website contains the dmalloc library, configuration scripts, and other useful items.

The last two statements in the installation log illustrate the implicit assumptions made by the author of the dmalloc software about the level of knowledge and experience of the software installer.

Please check-out Makefile and conf.h to make sure that sane configuration values were a result.
You may want to change values in settings.h before running 'make'.

The term "sane configuration values" can only be meaningful to an installer who is sufficiently familiar with Linux to determine if the configuration values are "sane." The same statement applies to anyone considering making changes to the file settings.h or to any other header file.

Compare this installation process to the installation process associated with most software for personal computers. A software package (known appropriately as an installer) can either be a multipurpose package available for general use or one designed specifically for this application. The installer software usually prompts the user for a choice of installation directories and performs the rest automatically.

This type of installation process is much simpler, at least when it works. The difficulty is the potential conflict with other software already residing on the system. Removal of such conflicts can be very difficult without both installation logs and knowledge of where critical files are stored.

The type of installation process chosen largely depends on the intended user. The three scenarios described for software delivery reflect the likely users and, as such, the degree of automation of the installation process.

There is one last point to make about installation procedures, especially for the personal computer system. Although they do require some detailed system knowledge, utilities such as the Setup installer for Microsoft Windows (various versions) do not require any special licensing arrangement in order to use them.

What about installation in the cloud? Cloud computing is, in many ways, ideal for the installation of software components, because in a well-designed cloud of the proper scale for the desired application, the installation process is not disruptive.

7.3 DOCUMENTATION

Even the best software with the most intuitive user interface needs some form of documentation. The documentation can be in any of the following forms:

- Internal documentation (embedded in the source code files)
- External documentation in the form of requirements and design documents

- Additional documentation explaining why design decisions were made, as in a design rationale
- Manuals, such as user, operations, or installation manuals
- Online help that is part of the software itself

Each of these forms of documentation has its own place within the software development process. Since these forms are so different, we will discuss each of them in a separate section.

There is one issue that applies to more than one type of documentation. Many long-lived systems have multiple releases. There is often a problem for documenters as to how to organize their materials in such cases. The choices are between having a single basis document, with many release notes indicating changes in the most recent version, and having entirely new documentation for each release. The first approach is less costly in terms of printing but can be a problem in terms of understanding. The second choice makes the documentation easier to understand but can easily fill up all available bookshelves in an office. Both approaches are used in practice. (Advice to beginning software engineers: avoid sharing an office with a person who has no room in bookcases or filing cabinets because of large stacks of paper documents. There will not be any room for the things you need for your work.)

Note that both the Internet and company-specific intranets can make some forms of documentation easier to disseminate and store, but require some organization to determine specific document versions. For example, the problem of multiple versions of documentation for different releases can be reduced by the use of some form of configuration management.

7.4 INTERNAL DOCUMENTATION

This section expands the discussion begun in Chapter 5, Section 5.2 when we considered coding styles. The term *internal documentation* refers to documentation that is included within the source code itself. Internal documentation may be limited to comments that are embedded within individual source code files. More commonly, it is also provided in readme files. Usually each source code directory and the parent directory of the entire system contain their own readme files.

Naturally, the coding standards of an organization take precedence over any general statement concerning internal documentation made in this book. You should keep in mind the discussion of coding standards that was given in Chapter 5. Nevertheless, we can make some general statements about internal documentation issues:

- Most organizations, especially if they are designing software that is to be reused or maintained, require extensive amounts of documentation.
- It is neither practical nor desirable to document every program statement with a comment.
- Good coding practices, such as naming conventions for files and subsystem directories, can aid in documentation.

The first point can cause some confusion for beginning software engineers. Many software standards require prologs or file headers for each source code file. The purpose is to indicate the basic functionality of the functions or procedures that are included in the file. This is followed by a detailed information header for each function or procedure included in the source code file. Certainly each major block of source code or each major branch of control will require additional commenting.

The second point to be made about documentation is that not every line needs to be commented. In general, meaningless documentation is not helpful. As an example, notice that nothing is added by the following comments:

```
i = i + 1;     /* add one to i */
temp = x;      /* assign value of x to temp */
x = y;         /* assign value of y to x */
y = temp;      /* assign value of temp to y */
```

A far better way to comment these lines is

```
i = i + 1;
/* swap x and y */
temp = x;
x = y;
y = temp;
```

The third point is the effect of good naming conventions for files and subsystems on documentation. Organization of a source code directory into three subdirectories named database, graphical user interface (GUI), and processing, makes it easy to determine the first step when trying to fix a problem with the graphical user interface. Even a cursory examination of directory and file names is often sufficient to eliminate several potential sources of problems.

As mentioned earlier, there is a last, informal rule of thumb that has been passed from programmer to programmer: If the documentation and source code do not agree, they are both wrong. Keep this in mind as you are documenting source code. Documentation is an integral part of software development activity. Attempting to document the code by "commenting" it at the last minute appears to be the greatest cause of this inconsistency problem. Besides, documenting the code as you develop it can make your development more efficient because you can tell what you last worked on.

7.5 EXTERNAL DOCUMENTATION

External documentation includes requirements and design documents, which are generally not given to potential users of a system, and training guides, which generally are.

The complete requirements and design documents should be produced and made available to the managers and technical monitors of software development projects. The documents may be in both written and electronic form. If electronic form is used, the documents may be stored in the format of the word processing program that produced them, or in a display-oriented format such as PostScript, Portable Document Format (PDF), or similar.

Often these external documents are signed off by a group known as the technical publications office, which is commonly known as "tech pubs." Tech pubs is responsible for creating and formatting external documentation that may have been started by others, proofreading the documentation, and checking it for technical consistency.

The most important members of a technical publications team are technical writers. They have experience in organizing documents and putting them into concise, standard English. It is important to have technical writers begin working on a software product's documentation well ahead of the deadline for release of a product, so that any major reorganization of documentation can be done if necessary.

External documentation of a software system is, by its nature, separated from the source code associated with the same system. Thus, there is a major problem with consistency between the different forms of documentation. One potential advantage of computer-aided software engineering (CASE) tools is the ability to have common views of software and its documentation.

7.6 DESIGN RATIONALES

Many large software systems have *design rationales* that attempt to explain the thinking behind several decisions that were made in the system's design. These are sometimes helpful in clarifying ambiguities that arise after a system has been put into use.

Perhaps the most widely available design rationale is *Rationale for the Design of the Ada Programming Language* by Ichbiah et al. (1986). This book is essential for an understanding of the principles used in the creation of the Ada programming language. Although not technically a design rationale, the book *Annotated C++ Reference Manual* by Margaret Ellis and Bjarne Stroustrup (1990) provides some related information for the C++ language.

Apart from their historical interest to researchers in the area of programming language evolution, these rationales can explain how to answer many questions about proper program design and library usage that cannot be answered by other language reference manuals.

On a personal note, I was recently involved with an effort at NASA to rehost a software system that was used for acquiring image data from a particular spacecraft, checking and organizing this data, and making the data available to widely scattered scientific researchers in a variety of formats. The original software system was developed in the 1970s. The software was to be rehosted from a collection of mixed FORTRAN and assembly language code on an obsolete mainframe to modern workstations. All paper copies of original requirements and design documents were lost. Only the source code, executable code, and operations manuals were still in existence. Reengineering the system to a modern environment would have been much easier if the other documentation had been available.

Our first task was to develop a rationale for certain previously made design and operational decisions so that it would be easy to determine which portions of the new software could be ported from the existing mainframe and which should be rewritten or discarded due to advances in technology that made porting them inappropriate. The rest of this work was more straightforward once the design rationale was developed. A cost savings of nearly 20 percent was obtained simply by making the initial effort to understand the naming conventions used in the design, thereby removing duplicates (Leach, 1998a).

7.7 INSTALLATION, USER, TRAINING, AND OPERATIONS MANUALS

The purpose of an installation manual should be clear from the discussion earlier in this chapter. Any system configuration information should be included there.

A user manual is intended for normal system use. Generally, such manuals are short and describe a few simple commands that can be used to provide most of the functionality of the system. Additional functionality may be documented in a second, longer manual. Many software systems come with a simple "getting started" manual and a more detailed manual for day-to-day operations.

When the software is sufficiently complex or controls mission-critical or safety-critical systems, separate manuals for training and software operation frequently will be necessary. Needless to say, training and operations manuals must be tested with actual people, preferably with the likely users themselves. Typically, a technical contract monitor will be responsible for ensuring that the manuals are adequate.

It should be noted that a large industry has evolved just to handle technical training issues for software packages.

7.8 ONLINE DOCUMENTATION

The term *online documentation* refers to assistance that is provided for users as part of the software itself. As such, the decision about including online documentation should have been made long before the software system is getting its final documentation, just to make sure that any links or references are placed properly.

The creator of an online facility must choose between developing an application-specific help system and using a general-purpose one with the information specific to the particular application inserted as necessary. The ability to use a general-purpose, online help system provides an enormous competitive advantage to makers of suites of related office software such as the Microsoft Office, Word Perfect, Apple Pages, or Lotus Smart Suite collections of word processors, spreadsheets, database systems, and electronic mail programs.

Online help for software generally can be classified as being one of four forms:

- A set of pages of documents (often linked in a hypertext format) can appear either from a menu that is always visible or from the use of a particular key.
- A set of well-thought-out FAQs (frequently asked questions) can be provided.
- An "intelligent agent" or "wizard" that interacts with the user can be invoked automatically whenever certain combinations of actions have occurred.
- One or more videos can be posted either on the software creator's website or, as is becoming more common, on a more general nonproprietary platform, such as YouTube.

In the first type of online help, the main difficulty is ensuring that the state of the user's computation or document is the same after selecting help. When the help menu is accessed, the user's work must be interrupted, saved (at least to a temporary location), and resumed when the use of the help menu is completed. This can be done most easily when the help information is provided in a separate web page. There often are security issues involved, especially if the user had to be logged on to the software via a password or other security control in order to access the help pages. You may recognize this type of problem as related to a state diagram pattern, which we have previously discussed.

Almost all computer platforms have some inexpensive or even free software that allows, say, an installation procedure to have the menu choices and mouse/pointer selections from various interactive sessions recorded and then placed in appropriate web pages.

I am reminded of an old joke about a software developer who creates a user manual for the software. The user manual is shown to two friends, both software engineers, who think it is very clear. The software developer then decides that the user manual would be unintelligible to most users and offers the job of creating a user manual to a technical writer.

The second form of online help is somewhat similar to the case of full documentation provided online. Indeed, the two may be connected. Of course, FAQs are only as useful as the information they convey and the manner in which it is organized. The same preceding joke also applies in this instance.

A similar situation holds for the case of an intelligent agent, which is the third form of online documentation discussed in this book. (This form of documentation is considered to be online, although it may be included on the software itself. Of course, in a world of ubiquitous cloud computing, we may not know where the software actually resides.) Here there is an additional complication: the agent is monitoring the user's progress, in addition to providing help. This greatly complicates program complexity and design, especially with potential security issues. Developing security using a state diagram pattern may be harder to apply than in the first form of online help. See Tecuci and Keeling (1998) for more information.

The fourth form of online documentation is in the form of video, which, as mentioned, can be hosted either on the software developer's website or on a commercial channel such as YouTube. Such a video requires a professional level of production, including lighting and sound. A poor-quality video reflects poorly on the software product, which is not the desired effect.

In all four cases, the most important part of the help mechanism is that it tells the user what he or she wants or needs to know.

7.9 READING LEVELS

How can you tell if the documentation is adequate? The most important thing is to test the documentation on people who are not already familiar with it. The idea is to have the tester of the software's documentation give feedback based on what he or she sees or reads, not on what he or she knows about the design of the system. Recall that we discussed different levels of user expertise in Section 3.8.

The reading level is a good metric to use for documentation. Generally speaking, lower reading levels are better, especially if the documentation is to be read by nondevelopers. A higher reading level is probably sufficient for documentation of a system's user interface.

Note, however, that there may be additional demands on even the most sophisticated user. For example, detailed knowledge of the application domain is generally required to understand more technical details of a potentially reusable system if it is to be reused elsewhere.

The Kincaid and other reading level tests are useful in this context. They are available from many sources and are included in the wwb (writer's workbench) package commonly available on AT&T System V UNIX and other classic UNIX variants. The wwb tool does what every modern, complete word processing system does: indicates incorrectly spelled words, double words, missing punctuation, and other ungrammatical constructions.

In addition, this software has many features not commonly available in commercial word processing systems. The metrics computed by the wwb tool generally include the reading level and an assessment of the values relative to some documents that are believed to be examples of clear writing. The wwb tool indicates unusual sentence complexity, such as too many compound-complex sentences, or the other extreme, too many simple declarative sentences, which make the document seem choppy. The wwb tool also flags documents with too much use of the passive voice.

A recent project on reading levels was carried out by the Harris Corporation for the U.S. Department of Defense. The project was directed toward reducing the gap between the average reading level of the recruits and the average reading level of the technical manuals. The solution to this problem was twofold: to improve the reading level of the recruits by specialized training and to decrease the reading level of the manuals by using certain linguistic approaches to lower the reading level.

You should be careful about using a single measure of reading level as the determining factor in document readability. There are many readability issues, including familiarity of the reader with the general subject matter; technical vocabulary; average length of words; number of simple, compound, or compound-complex sentences; and adjective-adverb ratio.

7.10 A MANAGER'S VIEW OF DELIVERY, INSTALLATION, AND DOCUMENTATION

In many ways, a typical manager's perspective on delivery and installation is similar to his or her perspective on most software life cycle activities: avoid risk. However, there is an additional factor that can occur at these stages: embarrassment.

The only surprises that can occur at this point are very unpleasant ones. Certainly the manager expects the entire delivery and installation process to have been checked in detail to detect any problems.

We note that many software development organizations use nontechnical personnel when testing installation materials. The purpose is to ensure that no unwarranted assumptions have been made by software developers who were intimately familiar with the software.

7.11 CASE STUDY OF DELIVERY IN AGILE DEVELOPMENT

Delivery is, perhaps, the easiest part of the agile process, because the agile development team is so familiar with the target application environment in which the software will be executed after installation. (I used the word "perhaps" because there is little actual data on the delivery costs and effort for projects that use an agile development process.)

It appears likely that most required components will already be in place in the target application environment, reducing installation effort. Think of an agile development process as showing the customer a succession of prototypes that execute in the appropriate environment and that have been carefully vetted by the "customer" and members of the agile development team who do most of the testing. The prototypes in this type of development have been refined until they meet the customer's requirements.

7.12 DELIVERY, INSTALLATION, AND DOCUMENTATION OF THE MAJOR SOFTWARE PROJECT

The first step for our major software project at this point is to deliver the software. As always, we must use the software requirements to determine what delivery format will be used.

How do we determine the proper delivery format? The answer is simple: read the system requirements. The requirements traceability matrix has several entries that are relevant to delivery. We include only the relevant items (Table 7.2).

It appears that the requirements were slightly vague on the matter of delivery. We believe, however, that the system is to be delivered on a single compact disk and that no compression utility was to have been used. Since there is no mention of a compiler, it is the responsibility of the developer to determine the environment and compilers that the customer expects to use. Not consulting the customer on this point can only lead to disaster.

As we did at each of the other milestones, we should also do a status check from a project management perspective. Were we ahead of schedule (unlikely), behind schedule (likely), or approximately on target? Were there any unpleasant surprises, any portions of the system that were more difficult than we expected? Did any portion of the system require extra attention, perhaps additional resources? Had technology or market pressures rendered any portion of the system obsolete?

In fact, we should do more. Delivery of a project is an excellent opportunity for a serious retrospective analysis. The lessons learned in this project should be shared with other

TABLE 7.2 A Portion of the Requirements Traceability Matrix for Our Major Continuing Software Project

Requirement	Design	Code	Test
1. Intel-based			
2. Windows 8.1			
3. Windows 8.1 UI			
4. Consistent with Excel 14.0			
5. System one size only			
6. One MB system			
7. 1 MB disk space			
8. One compact disk			
9. Includes installation			
10. No decompression utility			

project managers, as well as higher-level managers. If there is any hope of improving the way the organization does process development, data about the project's cost, schedule, and quality should be entered into a database for further analysis.

As was the case with Section 6.17 (Chapter 6) and several other sections, this section will close with the following two changes to the requirements for the major project. You will be asked to modify the delivery and installation processes in order to reflect the new requirements in the exercises. The first change is that the customer now wants a web-based user interface instead of one that is PC based. What changes need to be made to the delivery and installation processes? The second change is that the customer now wants to use a cloud for data storage instead of one that is PC based. What changes need to be made to the delivery and installation process?

SUMMARY

Software is useless unless it is delivered to the customer and can be installed on his or her computer. Delivery is different, depending on whether the software is to be delivered to a single customer, to multiple customers using multiple media, or over a distribution medium such as the Internet.

Regardless of the medium used for delivery, the following should be addressed in a predelivery review:

- Failure of the physical medium on which the software resides
- Incorrect placement of the software into directories, resulting in incorrect directory names
- Incorrect use of names of subdirectories, making subsystems inaccessible
- Incorrect access permissions, making files inaccessible
- Incorrect formatting of the software, such as using incorrect conversion tools for compressing or uncompressing files
- Incorrect documentation of the delivery instructions

Once the software has been delivered, it must be installed. Installation usually follows an installation script that is included with the software.

Software documentation can be in any one of several forms:

- Internal documentation (embedded in the source code files)
- External documentation in the form of requirements and design documents
- Additional documentation explaining why design decisions were made, as in a design rationale
- Manuals, such as user or installation manuals
- Online help that is part of the software itself

Regardless of the form of software documentation used, the documentation should be tested for accuracy and usability by persons not intimately familiar with the internal design of the software. Ideally, the persons testing the software's documentation will be chosen from a wide range of potential users, including both experienced and novice users.

KEYWORDS AND PHRASES

Delivery, installation, installation script, installation log, download, compressed software, documentation, internal documentation, external documentation, design rationale, hosting, rehosting

FURTHER READING

There are a few books on software documentation; the ones by Horton (1994) and Barker and Dragga (1997) are among the most accessible.

Two of the most interesting books on documentation are classics but still relevant. The rationale for the Ada programming language can be found in the book by Ichbiah et al. (1986). An annotated language reference manual for C++ can be found in the book by Ellis and Stroustrup (1990).

The recent book by Pawel Czarnu (2015) includes some interesting ideas about the installation of applications into the cloud.

EXERCISES

1. Select some software product that you have used recently. Describe how the software was delivered.

2. Install a software product on a computer system. Analyze the installation process, writing down each choice you made in response to prompts given during the installation. If it were your software, what changes would you make to the installation process?

3. Talk to a system administrator for the system or network that you use most often. Examine the log that he or she uses whenever new software is installed. What major decisions were made during the last software installation? What about the last software upgrade?

4. Programs written in C (and C++) often have performance problems when there are many calls to the memory allocator malloc() to allocate memory to arrays or structs in C and with malloc() underlying object creation and destruction in C++. Performance can be improved by fewer calls with larger amounts of memory allocated to malloc(), and having the programmer be responsible for allocation of memory by accessing this large memory directly as needed. Discuss this approach from the perspective of software maintenance. You might consider how some of your programs perform by compiling them with the –p option to turn on the compiler's profiler and executing them, then examining the profiler's output to determine where the program spent its execution time.

EXERCISES FOR THE MAJOR SOFTWARE PROJECT

1. Develop delivery guidelines for the major software project that we have been considering throughout this book.

2. Develop installation guidelines for the major software project that we have been considering throughout this book.

3. Examine a commercially available software package with which you are familiar. Which types of documentation were made available with the software? Evaluate the readability of the documentation.

4. Examine the source code for the large software project that we have been considering throughout this book. The only documentation provided is internal documentation, since it is restricted to the source code. Evaluate the readability of the documentation.

5. Write external documentation for the large software project that we have been considering throughout this book.

6. Write both user and operations manuals for the large software project that we have been considering throughout this book.

7. Write a design rationale for the large software project that we have been considering throughout this book.

8. The first change that the customer now wants is a web-based user interface instead of one that is PC based. What changes need to be made to the testing and integration processes?

9. The second change that the customer now wants is to use a cloud for data storage instead of one that is PC based. What changes need to be made to the testing and integration processes?

CHAPTER 8

Maintenance and Software Evolution

IN THIS CHAPTER, we will consider the part of the software engineering process that takes place after the software is delivered. You may be surprised to learn that for many long-lived systems, maintenance accounts for more than 75 percent of the total cost of the software during its lifetime. For these long-lived products, this tip-of-the-iceberg situation occurs during times of considerable change in supporting technology and changing user expectations, and causes considerable changes that are called *software evolution* if they are gradual and could be called *disruptive software* otherwise.

8.1 INTRODUCTION

At first glance, it seems strange to apply the term *maintenance* to software. After all, software does not wear out the way that a computer's on/off switch or other hardware might. The medium on which software is stored may change over time, with magnetic particles worn off the surface of a disk or tape, or even having a compact disk bent. The first situation can occur both on an individual computer and in the cloud. Careful system administrators will keep backup copies of all essential software (and user files), ideally in physically separated locations, in order to avoid problems with fire and flood.

Note also that software cannot rust nor can it fail because dirt gets in the middle of an electrical connection. Major changes to the power supplied to computers (e.g., 110 volt, 60 cycle alternating current in the United States; 220 volts in the United Kingdom) are not likely. Of course, power spikes or voltage reduction due to overtaxed electrical production facilities may have an effect, but these problems are well known. Prudent computer system administrators use a combination of surge suppressors, conditioned power sources, and uninterruptable power supplies to ensure the proper level of service. These problems are clearly hardware related and pose no particular problem for software engineers.

The reliability of hardware systems generally follows a pattern such as the one illustrated in Figure 8.1. The relatively high number of failures early in the lifetime of the hardware

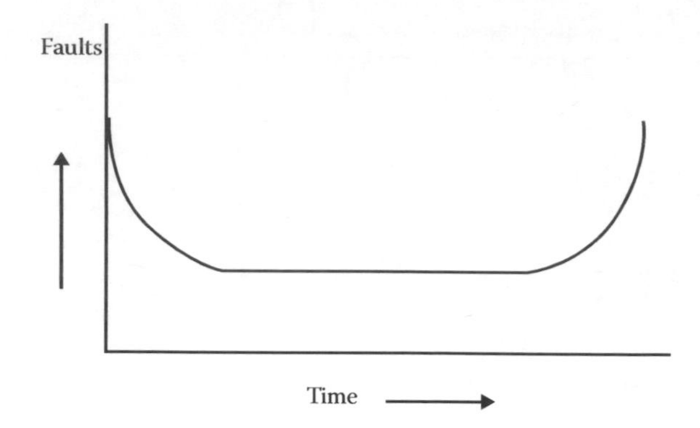

FIGURE 8.1 The typical bathtub curve for malfunctions of electrical and mechanical equipment.

is due primarily to three factors: faulty components, mistakes in the installation of these components, and improper usage by an inexperienced operator. The increased number of failures at the right-hand side of the graph represents failures that occur near the end of the hardware's useful lifetime and are due to actual component failures.

Hardware maintenance is different from hardware testing. For example, you should expect that a new car you recently purchased was assembled by experts using high-quality parts and had certain subsystems (such as those for electrical, steering, and braking operations) tested for safety. The car should also have received other basic checks when it was driven off the assembly line to a waiting cargo carrier and when it was driven from the cargo carrier to the dealer's showroom. These are all considered testing activities. If a new car needed to have its brakes or steering fixed during the warranty period, you would probably regard the car as being of poor quality and would certainly not recommend that a friend purchase a similar car.

You might also be very concerned if the plastic on your car dash split or the car rattled when going over bumps. On the other hand, if you had the same (non-safety-related) problems occur after 100,000 miles of operation, then you would probably consider fixing these problems as normal maintenance. You might choose to not fix these problems, because you might think that the remaining life of the car was so short that the maintenance expense was not worth it.

A personal note is appropriate here. With the average cost of a new car in the United States being well over $30,000, the average monthly payment for a five-year term would be in the neighborhood of $580 per month, assuming little or no down payment and a very low interest rate. My decision is based on an estimate of the value of the repairs and the utility of the car. If repairs mean that a car will be able to run safely for an additional two months, then I expect the repairs to cost no more than $1,160. Beyond that point, I have to ask if the car is worth it. The same would be true for a $7,000 repair, if I expect the car to run for at least a year. Thus, I apply an informal cost-benefit analysis to determine if the car should be maintained, if the car should be donated to a high school automotive repair program (to take a deduction on my income tax), traded to a dealer, or junked.

Proper maintenance of hardware involves keeping it clean, paying particular attention to moving parts, replacing faulty components, and generally planning to replace components that are old, obsolete, faulty, or at risk for imminent failure. A cost-benefit analysis should also be performed to determine if the maintenance effort is cost efficient, given the goals of the organization. Surprisingly, this approach is also a good way to view software maintenance.

Software maintenance could be described as the systematic process of changing software that is already in operation in order to prevent system failures and to improve performance. Software maintenance involves keeping software interfaces simple and standard, paying particular attention to troublesome modules, replacing faulty components, and generally planning to replace components that are old, obsolete, faulty, or at risk for imminent failure. The estimated remaining useful lifetime of the software must also be considered to determine if the maintenance effort is worth the added cost.

After this informal introduction to the reasons for software maintenance, we proceed to a more systematic discussion of the subject. There are several factors that require software to be maintained:

1. Hardware platforms change or become obsolete.
2. Operating systems change or become obsolete.
3. Compilers change or become obsolete.
4. Language standards change or become obsolete.
5. Communications standards change or become obsolete.
6. Graphical user interfaces change or become obsolete.
7. Related applications software packages change or become obsolete.
8. Relationships with developers of other applications or systems software have changed.
9. Software may have defects that become evident only after the customer has used the software. These defects must be corrected.
10. Customers demand new features. A recent newspaper article describes Workday, a product with about 600 business users that is working on its twenty-third release of a product first sold in 2007 (Hardy, 2014a). That is nearly a new release every four months!
11. Software needs upgrades to be competitive with new competition.
12. Existing software errors must be prevented from occurring in new releases of the software.

These factors may be classified as belonging to several distinct groups. Factors 1 through 8 may be classified as adaptive maintenance, since they are intended to help the software adapt to new technology. Factors 9 through 11 may be classified as corrective maintenance, since they attempt to make the software more correct, in the sense that it has fewer faults. The term *perfective maintenance* is also used in this case. Factor 12 may be classified as preventive maintenance, since it attempts to reduce the possibility of future software faults.

There is one common thread, however. An essential step in all types of software maintenance is understanding the software system to be maintained. Before a maintainer can change software in order to fix a problem or add a feature, he or she must understand the software as a whole and the particular modules that need to be modified. Many experiments have indicated that program understanding, which includes understanding of software requirements and designs, takes approximately half of all maintenance efforts.

Different approaches to software maintenance will be discussed in the remainder of this chapter. Keep in mind, however, that nothing in life is free and that any software maintenance activity must be budgeted for as part of the organization's software budget.

We will describe a typical approach to the software maintenance process that will consider the two actions of determining where the problem is and then fixing it. Many software problems can be traced to one or more of the following:

- Lack of adherence to software development standards. These problems include poor interfaces between modules, such as passing the wrong number or types of arguments to a function.
- Logical errors within program modules, such as loops, branches, or inconsistent program states.
- Nonatomic actions.
- Inconsistency between the requirements, design, code, and documentation for a system.

Students are often surprised by the amount of difficulty that they have understanding either commercial or government code. I have used actual source code from many software development organizations in my classes and the response is almost identical: the code is poorly documented and hard to understand. In reality, the students are faced with the problem of examining all or portions of systems that are too large to be understood in their entirety.

8.2 CORRECTIVE SOFTWARE MAINTENANCE

The first step in *corrective software maintenance* is to determine what needs to be maintained. In corrective software maintenance, all activity begins with the identification of a problem with the existing software. Determining the particular module that is the cause of a problem can be difficult. The basic approach is based on the technique of trial and error, guided by the maintainer's knowledge of the system, and his or her basic knowledge of computer science and software engineering.

Once problems have been determined, a decision can be made as to which, how, or when the problem will be fixed. This usually requires several related actions:

- Observation that a problem exists
- Documentation of the problem
- Determination of the importance of the problem

- Prioritizing the problems in order of importance of repair
- Determining which problems will not be repaired, due to insufficient resources
- Solution of the problem
- Testing the system to see if the fix to this problem can cause any faults in other portions of the software
- Documenting the solution as part of the source code
- Documenting the solution in other forms of the documentation if changes have been made to the original system design
- Updating a database of information about software errors

This may seem like a lot of work—it is. Software maintenance must, by its nature, require large amounts of paperwork to update documentation. After all, the developers of software systems in industry or government rarely stay on the same project forever. With high turnover of personnel, written documentation that traces any changes made must be left for future maintainers of a system.

Observation of a software problem can occur at several levels. In the most ideal situation, the observer is a person who is completely familiar with the system's requirements. In this case, the documentation of the problem is immediate, and the determination of the source of the problem can generally be easily determined.

A more common situation is for the persons observing the problem to be completely unfamiliar with the internal structure of the software. Indeed, the person observing the problem may even be unfamiliar with the software's expected operation. This is typical of the type of problem that is first observed by novice users who then call a toll-free number for technical support of their new word processing or similar application. The technically knowledgeable person working in technical support is responsible for guiding the caller through the steps that led up to this presumed problem, to look for problems in the setup and configuration of the software, and, in short, to determine if the problem is due to an error by the user or an actual fault in the software. There are essentially only two possibilities: the user has misunderstood the installation and operating instructions, or the software is in error.

Once the technical support person has determined that the user has misunderstood the documentation of either the system's installation or operation, he or she has two remaining tasks. The user must be led through the correct steps to fix his or her problem. In addition, the technical support person must document the problem so that the documentation and the user interface can be improved. This is often done in a future release of the product. The technical support person should fully describe the documentation problems and turn them over to a person responsible for new releases of the system so that the user community will consider the new releases to be improvements. We note in passing that most computer documentation could be improved by the employment of a technical writer.

In the case of a software fault, the user is not expected to provide detailed documentation of the details of the problem. The typical user does not have the technical ability and, in any event, it is not his or her job. The technical support person is responsible for taking the user's information and organizing it into a format that can be used in the maintenance process.

The goal of the person documenting a software problem is to be able to fill out a form as detailed as the one shown in Figure 8.2. This form is often called a software maintenance request form or software maintenance report form. Note that the form has places for complete descriptions of the problem and for an assessment of the relative importance of the problem. A sensible software manager generally will prefer to set priorities for the deployment of maintenance personnel based on an assessment of the relative importance of the problems encountered. A realistic estimate of the difficulty of fixing an incorrect module is also useful in assigning priorities. Some minor problems might never be fixed.

After the problem has been fixed, another form should be filled out. This second form indicates the response to the software maintenance form and, as such, is often called a software maintenance response form. Figure 8.3 illustrates a typical software maintenance

Maintenance request number________________

System________ **Version** ____________**Date**______________

Related maintenance request numbers__________________________

Environment
Computer hardware:

Operating system______________________**Version**____________

Related system software: GUI ______________________________

Related application software ____________________________

Description of problem

Severity of problem
Critical_____**Urgent**_____**Important**_____**Routine**_____

Classification: Hardware_____**Software**_____ **Documentation**_______
Person making report

Person verifying report

Date problem verified

Organization responsible for fixing problem

FIGURE 8.2 A typical software maintenance request form.

Maintenance repair number__________________

In response to maintenance request number__________________

Related maintenance request numbers______________________________

System__________ Version _____________ Date_____________

Environment

Computer hardware:

Operating system ________________________________ Version_________

Related system software: GUI _______________________________________

Related application software _______________________________________

Severity of problem
Critical Urgent Important Routine_____

Classification: Hardware Software_____ Documentation_______
Person making report

Person verifying report________________________ Date__________

Person fixing problem_____________________________

Resolution of problem

Person verifying solution__________________________

Modules affected

Time needed to fix problem

FIGURE 8.3 A typical software maintenance response form.

response form. Many organizations combine these request and response forms into a single form as shown in Figure 8.4. Note that the response form will always indicate a list of affected modules and the time needed to fix the problem.

Note the importance of the extra documentation on these forms. The ability to attach hard copies of screen dumps is often very helpful to maintainers. If no screen dump facility is available in the software or operating system, a photograph of the screen can suffice. Obviously, a digital image is best.

Maintenance request number_______________
Related maintenance request numbers_________________________

System_________ Version ____________ Date______________

Environment

Computer hardware:
Operating system __________________________Version________

Related system software: GUI ______________________________

Related application software ______________________________

Description of problem

Severity of problem
Critical_____Urgent_____Important_____Routine_____
Classification: Hardware_____Software_____ Documentation_______
Person making report____________________

Person verifying report____________________
Person fixing problem____________________
Person verifying solution__________________
Resolution of problem

Modules affected

Time to fix problem

FIGURE 8.4 A typical combined software maintenance report and response form.

You may be concerned about the proliferation of paperwork required by these forms. Why have all this extra information? Who needs it? An actual example indicates the need for these activities.

Suppose that you were examining a software system that had to be of highest quality because it was to be used without modification as the core of many other software products in a safety-critical application. Primarily because of its design and adherence to essential interface standards, this software had a much lower defect ratio (faults per 1,000 lines of code [KLOC]) than did most other software in the organization. However, high quality was not enough. Since this reusable core had to be reused in so many applications, it was important to check to determine where software errors tended to occur.

An examination of a database of maintenance requests and responses indicated that 44 percent of identifiable software faults were traced to a single set of modules used to interface to an old locally developed command language. It was decided to scrap this command language and use a more modern one with a grammar that could be parsed easily using standard grammatical parsing tools such as flex and bison (the successors to the classical tools lex and yacc). The identification of the command language parsing tools as major sources of software maintenance costs certainly influenced this decision.

Of course, there is no need to have the forms written on paper. The forms themselves can be stored electronically, with automatic entry of data into a database. Ideally, a software package that supports maintenance activities with automatic forms-generation and storage of information into a database would be available to maintainers. In many software environments, the lack of forms-generation capability is the most critical problem.

The simplest solution is to enter the maintenance data in text format directly into a spreadsheet or database. We prefer to use the lowest level of technology in this book in order to be able to work in nearly all situations.

8.3 ADAPTIVE SOFTWARE MAINTENANCE

Adaptive software maintenance is the process of changing software to meet new trends in technology or the marketplace. Expected changes in hardware, systems, interoperable application packages, and new features provided by the competition are all important in determining the need for adaptive maintenance of software.

Consider the case of apps written for smartphones or tablets. An app designer focusing on devices running on a number of phones, each of which is running on multiple versions of the Android operating system, may have to change the apps to accommodate the most current versions or may have to settle for being consistent with the state of a previous operating system version. The layout of apps might have to change to have the proper look and feel for various screen dimensions.

A designer for the Apple iOS devices may have fewer changes to consider, since there are fewer versions of iOS. However, the change of programming languages to Swift from Objective C may cause many problems if any new features are added to an existing app.

Here is another example that illustrates the effect of changes in application development environments on software development, Ruby on Rails, especially with the need to have software be installed campuswide by professionals as well as by individual students.

Notice the reference to Git, an open source configuration management system, as well as to Ruby-specific tools. The key concerns are, clearly, the need for the latest version of Rails and a seamless integration package. The information was posted to the SIGCSE (Computer Science Education Special Interest Group) mailing list by Professor Steven Bogaerts of DePauw University on May 2, 2014.

> Hi everyone – I'll be teaching a course that uses Ruby on Rails in the fall, and I need to decide how to request our IT staff to install everything we need on campus machines. Students will also want to install this on personal machines, so ease of installation is important.
>
> In the past I've used http://railsinstaller.org, which has been very nice. It includes not just Ruby and Rails but also Bundler, Git, SQLite, etc, all integrated properly from the start.
>
> The problem is it uses Rails 3.2. I'd really like to be using Rails 4.x if possible. I did find http://railsftw.bryanbibat.net/, which looks similar to railsinstaller, only with 4.0.3. But it seems not to include quite as much as railsinstaller (I'm not sure of the consequences of this right now), and it's also Windows-only. That's not an insurmountable problem, but it is a slight complication.
>
> So I'm wondering how others have dealt with this problem:
>
> - Simplifying student installation procedures
> - Enabling use of Rails 4.x and various tools (as in railsinstaller) with no/minimal configuration required
>
> Has anyone had success with railsftw? Or a piecemeal manual configuration with the same tools that railsinstaller provides? I appreciate very much any suggestions you could provide, thank you!

As another example, consider the case of Microsoft Word, which has gone through many versions since its inception. Some of these changes involve the user interface and usability. Other changes provided additional features and common file formats. Frequently, documents opened in one version of later versions of Word either cannot be read or have page formats changed by an earlier version of the same software. (This problem is not unique to Microsoft. The same statement applies to Corel WordPerfect, among others.)

For some users, the advantage of a common file format in which documents can be read on both PCs and Macs is outweighed by having larger file sizes used for documents. For others, the additional features provided by macros may be less important than the additional security requirements to treat macro viruses, which may migrate from PCs to Macs and vice versa.

As with corrective maintenance, considerable testing of software is necessary, especially regression testing.

It is appropriate at this point to indicate some problems that cannot reasonably be expected to be uncovered during even the most exhaustive software testing.

Unfortunately, standard testing procedures generally cannot detect erroneous events such as race conditions, which will occur very infrequently. The term *race condition* refers to a situation where the result of a computation (or even the continuation of system operation) depends upon the order in which distinct concurrently executing processes get access to the CPU.

An example of a race condition is shown in Figure 8.5. The value of the variable X at the end of the asynchronous computation depends on the order in which the concurrently running processes execute.

There is another equally important problem that can cause problems for maintainers of concurrent systems, especially real-time systems. The maintainer of such systems must pay special attention to functions that make operating system calls but are not themselves atomic. (An "atomic instruction" is one that cannot be interrupted by an external event and, therefore, such instructions will always complete once their execution has started.)

Scheduling of processes on one or more CPUs cannot be predicted and thus race conditions and nonatomic system calls can be disastrous.

These two problems can occur in software systems that must execute concurrently with other processes. Both nonatomic function calls and race conditions are extremely dangerous for real-time or safety-critical systems, because the system cannot be guaranteed to operate successfully and because system run-time performance cannot be predicted. Even worse, testing cannot provide guarantees of the necessary system quality.

In the case of nonatomic functions used to synchronize access to system resources, even the most exhaustive testing of a software system with concurrently executing processes will be inadequate, because the likelihood of a system failure during testing is small. Figure 8.6 illustrates the issue.

The problem will occur because the synchronization depends on the values of two distinct variables, X and Y, which are tested in two distinct steps. However, if the processes get access to the CPU in such a way that the third instruction of process 2 executes before the third instruction of process 1, the synchronization condition may not hold, and access to the sharable resource may be granted (or denied) incorrectly.

Another problem is particularly vexing for software maintainers. Many software engineers, especially those whose primary experience is with computers with large amounts of physical memory, do not think of their systems as having any limitations. However, memory can be exhausted, especially if there are commercial off-the-shelf (COTS) products whose memory demands are unknown.

Process 1	Process 2
X = 1;	X = 1;
Y = 7;	Y = 7;
X = X + Y;	X = X – Y;

FIGURE 8.5 A race condition with different results of a computation, depending on the order in which instructions in a process are executed by the CPU.

Process 1	Process 2
X = 1;	X = 1;
Y = 7;	y = 7;
IF (X == 1) AND (Y ==7)	X = X - Y
ACCESS shared resource;	Y = X;
X = X + Y;	
Y = 1;	

Shared Resource

FIGURE 8.6 Access to a shared resource controlled by a nonatomic function.

An equally difficult problem for maintainers is the use of excessive numbers of other restricted system resources. For example, two different software packages may each write data to the same socket (which may be implemented as a "winsoock" or a "socket descriptor," depending on the operating system) for network communications, and this may cause unexpected conflicts.

In another more complex example, the number of files that can be open at any one time for a user's process can range from eight in programs using older versions of C to twenty for ANSI standard C to sixty-four on some UNIX systems (the "soft limit," which can be increased by a system administrator) to the maximum allowable (the "hard limit," often 100 on UNIX and Linux systems).

You should be aware that the situation is especially complex if the software to be maintained must interact with many COTS products whose internal structure is not known.

It is clear that a software maintainer may need many skills beyond simple programming experience.

8.4 HOW TO READ REQUIREMENTS, DESIGNS, AND SOURCE CODE

Once a maintenance problem has been given to a maintainer, the source code must be modified. However, many software systems consist of millions of lines of source code. Each line of source code has the potential to be the place where a problem occurs. Of course, not all problems in the user interface, for example, need to be checked. But where do we begin?

A software maintainer will look at all available information: requirements, design, manuals and other documentation and, of course, the source code itself. The goal is to understand the likely effects of any changes in the software before the changes are made. Everyone who has written software is familiar with having a change in one portion of the software affect performance in some other portion. Thus, a systematic approach is necessary.

Keep in mind the old adage: If the source code and any of the requirements, design, or documentation do not agree, they are all wrong. In this case, the source code is the only certain guide, because it reflects the system as it currently works, not as it was intended to work. The available source code is always read as part of the maintenance process.

Let us suppose at this point that we are only considering source code. What do we look for first?

The answer is composed of several parts. We would start by looking at the maintenance reports for related situations in the software to seek out any suggestions on where to look for problems. We would also make use of any available integrated development environment (IDE) or computer-aided software engineering (CASE) tool to analyze the source code and show the program's structure. Be aware that, even if the original IDE or CASE tool in which the software was created is no longer available, perhaps due to obsolescence or to expiration of a license, it still may be possible to obtain information on the program's structure using either newer IDEs or CASE tools available to the maintenance project, or even to freeware or shareware tools to analyze program structure.

After we had obtained enough information to restrict our attention to a limited number of modules, each of them would be inspected at essentially the same level of detail as it would be in the type of code review discussed in Chapter 5.

8.5 A MANAGER'S PERSPECTIVE ON SOFTWARE MAINTENANCE

Although it may be surprising to students, software maintenance is one of the most costly items in the software life cycle. One thing that managers like is reducing costs. Senior managers recognize that software maintenance does not usually increase market share and cannot, therefore, aid in improving the organization's bottom line. You can often help your manager by suggesting some ways that maintenance costs can be reduced.

A database of maintenance requests and actions can be very helpful. For example, the system described in Section 8.2 had a maintenance database in which a simple query indicated that one particular software module was causing 44 percent of the traceable software failures. The problems were traced to new dialects of a command language, which were created with every new version of the software. Freezing the command language caused little functionality loss in the opinion of the users, but it reduced maintenance costs by a considerable amount. Managers like this type of information. It makes their job easier, especially in times of increased budget pressure, where reducing costs with little or no decrease in functionality or quality is a paramount goal.

This kind of analysis of maintenance effort is extremely important to managers at all levels. For most commercial organizations, little revenue is brought in during maintenance unless new releases of the software are planned. Thus, maintenance is viewed as a cost to be minimized whenever possible.

Of course, it is always essential to follow your organization's standards-and-practices manual (if such a manual exists) for documentation procedures.

8.6 MAINTENANCE COSTS, SOFTWARE EVOLUTION, AND THE DECISION TO BUY VERSUS BUILD VERSUS REUSE VERSUS REENGINEER

As stated in Chapter 1, many software systems have very long useful lives. Think of the software used in the United States for providing checks for the Social Security Administration. The software originally provided for the printing of checks that were mailed to elderly and disabled people, as well as to the surviving parents or guardians of children under the age of 18, one or both of whose parents had died. Now the dispersal of funds is almost entirely Internet based, with funds transmitted to recipient's bank accounts. The underlying software, used to analyze large databases to make this disbursement possible, runs on large mainframe computers, with very old computer languages such as COBOL and Codasyl-enabled databases that allow efficient, batch mode processing that is necessary for delivery of funds without delays.

It is impossible to buy such software on the open market, so changes (which often occur due to changes in U.S. laws and regulations) are often done by a combination of building, reusing, and reengineering. Reengineering can be very difficult to understand for students and beginning programmers (Leach et al., 2008).

Another example is Microsoft Word, which was a stand-alone word processor running on 16-bit microcomputers running an operating system with a text-based command-line interface in its first inception. Now, Microsoft Word is part of Microsoft Office, an integrated suite of tools running on an operating system that supports multiple windows, with a mouse as a device for pointing and selecting. Recently, Microsoft Office has been ported to tablet devices, with their own user interfaces depending on the use of a single touch of a finger for selection, sweep actions for selection of text, and pushing two finders together or pulling them apart to indicate changes in size, among other things. Rentals are available through the Office 360 program. This is more a case of reengineer, build, and reuse.

Clearly, such systems *evolve*, based on factors such as changing technology, competitive pressures, and the desire to reengineer to make software development more efficient.

I think the best way to understand the issues is to consider a real-life example from the aerospace industry. (We have previously discussed this example in a simpler form.) Many satellites continued producing useful research data about the earth's magnetic and atmospheric condition for many years beyond their expected useful life. The Interplanetary Monitoring Platform, hereinafter referred to as IMP-8, was launched on October 26, 1973, and data transmission was stopped on October 7, 2006, a term of 33 years. (The web portal for this project is http://spdf.gsfc.nasa.gov/imp8/project.html.)

Because of the long lead times for planning, building, and launching many satellites, many of the hardware and software computer systems used to control the satellites and process the data had long been obsolete. Not surprisingly, hardware and software companies were deciding to not support their systems any more.

In order to focus on its core business, Unisys decided to raise its annual charge for hardware and software maintenance of just its dedicated mainframe to the level of the entire budget for maintenance, operation, ground-based control, data storage, and personnel costs. The decision by Unisys was announced in 1997, nearly nine years before the IMP-8 project finally terminated, so there was a lot of useful data still to be collected.

The decision was made to use entirely new computer hardware for ground control of the satellite and to simultaneously reengineer the software. The new computer hardware chosen was an HP workstation running HP-UX, a UNIX-variant, which was, essentially, a standard PC far more powerful than a 1970-vintage mainframe.

The software was another matter, consisting of 33 files, containing a total of 108,161 lines of code. Most files contained both FORTRAN code and Unisys assembly language. The FORTRAN was ported to the most common dialect, FORTRAN 77, and the assembly code was rewritten in FORTRAN. csplit, a standard UNIX utility, was used to separate the 33 files into 584 separate files, each containing a single FORTRAN or Unisys assembly module. Freeware programs such as FORTRAN Source Code Analyzer (fsca) previously developed at Howard University, floppy, and flow, were used to create program control graphs. These metrics were used to identify 2 percent of the modules as having a complex internal structure and, therefore, requiring additional testing effort. By examining the files pairwise using the standard UNIX utility diff, it was determined that naming conventions had not been used but that the same source code was copied into other locations. With proper naming conventions, it was found that 19.4 percent of the files were duplicates, reducing potential maintenance effort by this amount. It was clearly much cheaper to reengineer than to buy or build this software.

At times, software maintenance also provides an opportunity for code improvement. Here is a simple example. As stated in Chapter 6, the technique of multiple inheritance has fallen into disfavor because of the possibility that a portion of a class may be inherited along different inheritance paths. In some cases, if a portion of the code is newly written, it may be better for the test team to ask the coders to recode any software with multiple inheritance to avoid this problem. If the code has been used without incident for many years, the best choice may be to leave the code alone.

We briefly turn our attention to software reuse. The most extreme case of software reuse is when an entire software system, such as a COTS product, is reused without change. The question we address is when is the reuse not cost-effective, due to a large number of evolutionary changes in COTS software that cause too many changes to systems, due to the hidden costs of configuration and testing of interfaces to systems for which the internal structure is not known, as is the case with most COTS products. A discussion generated by an important paper of Jeffrey Voas (1999) and followed up by research of Greg Saunders (2003) and myself (Leach, 2005) leads to the following suggestions:

- Collect data on how your project spent its money.
- Map this data over time to try to get a graphical image of how costs increased when new releases of COTS products occurred and had to be incorporated into a system.
- Keep data on how much the process of upfront systems analysis and COTS evaluation costs and how long it takes. Consider the use of some less expensive, junior-level personnel as part of the COTS evaluation team, at least for the second or third iteration of the COTS-based system.
- Consider ways to reduce the number of senior-level people kept on long-term projects, by phasing in some of the more junior personnel. You will probably have to do it anyway, because talented people want professional growth!
- Make sure there are at least a few domain experts and senior software engineers always available, even if you have to pay more. Consultants may be of assistance in filling some portion, but not all, of this role.
- Learn from your mistakes!

8.7 MAINTENANCE IN AGILE DEVELOPMENT AND THE TOTAL LIFE CYCLE COSTS

This is an easy section to write for a very simple reason: Agile development emphasizes rapid development of well-tested software in cooperation with a "customer." It is highly unlikely that any significant portion of the agile development team will be around the organization, much less be willing to work on what is almost always a much less productive activity such as software maintenance, which requires an enormous effort in program understanding before a system can be maintained.

Unfortunately, due to the volatility of agile software development teams, there is essentially no data that describes the amount of maintenance time and effort needed for the systems the agile development team has created, at least in the level of detail necessary to guide practitioners and managers.

8.8 MAINTENANCE OF THE MAJOR SOFTWARE PROJECT

One view of software maintenance would suggest that the system was already under maintenance when the requirements were changed to use the Internet. We will ignore the previous requirements change and instead consider some possible changes after delivery.

The first type of maintenance to consider is corrective. Does the software behave as it should? After we have considered correctness, we must worry about the user interface and the potential for enhancements. Of course, performance issues can always arise if the response time needs to be improved.

What about the human–computer interface? The original system has a user interface developed for a Windows-based PC. What changes are necessary for the Microsoft Windows, version 8.1? The maintenance effort might be focused on updating the software to work on later operating systems, with new language versions and application toolkits.

A system intended for use in a single computer also could be ported to an Apple Mac computer and use the Apple user interface. It could also use the Linux-based operating system underlying the Apple user interface. (This Linux-based system is called Darwin.) What additional maintenance efforts must be required to port from a Windows-based computer to an Apple Mac?

Could the system be reengineered to become web-based, with the metrics computations performed on servers or, perhaps, a set of individual servers each hosting many source code files? Could the user interface be based primarily on tablets, such as, say, a banking app with its connection to a remote database? Could it be created using only open source software?

We will leave these issues to the exercises.

SUMMARY

Much of the effort for software projects occurs after the software is delivered and installed. This is called the maintenance phase.

There are three major types of software maintenance:

1. Corrective maintenance
2. Adaptive maintenance
3. Preventive maintenance

The Y2K problem is an example of preventive maintenance until the date changed and the problem occurred. It then became a corrective maintenance problem.

There is an adage that describes the skepticism a maintainer must have: If the source code and any of the requirements, design, or documentation do not agree, they are all wrong. One of the major issues to consider during software maintenance and evolution is the decision whether to buy versus build versus reuse versus reengineer a complex, existing software system.

KEYWORDS AND PHRASES

Software maintenance, software evolution, reengineering, corrective maintenance, adaptive maintenance, preventive maintenance, COTS

FURTHER READING

Perhaps the best source of information for a beginning software engineer is a company's software development standards and practices manual. Forms such as the ones given in this text are usually available from these documents.

There are some general references on software maintenance available. The book *Software Change Impact Analysis*, edited by Arnold and Bohner (1997), provides an excellent overview of the state of the art in assessing the impact of maintenance decisions. The book by Pigoski (1997) is also useful.

See the articles by Voas (1999), Saunders (2003), and Leach (2005) for perspectives on COTS products during software evolution.

There are several important articles on the important subject of software maintenance, and you should check some of the conferences that are more practitioner oriented. There is even a journal related to maintenance issues: the *Journal of Software Maintenance*, which was renamed to the *Journal of Software Maintenance and Evolution: Research and Practice*, and now is known as the *Journal of Software: Evolution and Process.*

EXERCISES

1. List at least ten software packages that will have to be changed to accommodate an increase in the number of characters used for Social Security numbers.

2. List at least ten software packages that will have to be changed to accommodate changes in the number of digits used for phone numbers.

3. List at least ten software packages that will have to be changed to accommodate changes in the UNIX date and time computation. Be sure to discuss the use of the make utility.

4. Consider a popular application for personal computers. Describe major changes between successive releases. List them in the order of estimated likely costs.

EXERCISES FOR THE MAJOR SOFTWARE PROJECT

1. Examine the dialogue we presented in Chapter 3, when we developed the requirements for the large software project that we have discussed throughout this book. Based on that dialogue, what changes do you expect to be made to the software during its useful lifetime?

2. Compare the changes made in the software in the major project with the ones you might have predicted in the previous question.

3. Change the major software project to allow the separation of the computational portion from the user interface. Then incorporate a client–server relationship, with the computation done on a server and the user interface on a client machine.

4. Change the user interface for the current system in the major software project to run on Windows 8.1.

5. Change the user interface for the current system in the major software project to use the Apple user interface.

6. Change the user interface for the current system in the major software project to use the Linux-based operating system underlying the Apple user interface. (This Linux-based system is called Darwin.)

7. Change the user interface for the current system in the major software project to become web-based, with the metrics computations performed on servers or, perhaps, a set of individual servers each hosting many source code files.

8. Change the user interface for the current system in the major software project to run its user interface based primarily on tablets, such as, say, a banking app with its connection to a remote database.

9. The first change is that the customer now wants a web-based user interface instead of one that is PC based. What changes need to be made to the maintenance processes? Is a new maintenance plan needed?

10. The second change is that the customer now wants to use a cloud for data storage instead of one that is PC based. What changes need to be made to the maintenance process? Is a new maintenance plan needed?

CHAPTER 9

Research Issues in Software Engineering

THE NATURE OF SOFTWARE ENGINEERING has changed considerably in recent years. The concept of a program is being replaced by the concept of a system, which is comprised of many different communicating and cooperating components. Different components may be fully or partially distributed, executing on different computers. This approach to software development requires planning, resource allocation, coordinated scheduling, requirements engineering, risk assessment, modular designs, testing, quality assessment, flexible and repeatable development processes, and proper reviews. In short, this approach has all the aspects of an engineered process.

Market pressures force organizations to produce high-quality, well-documented systems with many new features with technically correct performance in a timely manner and with efficient use of resources. Competitive pressures often make it difficult for many smaller companies, especially those without research and development groups, to spend much time improving their processes, particularly since they are often focused on improving their use of current and future technology. Not keeping up with current technology can cause a company to fail. (So can marketing and distribution, but we will not discuss them here.)

However, the cost of not improving the software engineering process can be equally disastrous, since those companies that produce high-quality software efficiently will often have an enormous technical advantage based on their efficiency.

This brief chapter describes how the software process can be improved. Section 9.1 lists some research issues that are considered to be important now and are likely to be in the foreseeable future. Section 9.2 provides guidance on how to interpret the software engineering research literature.

The purpose of this chapter is to enable you to read and interpret the vast, ever-increasing literature in software engineering. Even if you never intend to be a researcher in this area, the information presented here in the references will be important when you have to assess what happened during a project postmortem.

9.1 SOME IMPORTANT RESEARCH PROBLEMS IN SOFTWARE ENGINEERING

In this section, I present my views on some of the most important research areas in software engineering, together with a few important research questions in each area.

9.1.1 The Fundamental Question

Buy versus build versus reuse versus reengineer—This decision is arguably the most important choice that must be made in a software development project. Devise a series of case studies, comparative studies, and controlled experiments to determine a systematic process for making this decision in some nontrivial application domain. Note that many of the research questions posed later in this section will address portions of this question. (We will discuss the importance of case studies, comparative studies, and controlled experiments in software engineering research in Section 9.2.)

9.1.2 Requirements

Requirements representation—Devise and implement an experiment to determine which is the most effective method for writing requirements in a given application domain. Such a determination would consider plain text, graphical descriptions, and formal methods.

Requirements reviews and inspections—What is the most effective way to hold a requirements review or inspection?

Requirements aggregation—How should a set of requirements be aggregated in order to encourage software reuse?

Procedurally oriented requirements—What is the most efficient method of determining requirements for a procedurally oriented system in a particular application domain?

Object-oriented requirements—What is the most efficient method of determining requirements for an object-oriented system in a particular application domain?

Hybrid requirements—What is the most efficient method of determining requirements for a system that will have both procedurally and object-oriented components in a particular application domain?

Requirements generation tools—Create a tool (or set of tools) to generate detailed requirements from a particular set of high-level requirements.

Validation of requirements generation—Develop a method to determine the validity of the requirements generation tool (or set of tools) as an improvement to the requirements process.

Comparison of requirements generation and high-level languages—Devise and carry out an experiment to determine if the requirements generation tool (or tools) described earlier is more efficient than a fourth- or fifth-generation language in improving the efficiency of the software development process.

Determine the most efficient way to develop requirements in a particular application domain where the entire system will be written in a functional programming language such as LISP.

Determine the most efficient way to develop requirements in a particular application domain where the entire system will be written in a logic programming languages such as Prolog.

Determine the most efficient way to develop requirements in a particular application domain where the system will be written in multiple programming languages, including procedural, object-oriented, functional, logic, or higher-level languages.

9.1.3 Design

Design representations for procedurally oriented systems—What is the most effective method for representing designs for a procedurally designed system?

Design representations for object-oriented systems—What is the most effective method for representing designs for a object-designed system?

Design representations for hybrid systems that have both procedurally and object-oriented components—What is the most effective method for representing designs for a hybrid of procedurally and object-oriented component subsystems?

Graphical design representations—Are graphical representations more effective than text-based ones?

Design reviews and inspections—What is the most effective way to hold a design review or inspection?

9.1.4 Coding

Coding standards—Devise and implement an experiment to determine which, if any, coding standards improve the efficiency of writing source code.

9.1.5 Testing

Efficiency of testing methods—Develop a methodology to determine which testing method is most likely to be more effective in a particular application domain: white-box or black-box testing.

White-box testing—Develop a method to determine the most efficient way to perform white-box testing in a particular class of application domains.

Black-box testing—Develop a method to determine the most efficient way to perform black-box testing in a particular class of application domains.

Object-oriented testing—Develop a method to determine the most efficient way to perform object-oriented testing in a particular class of application domains.

Performance testing—Develop a method of software performance testing that provides early feedback about performance issues that may delay software deployment.

Stress testing—Develop a method of software system stress testing that provides early feedback about system limitation issues that may delay software deployment.

Fault seeding and testing—Determine a methodology for determining the effectiveness of fault seeding as a way to measure the efficiency of defect removal during testing.

Intelligent agents—Develop a method for testing intelligent, autonomous agents that may move around the nodes of a network.

Testing of functional systems—Develop a method for systematically testing software that is written in a functional programming language such as LISP.

Testing of logic systems—Develop a method for systematically testing software that is written in a logic programming language such as Prolog.

Testing of artificial neural networks—Develop a method for systematically testing software that is written in the form of an artificial neural network. (This testing is presumed to follow the normal training of an artificial neural network.)

9.1.6 Integration

Top-down integration—Devise and implement an experiment to determine the efficiency of top-down integration in a realistic application domain with proper integration tools.

Bottom-up integration—Devise and implement an experiment to determine the efficiency of bottom-up integration in a realistic application domain with proper integration tools.

Plug and play integration—Devise and implement an experiment to determine the efficiency of plug-and-play integration in a realistic application domain with proper integration tools. (Plug-and-play integration involves the insertion of components into a running system.)

Big-bang integration—Devise and implement an experiment to determine the efficiency of big-bang integration in a realistic application domain with proper integration tools.

9.1.7 Maintenance

Coding standards—Devise and implement an experiment to determine which, if any, coding standards improve the efficiency of maintaining source code.

COTS maintenance—Develop a process for maintaining interfaces to commercial off-the-shelf (COTS) products during the lifetime of the project that includes the COTS product. Make sure that the process includes a mechanism for treating the case where there are multiple releases of the COTS product.

9.1.8 Cost Estimation

Cost estimation and the Internet—Create both a process and a model for estimating the cost of software projects that make extensive use of the Internet. Devise and implement a series of experiments to validate your model and process.

Cost estimation, COTS, and reuse—Create both a process and a model for estimating the cost of software projects that make extensive use of COTS and reuse. Devise and implement a series of experiments to validate your model and process.

9.1.9 Software Reuse

Management—What is the most efficient organizational or management structure from the perspective of encouraging enterprise-wide systematic software reuse in such a way as to greatly reduce the organization's total software development and maintenance costs?

Domain engineering—What is the most effective method of organizing and classifying the software artifacts within a domain so as to support systematic software reuse?

Reuse library search—What is the most effective method for determining the software artifact (or artifacts) within a reuse library that are closest to the desired artifact?

Library organization—What is the most efficient method of reuse library organization in order to allow artifacts at many different levels of the software life cycle to be stored in such a way that we can retrieve them according to their functionality, the interface standards the artifacts meet, or the computer-aided software engineering (CASE) tool that created them?

FAST in a single application domain—The family-oriented abstraction, specification, and translation (FAST) approach involves concurrent analysis of both the reusable components within a domain and the development of applications using these reusable components. For which domains can the FAST process work efficiently? The FAST process was developed at Lucent Technologies and is best described in a book by Weiss and Lai (1998).

FAST deployment—How can the FAST approach be used efficiently within different application domains?

Generative reuse—Generative reuse is software reuse using highly configurable frameworks and application generators. Develop a method to determine the applicability of generative reuse for particular application domains and the likelihood of this approach being successful.

Component reuse—Component reuse is software reuse using existing software components. Develop a method to determine the applicability of component reuse for particular application domains and the likelihood of this approach being successful.

Generative and component reuse—Develop a methodology for predicting whether generative reuse or component reuse is more likely to be successful in a given application domain.

Reuse cost accounting—Determine an efficient method of cost estimation and accounting for systematic reuse programs that extend across multiple organizations.

Reuse and design representations—There are many design representations (flowcharts, dataflow diagrams, statecharts, UML, etc.). Which is the most effective in promoting software reuse?

Determine the most efficient way to implement a systematic practice of software reuse where all components are written in a functional programming language such as LISP.

Determine the most efficient way to implement a systematic practice of software reuse where all components are written in a logic programming language such as Prolog.

9.1.10 Fault Tolerance

Recovery blocks and fault tolerance—Determine the class of applications for which rollback of a system to a recovery block is an effective technique of improving software fault tolerance.

Recovery blocks, fault tolerance, and system correctness—Develop a method to develop recovery blocks that are error free and, thus, whose incorporation into a program improves, rather than degrades, system fault tolerance.

Exceptions and fault tolerance—Develop a formal method to prove the correctness of programs that use exceptions to control program execution after abnormal situations are detected.

Assertions and fault tolerance—Develop a formal method to prove the correctness of programs that use assertions to control program execution after abnormal situations are detected.

Fault tolerance and replication (N-version, or multiversion, programming)—Determine the class of applications for which a combination of replication of major components of a system and a "voter controller" to determine system correctness by majority rule is an effective technique of improving software fault tolerance.

Intelligent agents and fault tolerance—Devise and implement an experiment to determine the effectiveness of intelligent agents to detect, and perhaps correct, run-time faults that occur in a system.

9.1.11 Metrics

Process metrics—The general problem is to determine which metrics describe the state of a software development process with sufficient accuracy to be able to detect inefficiencies.

Descriptive process metrics—Determine a set of metrics that can be used to determine if a project is late or over budget at a specified milestone in the development.

Predictive process metrics—Determine a set of metrics that can be used to predict if a project will be late or over budget.

Product metrics—The general problem is to determine which metrics describe the quality of a software product with sufficient accuracy to be able to detect inefficiencies.

Metrics for procedurally oriented system requirements—Determine which metrics reflect the complexity of the requirements of procedurally oriented programs.

Metrics for procedurally oriented system designs—Determine which metrics reflect the complexity of the designs of procedurally oriented systems.

Metrics for procedurally oriented systems—Determine which metrics reflect the complexity of procedurally oriented systems.

Metrics for object-oriented system requirements—Determine which metrics reflect the complexity of the requirements of object-oriented programs.

Metrics for object-oriented system designs—Determine which metrics reflect the complexity of the designs of object-oriented systems.

Metrics for object-oriented systems—Determine which metrics reflect the complexity of object-oriented systems.

Metrics for hybrid system requirements—Determine which metrics reflect the complexity of the requirements of systems that are hybrids of procedurally and object-oriented subsystems.

Metrics for hybrid system designs—Determine which metrics reflect the complexity of the designs of systems that are hybrids of procedurally and object-oriented subsystems.

Metrics for hybrid systems—Determine which metrics reflect the complexity of systems that are hybrids of procedurally and object-oriented subsystems.

9.1.12 Languages and Efficiency

Java efficiency—Determine the best techniques for grouping of Java language software components into "jars" in such a way as to improve program efficiency.

C efficiency—Determine the best techniques for grouping of C language software components into files and directories in such a way as to improve program efficiency.

C++ efficiency—Determine the best techniques for grouping of C++ language software components into files and directories in such a way as to improve program efficiency.

Ada efficiency—Determine the best techniques for grouping of Ada language software components into packages in such a way as to improve program efficiency.

Smalltalk efficiency—Determine the best techniques for grouping of Smalltalk language software components into files and directories in such a way as to improve program efficiency.

Functional languages efficiency—Determine the best techniques for grouping of software components written in functional languages such as LISP into files and directories in such a way as to improve program efficiency.

Logic languages efficiency—Determine the best techniques for grouping of software components written in logic languages such as Prolog into files and directories in such a way as to improve program efficiency.

Interfaces between languages—Determine the best way to interface between language components written in multiple languages. Write an experiment comparing Interface Definition Language (IDL), CORBA, COM, Java, and HTML in a particular application domain.

9.1.13 Language Generators

Lexical analyzers—Devise and implement an experiment to determine the relative efficiency of lexical analyzers written using an analysis tool such as lex or flex.

Devise a method for testing the code produced by a lexical analyzer such as lex or flex at a high level before the source code is generated.

Parser generators—Devise and implement an experiment to determine the relative efficiency of parsers written using tools such as yacc or bison.

Devise a method for testing the code produced by a parser generator such as yacc or bison at a high level before the source code is generated.

Application generators—Devise and implement an experiment to determine the relative efficiency of software written using domain-specific parser generators.

Devise a method for testing the code produced by a parser generator at a high level before the source code is generated.

9.1.14 Inspections and Reviews

General question—What is the most efficient method of running inspections and reviews?

Inspections by groups—How should inspection groups be organized and inspections be carried out to discover errors efficiently?

False positives and inspections by groups—How should inspection groups be organized and inspections be carried out to reduce the number of false positives but still discover errors efficiently?

Inspections by individuals—How should inspections be carried out to discover errors efficiently?

False positives and inspections by individuals—How should inspections be carried out to reduce the number of false positives but still discover errors efficiently?

The open source software approach, such as the one used for coding and maintenance of the Linux operating system, depends on detailed inspections of code by multiple reviewers using the Internet (McConnell, 1999). How efficient is this process in terms of reducing software faults per unit of programmer time?

This question also concerns the open source approach (McConnell, 1999). For which types of projects is the open source approach suited?

9.1.15 Distributed Systems

Client–server systems—What is the best way to partition a software system into a server and a client?

Remote procedures—Devise and implement an experiment to compare remote procedure call and Java RMI.

9.1.16 Software Project Management

Software project management—Determine the most effective way to manage a project.

Self-organizing teams—Devise and implement an experiment to determine if allowing a team to organize itself without much direction from a manager is efficient for small projects.

9.1.17 Formal Methods

Appropriate application domains for formal methods—Determine a class of application domains in which formal methods work efficiently.

Verifying a verifier—Develop a formal method of verifying the theorem provers used to prove the correctness of certain formally defined systems. (Theoretical note: It is impossible to completely verify a "universal" theorem prover.)

9.1.18 Processes

General problem—Determine the efficiency of several different software development processes.

Return on investment of attaining different Software Engineering Institute (SEI) levels—Devise and implement an experiment to determine the return on investment on the efficiency resulting from attaining SEI levels 2, 3, 4, or 5.

Return on investment of attaining different International Organization for Standardization (ISO) levels—Devise and implement an experiment to determine the return on investment on the efficiency resulting from attaining ISO 9000 or 9001.

Return on investment of attaining Department of Defense (DOD) 2167A compliance—Devise and implement an experiment to determine the return on investment on the efficiency resulting from following DOD 2167A. (The same can be done for other standards.)

9.1.19 Risk Management

General problem—Develop a precise way to determine the risk of any new technology.

Risk assessment in the life cycle—Incorporate risk assessment into the classical waterfall and rapid prototyping software development models.

9.1.20 Quality Assurance

Efficient teams—Devise and implement an experiment to determine the efficient allocation of personnel to quality assurance teams.

Return on investment in quality assurance—Determine the return on investment in a quality assurance team.

9.1.21 Configuration Management

Configuration management for distributed development—Develop a flexible process to be used in configuration management for very large software systems that have software developers working in several geographically separate places, perhaps even different time zones.

Configuration management throughout the life cycle—Develop a flexible process that allows configuration management to be applied to artifacts created at various stages of the life cycle.

9.1.22 Crystal Ball

Determine the software engineering tool, approach, process, or environment that will be most important in twenty years. Keep in mind that doing this twenty years ago would have required you to discern the importance of object-oriented programming, Java, and the Internet. Ten years ago you would have had to discern the importance of the cloud, agile processes, and app stores.

9.2 HOW TO READ THE SOFTWARE ENGINEERING RESEARCH LITERATURE

Many research papers in software engineering either describe a theory, a tool, a process, or a software system. Others describe an attempt to show that something works particularly well by means of some sort of numerical evaluation of the performance of the tool, process, or software system. In this section we will consider only the second type. Of course, there will be a considerable difference in the amount of technical detail, especially proprietary detail, between something written for use internally within an organization and something intended for the general literature. (We will ignore anything written by the sales or marketing departments of an organization.)

Generally speaking, any report describing research that attempts to quantify the utility of a tool, a process, or software system can be classified as falling into one of three categories:

1. A case study
2. A comparative study
3. An experiment

Understanding the proper use of these three categories of research can help you to understand the applicability to your particular environment and avoid problems by misusing the research results in an application environment for which the research is not appropriate. We will describe each of these types of research in what follows.

A *case study* describes a particular situation in considerable detail. A good paper presenting a case study will include a description of the environment in which the tool, process, or software system was created. It will describe how the work was done and what conclusions were drawn, such as the following:

"This tool reduced logical software faults by 20%."

"Going directly from requirements to code without creating a system design increased costs by 30%."

"Demanding rigorous inspections reduced testing time by 10% and reduced testing costs by 25%."

"Use of the proprietary cost estimation software made us underestimate cost by 50%."

As a reader of the case-study type of research, you should look for information that suggests that similar results will hold for your organization. This means that you must determine if you have a similar environment and application domain, and if you are able do the same thing that was described in the case study. This, in turn, implies that the paper presents the development environment and infrastructure in enough detail for you to compare it to your own environment and infrastructure.

If the detail in the paper is not sufficient, you should look for some additional detail before you try to implement the tool, process, or software system in your organization. If you cannot get the additional information you need, you should not attempt to do the same thing without being willing to risk complete failure for the recommended tool, process, or software system.

Even if you have sufficient detailed information to determine if the environment, infrastructure, and application are similar to your situation, you must still be careful. Unfortunately, the problem is due to the very nature of the case study approach.

The primary problem with even the most detailed case study is the lack of controls. There may be factors other than those reported in the case study that may have made major contributions to the results obtained. On the other hand, case studies often describe large systems.

Another type of research approach is the *comparative study.* A comparative study also provides an assessment of two or more instances of a tool, process, or software system. It has an advantage over a case study in that there is a framework for comparison. A good comparative study often describes the environment and application domain, but there is frequently much less detail than in a case study because of space limitations. Comparative studies often describe realistic situations.

A comparative study will often have statements such as the following:

"System A had 20% fewer logical software faults than system B."

"Tool A was used by fewer than 50% of the developers, while tool B was used by more than 85%."

"System A was three months behind schedule. System B was cancelled after missing its deadline by two years."

"There was no discernible difference between the two systems in terms of quality or speed of execution. However, system B cost 10% less than system A."

As with a case study, there are few controls in a comparative study and results can only be suggested, not proved. There may be many factors in addition to those presented in the paper that account for the results reported.

When reading a comparative study, watch that the systems compared in the paper are described in sufficient detail and convince yourself that the comparison is fair. The most common problem with comparative studies is that the comparison is done along lines that may not be fair in the sense that one of the tools, processes, or software systems being compared is used in a way that was not intended.

The last form of software engineering research that we describe is the *controlled experiment.* Controlled experiments offer great promise in the sense that well-designed and implemented ones can clearly indicate that a particular tool, process, or software system is effective in some instance. A good experiment will have a clear design and the design will be reported in detail in the research report.

Unfortunately, controlled experiments are usually done in academic institutions, where the students are relative novices. In many experiments with large numbers of subjects, the subjects are usually chosen from beginning programming classes. As such, you should be suspicious that the results will scale up to large development efforts with experienced personnel.

Which should you believe most: a case study, comparative study, or controlled experiment? The answer is all three, but you should only believe them a little until there is a body of supporting evidence.

The approach of the Software Engineering Laboratory (SEL) at NASA/Goddard Space Flight Center is typical. Under the direction of Victor Basili and Marvin Zelkowitz of the University of Maryland and several colleagues from Goddard, a large body of knowledge was created to support or refute several commonly held expert opinions and industry practices in software engineering. The knowledge base includes all three forms—case study, comparative study, and controlled experiment—and each builds upon knowledge for the others. For example, case studies suggest comparative studies, which lead to experiments. Analyses of experimental results suggest new case studies. Comparative studies suggest new case studies, and so on. Even a cursory examination of the titles of the references provided in this book suggests that many of the problems of software engineering have been revisited many times, sometimes by case study, sometimes by comparative study, and sometimes by controlled experiment.

FURTHER READING

Reading the following papers is a must for the software engineering researcher wishing to study empirical research in the discipline: Fenton and Pfleeger (1996), Fenton and Bieman (2014), Tichy et al. (1995), and Kitchenham, Pfleeger, and Fenton (1995). Nearly every issue of *IEEE Software, IEEE Computer, Communications of the ACM, ACM Transactions on Software Engineering and Methodology, IEEE Transactions on Software Engineering, Journal of Systems and Software, Software: Practice and Experience*, and *Experimental Computer Science* contain important and thought-provoking articles on software engineering. You should try to attend a meeting of the International Conference on Software Engineering. There is also a conference devoted to empirical work: the annual Empirical Assessment in Software Engineering (EASE) conference.

EXERCISES

1. (For those of you who are practicing software engineers.) Choose one of the first 123 research problems listed in this chapter and determine how the issue affects your organization.

2. (For those of you who are considering graduate school.) Choose one of the first 99 research problems listed in this chapter and start working on a PhD in computer science.

3. Read a recent issue of one of the following journals: *IEEE Software, IEEE Computer, Communications of the ACM, ACM Transactions on Software Engineering and Methodology, IEEE Transactions on Software Engineering, Journal of Systems and Software, Software: Practice and Experience, Experimental Computer Science*, or a recent EASE conference proceedings. How many of the articles describe case studies? How many describe comparative studies? How many describe controlled experiments?

4. Read a company internal report that describes a project. How would you classify the report: case study, comparative study, or controlled experiment? Write a critique of the report, using the ideas given in this chapter.

Appendix A: An Interesting Software Patent

THIS APPENDIX BRIEFLY COVERS a United States patent named "Automatically enabling private browsing of a web page, and applications thereof" that was invented by Michael David Smith and was assigned to Google on January 13, 2015. The patent application, available online, contains four diagrams, one of which is included here. From the application:

> FIG. 1 illustrates an example web browser system architecture, according to an embodiment.
>
> FIG. 2 depicts a more detailed view of an example privacy mode enabler in an example web browser system architecture, according to an embodiment.
>
> FIG. 3 is a flowchart illustrating an example aspect of operation, according to an embodiment.
>
> FIG. 4 is a system document that can be used to implement embodiments described herein.

I believe that the figure of highest educational value to readers of this book is the one the patent labeled "FIG. 3," which, as described in the patent application, is a flowchart. I think the existence of this diagram in this patent makes it clear that even though older design representations are rarely used in modern software, an understanding of them can be very helpful, even for use with cutting-edge products from first-rank technology companies.

For simplicity, only one of the diagrams that comprise the application submitted to obtain the patent is shown (Figure A.1). Notice that UML was not used to convey the information in the application.

For more information on this particular patent, see the entire application online at the United States Patent Office. An easy way to do this is to use the URL http://patft.uspto.gov/netahtml/PTO/srchnum.htm and search for the patent number 8,935,798. Selecting the images option allows the images to be easily located.

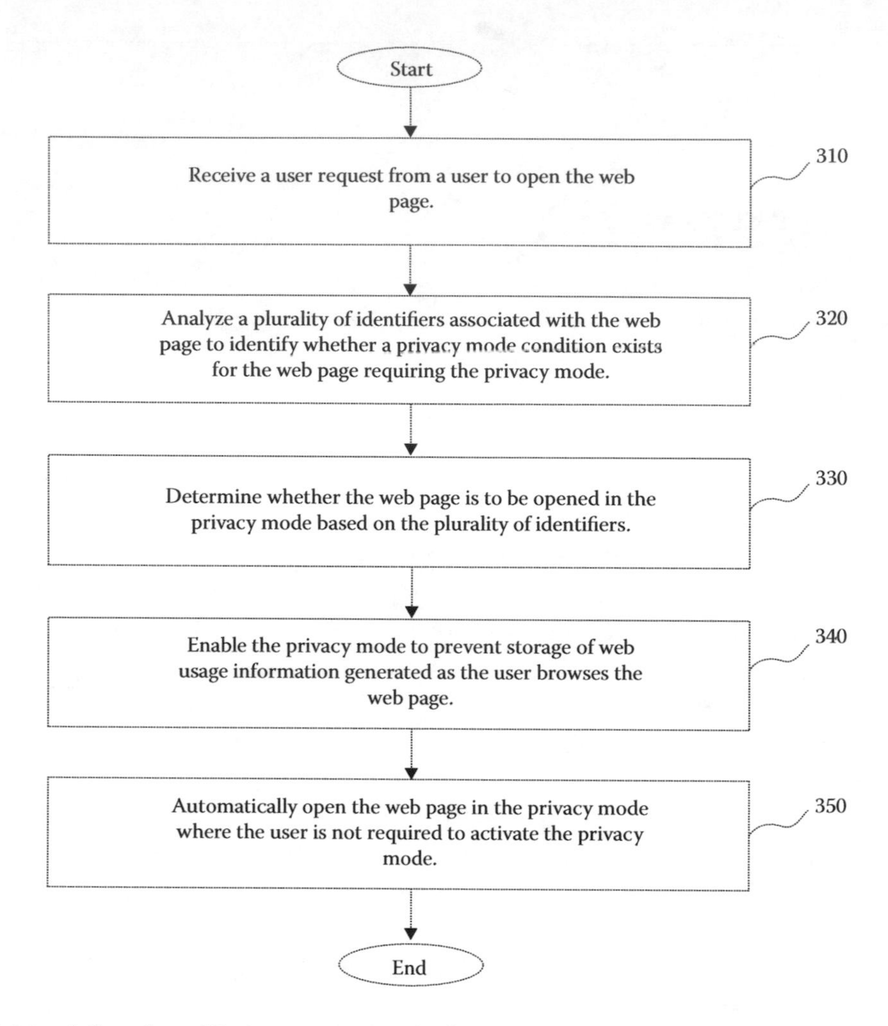

FIGURE A.1 A flowchart-like image as part of a figure in a recent Google patent.

There is one final observation that should be made about patents. The decision to patent a new piece of software, whether it be an entire system or a new smaller unit of code, is not one to be taken lightly. Even a few hours spent browsing the patent database can provide interesting insights. At the very least, these hours spent browsing patents can provide a useful perspective on the social and ethical implications of computing.

Appendix B: Command-Line Arguments

THE FUNCTION MAIN() IN a C or C++ program can interact with its own environment—the operating system of the computer. The communication between the program and the operating system can go in two directions. If the program has an error such as division by 0, then the program communicates with the operating system and the program is halted. Communication in the other direction is done by means of command-line arguments.

Example B.1 presents an example of a program that has command-line arguments. Notice the form of the arguments to main(). These arguments are character strings that are put into the program by an interface between the command shell, which is part of the operating system, and the running program.

Example B.1

```
#include <stdio.h>

int main(int argc, char *argv[ ])
{
  int i;
  puts("The arguments to this function are:");
  for (i = 0; i < argc; i++)
     printf("%s\n",argv[i]);
   printf("\n");
}
```

Type in this program and run it. If your executable program was named a.out (which is the default name for executable files under the UNIX operating system), then your output would be the character string

a.out

on a line by itself. (The corresponding output in computers running under MS/DOS would be the character string

```
progname
```

assuming that the executable file was named progname.) For the input

```
a.out Fred Marie Harry Computer Science types
```

the output would be

```
a.out
Fred
Marie
Harry
Computer
Science
types
```

with each string of characters printed on a line by itself. The command-line arguments in the first run of the program were

```
a.out
```

In the second run of the program, there were the seven command-line arguments previously listed. The value of argc was 1 in the first run and 7 in the second run. The value of argc is always at least 1 since the name of the executable file (a.out in this example) is always counted as a command-line argument. The other command-line arguments are stored in argv[]. (On systems running under MS/DOS, the output would be identical except for the name of the executable file.)

This example shows that the command line, which included the name of the executable file and other character strings, was processed by the operating system and the information was given to the program. This is useful when we wish to pass arguments to a C program.

Example B.2 shows how we can read a command-line argument as a string of characters (the only way to get commands from the operating system) and change it into an integer.

Example B.2

```
#include <stdio.h>

int main(int argc, char *argv[ ])
{
int i,lim;
lim = atoi(argv[1]);
for(i=1;i <= lim;i++)
  printf("%d\n",i);
}
```

Type in this program and test it with several different arguments. It gives the correct answer if you type in a command line such as

```
a.out 2
a.out 23
```

It also gives a correct output (nothing) if the command line is

```
a.out c
```

where c can be any character. However, there is a serious error if you simply type the command line

```
a.out
```

The program terminates ungracefully with a core dump. The program is not well written from the standpoint of defensive programming since it can fail unpleasantly when presented with only minor errors in input. The example illustrates an extremely important principle of programming (especially C programming): Be sure that possible errors in interfaces do not cause the program to crash. This warning applies both to interfaces to a human user entering data and to regular processing in the program.

The best way to write this program is to insert some defensive code immediately after the type declarations. Since the variable argc keeps track of the number of command-line arguments, we can use the argument count to exit if there are not enough arguments. This allows the program to terminate gracefully rather than dump the in-core memory into a file.

The defensive code is

```
if (argc == 1)
  {
  printf("Error - not enough arguments\n");
  exit(1);
  }
```

This code allows the termination of the program by using the exit() function, which is called with an argument of 0 to indicate that no abnormal action should be taken by the operating system. A call to exit() always terminates the program.

Appendix C: Flowcharts

There are many symbols that can be used in flowcharts. It is helpful that the drawing feature of Microsoft Word has twenty-eight predefined icons that represent the different symbols most commonly used within flowcharts. Five of these common symbols are shown in Figure C.1.

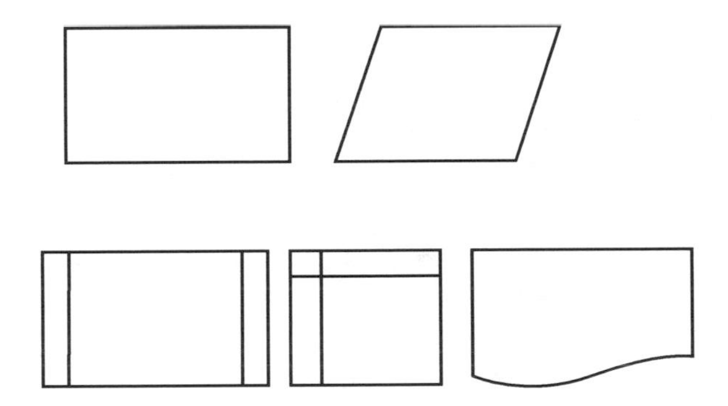

FIGURE C.1 Some symbols commonly used in flowcharts. Reading clockwise from upper left, the symbols are process, data, document, predefined process (subroutine), and internal data storage.

References

Ada, *Reference Manual for the Ada Programming Language*, ANSI-MIL-STD-1815A, 1983.

Ada, *Reference Manual for the Ada Programming Language*, ANSI-MIL-STD-1815B, 1995.

Ada, *2005 Language Reference Manual*, 2005, http://www.adaic.org/ada-resources/standards/ada05/.

Ada, *2012 Language Reference Manual*, 2012, http://www.ada-auth.org/standards/ada12.html.

Albrecht, A. J., Measuring application development productivity, *Proceedings of the IBM Applications Development Joint SHARE/GUIDE Symposium*, 83–92, Monterey, California, 1979.

Albrecht, A. J., and Gaffney, Jr., J. E., Source lines of code, and development effort prediction: A software science validation, *IEEE Transactions on Software Engineering*, vol. SE-9, 639–648, 1983.

American National Standards Institute, *Standard Glossary of Software Engineering Terminology*, ANSI/IEEE Standard 729–1991.

American National Standards Institute, *Recommended Practice for Software Reliability*, ANSI Standard R-012-1992.

Armour, P. G., The business of software: When faster is slower, *Communications of the ACM*, vol. 56, no. 10, 30–32, October 2013.

Arnold, R. S., ed., *Software Reengineering*, IEEE Press, Los Alamitos, California, 1992.

Arnold, R. S., and Bohner, S., *Software Change Impact Analysis*, IEEE Press, Los Alamitos, California, 1997.

Arnold, T. R., and Fuson, W. A., Testing "in a perfect world," *Communications of the ACM*, vol. 37, no. 9, 78–86, September 1994.

Atkinson, C., *Object-Oriented Reuse, Concurrency, and Distribution*, ACM Press, New York, 1991.

Babar, M. A., Kitchenham, B., Zhu, L., and Jeffery, R., An exploratory study of groupware support for distributed software architecture evaluation process, *Proceedings of the 11th Asia-Pacific Software Engineering Conference (APSEC '04)*, 222–229, November 30–December 3, 2004.

Baker, A. L., and Zweben, S. H., The use of software science in evaluating modularity, *IEEE Transactions on Software Engineering*, vol. SE-5, no. 3, 110–120, 1979.

Barbey, S., and Strohmeier, A., The problematics of testing object-oriented software, *Proceedings of the Second Conference on Software Quality Management (SQM '94)*, Edinburgh, Scotland, vol. 2, 411–426, July 26–28, 1994.

Barker, T. T., and Dragga, S., *Writing Software Documentation: A Task-Oriented Approach*, Allyn & Bacon, New York, 1997.

Barnard, J., and Price, A., Managing code inspection information, *IEEE Software*, vol. 11, no. 2, 59–69, March 1994.

Basili, V. R., and Selby, R. W., Comparing the effectiveness of software testing strategies, University of Maryland at College Park, Department of Computer Science, Technical Report Number TR-1501, 1985.

Basili, V. R., and Rombach, H. D., The TAME project: Towards improvement-oriented software development, *IEEE Transactions on Software Engineering*, vol. SE-14, no. 6, 758–773, 1988.

Basili, V. R., Viewing maintenance as reuse-oriented software development, *IEEE Software*, vol. 7, no. 1, 19–25, January 1990.

Basili, V. R., Caldiera, G., and Cantone, G., A reference architecture for the component factory, *ACM Transactions on Software Engineering and Methodology*, vol. 1, no. 1, 53–80, January 1992.

Batory, D., and O'Malley, S., The design and implementation of hierarchical software systems with reusable components, *ACM Transactions on Software Engineering and Methodology*, October 1992.

Beck, K., *Test Driven Development: By Example*, Addison-Wesley, Boston, 2003.

Behrens, C. A., Measuring the productivity of computer systems development activities with function points, *IEEE Transactions on Software Engineering*, vol. SE-13, no. 1, 311–323, 1987.

Beizer, B., *Software Testing Techniques*, Van Nostrand Reinhold, New York, 1983.

Beizer, B., *Software Testing Techniques*, 2nd ed., Van Nostrand Reinhold, New York, 1990.

Bell Communications Research Technical Standard for Software Engineering, 1988.

Berard, E. V., *Essays on Object-Oriented Software Engineering*, vol. I, Prentice-Hall, Englewood Cliffs, New Jersey, 1993.

Bergin, J., *Polymorphism: As It Is Played*, Slant Flying Press, New York, 2015.

Berry, D. M., The importance of ignorance in requirements engineering, *Journal of Systems and Software*, vol. 28, no. 2, 179–184, February 1995.

Bethea, R. M., Duran, B. S., and Boullon, T. L., *Statistical Methods for Engineers and Scientists*, 2nd ed., Marcel Dekker, New York, 1985.

Bieman, J. M., and Zhao, J. X., Reuse through inheritance: A quantitative study of C++ software, *Proceedings of the ACM-SIGSOFT Symposium on Software Reusability*, Seattle, Washington, April 28–30, 1995.

Bieman, J. M., and Karunanithi, S., Measurement of language-supported reuse in object-oriented and object-based software, *Journal of Systems and Software*, vol. 30, no. 3, 271–293, September 1995.

Biggerstaff, T., and Richter, C., Reusability framework, assessment, and directions, *IEEE Software*, vol. 4, no. 2, 41–49, March 1987.

Biggerstaff, T., and A. Perlis, A., eds., *Software Reusability*, vols. 1 and 2, ACM Press, New York, 1989.

Binder, R. V., Design for testability in object-oriented systems, *Communications of the ACM*, vol. 37, no. 9, 87–101, September 1994.

Bleser, T. W., *Human-Computer Interface Guidelines Evaluation*, Century Computing, Laurel, Maryland, 1994.

Boehm, B., Brown, J. R., MacLeod, G. J., and Merritt, M. J., *Characteristics of Software Quality*, Elsevier, North-Holland, 1978.

Boehm, B., *Software Engineering Economics*, Prentice Hall, Englewood Cliffs, New Jersey, 1981.

Boehm, B., A spiral model of software development and enhancement, *IEEE Computer*, vol. 21, no. 2, 61–72, February 1988.

Boehm, B., Egyed, A., Kwan, J., Port, D., Shah, A., and Madachy, R., Using the WinWin spiral model: A case study, *IEEE Computer*, vol. 31, no. 7, 33–44, July 1998.

Boehm, B., and In, H., Identifying quality-requirement conflicts, *Proceedings of the Second Annual Conference on Requirements Engineering*, Colorado Springs, Colorado, April 15–18, 1996. (Most Influential Paper Award given 2006.)

Bollinger, T., and Pfleeger, S. L., Economics of reuse: Issues and alternatives, *Information and Software Technology*, vol. 32, no. 10, 643–652, December 1990.

Booch, G., *Software Engineering with Ada*, 3rd ed., (with D. Bryan and C. G. Petersen), Benjamin Cummings, Redwood City, California, 1991.

Booch, G., *Object-Oriented Analysis and Design with Applications*, Benjamin Cummings, Redwood City, California, 1993.

Box, G. E. P., Hunter, W. J., and Hunter, J. S., *Statistics for Experimenters*, John Wiley & Sons, New York, 1978.

Bracken, M. R., Hoge, S. L., Sary, C., Rashkin, R., Pendley, R. D., and Werking, R. D., IMACCS: An operational COTS-based ground system proof-of-concept project, *Proceedings of the European Space Conference*, pp. 37.1–37.8, 1995.

Brooks, F., *The Mythical Man-Month*, Addison-Wesley, Reading, Massachusetts, 1975.

Budd, T., *Classic Data Structures in C++*, Addison-Wesley, Reading, Massachusetts, 1994.

Burnstein, I., *Practical Software Testing*, Springer-Verlag, New York, 2003.

Burton, B. A. et al., The reusable software library, *IEEE Software*, vol. 4, no. 4, 25–33, July 1987.

Campbell, R., *Client-Server Systems User Interface Style Guide* (DSTL-95-DRAFT), NASA/Goddard Space Flight Center, Greenbelt, Maryland, 1994.

Card, S. K., User perceptual mechanisms in the search of computer command menus, *Proceedings of the Human Factors Conference*, Washington, D.C., 1982.

Card, D., Church, V. E., and Agresti, W. W., An empirical study of software design practices, *IEEE Transactions on Software Engineering*, vol. SE-12, 264–271, 1986.

Chikofsky, E., and Cross, J., Reverse engineering and design recovery: A taxonomy, *IEEE Software*, vol. 7, no. 1, January 1990.

Chisnall, D., The challenge of cross-language interoperability, *Communications of ACM*, vol. 56, no. 12, 50–56, December 2013.

Coad, P., and Yourdon, E., *Object-Oriented Analysis*, 2nd ed., Prentice-Hall, Englewood Cliffs, New Jersey, 1990.

Cohen, D., Lindvall, M., and Costa, P., An introduction to agile methods. In *Advances in Computers*, 1–66. Elsevier Science, New York, 2004.

Computer Science Corporation (CSC), Systems Engineering and Analysis Support Contract, NASA Goddard Space Flight Center, Greenbelt, Maryland, 1991.

Conte, S. D., Dunsmore, H. E., and Shen, V. Y., *Software Engineering Metrics and Models*, Benjamin Cummings, Menlo Park, California, 1986.

Coplien, J., Hoffman, D., and Weiss, D., Commonality and variability in software engineering, *IEEE Software*, 37–45, November/December 1999.

CORBA (Common Object-Oriented Request Broker Architecture), Object Management Group, http://www.omg.org, 1997.

Cusumano, M. A., and Selby, R. W., *Microsoft Secrets: How the World's Most Powerful Software Company Creates Technology, Shapes Markets and Manages People*, Touchstone, New York, 1995.

Cusumano, M. A., and Selby, R. W., How Microsoft builds software, *Communications of ACM*, vol. 40, no. 6, 53–61, June 1997.

Czarnul, P., *Integration of Services into Workflow Applications*, Chapman & Hall/CRC, Boca Raton, Florida, 2015.

Date, C. J., *An Introduction to Database Systems*, 6th ed., Addison-Wesley, Reading, Massachusetts, 1995.

Davis, A. M., *Software Requirements: Analysis and Specification*, Prentice-Hall, Englewood Cliffs, New Jersey, 1990.

Davis, M. J., Adaptable, reusable code, *Proceedings of the Symposium on Software Reusability*, SSR '95, Seattle, Washington, 38–46, April 28–30, 1995.

Deitel, H. M., and Deitel, P. J., *C++ How to Program*, Prentice-Hall, Englewood Cliffs, New Jersey, 1994.

DeMarco, T., *Structured Analysis and System Specification*, Prentice Hall, Englewood Cliffs, New Jersey, 1979.

DeMillo, R. A., McCracken, W. M., Martin, R. J., and Passafiume, J. F., *Software Testing and Evaluation*, Benjamin Cummings, Menlo Park, California, 1987.

Denning, P. J., The profession of IT: Design thinking, *Communications of the ACM*, vol. 56, no. 12, 29–31, December 2012.

Department of Defense Architectural Framework (DoDAF), version 2.0, dodcio.defense.gov/Portals/0/Documents/DODAF/DoDAF_v2-02_web.pdf.

Dhama, H., Quantitative models of cohesion and coupling in software, *Journal of Systems and Software*, vol. 29, 65–74, 1995.

Dingsøyr, T., Dybå, T., and Moe, N. B., eds., *Agile Software Development: Current Research and Future Directions*, Springer, Berlin, 2010.

Dijkstra, E. W., Guarded commands, nondeterminism, and formal derivation of programs, *Communications of the ACM*, vol. 21, no. 8, 453–457, 1990.

DISA, Domain analysis and design process, Defense Information Systems Agency Center for Information Management (CIM) Software Reuse Office, Document #1222-04-210/30.1, 1993.

Donat, M., Husein, J., and Coatta, T., Automated QA testing at Electronic Arts: A discussion with Michael Donat, Jafar Husein, and Terry Coatta, *Communications of the ACM*, vol. 57, no. 7, 50–57, July 2014.

Dougherty, D., *sed & awk*, O'Reilly & Associates, Sebastopol, California, 1992.

Dromey, R. G., A model for software product quality, *IEEE Transactions on Software Engineering*, vol. SE-21, no. 2, 146–162, February 1995.

Ellis, M., and Stroustrup, B., *The Annotated C++ Reference Manual*, Addison-Wesley, Reading, Massachusetts, 1990.

Ellis, T., COTS integration in software solutions—A cost model, *Systems Engineering in the Global Marketplace, NCOSE International Symposium*, St. Louis, Missouri, July 24–26, 1995.

Erl, T., Carlyle, B., Pautasso, C., and Balasubramanian, R., "5.1," in Thomas Erl, *REST Constraints and Goals*, Prentice Hall, 2013.

Ezran, M., Morisio, M., and Tully, C., Failure and success patterns in reuse programs: A synthesis of industrial experiences, *Tutorial TTF3, Twenty-First International Conference on Software Engineering*, ICSE '99, Los Angeles, May 17–21, 1999.

Fagan, M., Design and code inspections to reduce errors in program development, *IBM Systems Journal*, vol. 15, no. 3, 182–211, 1976.

Farchamps, D., Organizational factors and reuse, *IEEE Software*, vol. 11, no. 5, 31–41, September 1994.

Fenton, N. E., and Pfleeger, S. L., *Software Metrics: A Rigorous Approach*, 2nd ed., International Thomson Press, London, 1996.

Fenton, N. E., and Bieman, J., *Software Metrics: A Rigorous Approach*, 3rd ed., Chapman & Hall/CRC Press, London, 2014.

Finelli, G. B., NASA software failure characterization experiments, in *Software Reliability and Safety*, B. Littlewood and D. R. Miller, eds., Elsevier Science, London, 1991.

Fowler, G. S., Korn, D. G., and Vo, K.-P., Principles for writing reusable libraries, *Proceedings of the Symposium on Software Reusability*, SSR '95, Seattle, Washington, 150–159, April 28–30, 1995.

Fowler, M., Is design dead?, in *Extreme Programming Explained*, G. Succi and M. Marchesi, eds., Addison-Wesley, Boston, 2001.

Frakes, W. B., and Gandel, P. B., Representing reusable software, *Information and Software Technology*, vol. 32, no. 10, 653–661, December 1990.

Frakes, W. B., ed., *Advances in Software Reuse: Selected Papers from the Second International Workshop on Software Reusability*, Luccia, Italy, March 24–26, 1993.

Frakes, W. B., ed., *Advances in Software Reuse: Selected Papers from the Third International Workshop on Software Reusability*, Rio de Janeiro, Brazil, November 1–4, 1994.

Frakes, W. B., and Isoda, S., Success factors for systematic reuse, *IEEE Software*, vol. 11, no. 5, 14–22, September 1994.

Frakes, W. B., and Pole, T. P., An empirical study of representation methods for reusable software components, *IEEE Transactions on Software Engineering*, vol. SE-20, no. 8, 617–630, August 1994.

Frakes, W. B., and Fox, C. J., Modeling reuse across the software life cycle, *Journal of Systems and Software*, vol. 30, no. 3, 295–301, September 1995.

Frakes, W. B., and Fox, C. J., Sixteen questions about software reuse, *Communications of the ACM*, vol. 38, no. 6, 75–87, June 1995.

Freeman, P., Reusable software engineering: Concepts and research directions, *ITT Proceedings of the Workshop on Reusability in Programming*, 129–137, 1983.

Freeman, P., Reusable software engineering: Concepts and research directions, *Tutorial: Software Reusability*, IEEE Computer Society, Los Alamitos, California, 10–23, 1987.

Freeman, S., and Pryce, N., *Growing Object-Oriented Software, Guided by Tests*, Addison-Wesley, Upper Saddle River, New Jersey, 2009.

Gaffney, J., and Cruickshank, R. D., A general economics model for software reuse, Software Productivity Consortium, 1991.

Galorath, D. D., and Evans, M. W., *Software Sizing, Estimation, and Risk Management*, Auerbach, Boca Raton, Florida, 2006.

Gamma, E., Helm, R., Johnson, R., and Vlissides, J., *Design Patterns: Elements of Reusable Object-Oriented Software*, Addison-Wesley Professional, Boston, 1998.

General Accounting Office, *Software Reuse—Major Issues Need to Be Resolved Before Benefits Can Be Achieved*, GAO Report Number GAO/IMTEC-93-16, Washington, D.C., 1993.

Ghezzi, C., Jazayeri, M., and Mandrioli, D., *Fundamentals of Software Engineering*, Pearson Prentice Hall, Upper Saddle River, New Jersey, 2002.

Gilb, T., *Principles of Software Engineering Management*, Addison-Wesley, Reading, Massachusetts, 1987.

Goguen, J. A., Reusing and interconnecting software components, *IEEE Computer*, 16–28, February 1986.

Gold-Bernstein, B., and Ruh, W. A., *Enterprise Integration: The Essential Guide to Integration Solutions*, Addison-Wesley, Reading, Massachusetts, 2005.

Goldberg, A., and Rubin, K. S., Taming object-oriented technology, *Computer Language*, vol. 7, no. 10, 34–35, October 1990.

Gomaa, H., Methods and tools for domain-specific software architectures, *Proceedings of the Fifth Workshop on Software Reuse*, WISR-5, 1992.

Goodman, M. A., Goyal, M., and Massoudi, R. A., *Solaris Porting Guide*, SunSoft Press, Mountain View, California, 1993.

Gotel, O. C. Z., and Finkelstein, A. C. W., An analysis of the requirements traceability problem, Requirements Engineering, 1994, *Proceedings of the First Annual Conference on Requirements Engineering*, Colorado Springs, Colorado, 94–101, April 18–22, 1994. (Most Influential Paper Award given 2004.)

Gothelf, J., How we finally made agile development work, *Harvard Business Review*, October 11, 2012. Accessed December 2014 from https://hbr.org/2012/10/how-we-finally-made-agile-development-work.

Goulde, M., Developing a reuse strategy, *Open Computing*, vol. 12, no. 8, 29, August 1, 1995.

Grady, R. B., and Caswell, D. L., *Software Metrics: Establishing a Company-Wide Policy*, Prentice-Hall, Englewood Cliffs, New Jersey, 1987.

Grady, R. B., Successfully applying software metrics, *IEEE Computer*, vol. 27, no. 9, 28–26, September 1994.

Green, M., University of Alberta user interface management system, *Proceedings SIGGRAPH Conference—Computer Graphics (ACM)*, vol. 19, no. 3, 205–213, July 1985.

Griss, M. L., A multi-disciplinary software reuse research program, *Proceedings of the Fifth Workshop on Software Reuse*, WISR-5, 1992.

Guerrieri, E., Case study: Digital's application generator, *IEEE Software*, vol. 11, no. 5, 95–96, September 1994.

Hall, P. A. V., *Software Reuse and Reverse Engineering in Practice*, Chapman & Hall, London, 1992.

Halstead, M. H., *Elements of Software Science*, Elsevier, North-Holland, 1977.

Hardy, Q., Business software keeps evolving quickly, *The New York Times*, May 12, 2014a.

Hardy, Q., Free software's big challenge, *The New York Times*, July 28, 2014b.

Harms, D. E., and Weide, B. W., Copying and swapping: Influence on the design of reusable software components, *IEEE Transactions on Software Engineering*, vol. SE-17, no. 5, 424–435, May 1991.

Harrold, M. J., McGregor, J. D., and Fitzpatrick, K. J., Incremental testing of object-oriented class structures, *Proceedings of the 14th International Conference on Software Engineering*, Melbourne, Australia, 68–79, May 11–15, 1992.

Hartson, H. R., and Hix, D., Human-computer interface development concepts and systems for its management, *ACM Computing Surveys*, vol. 21, no. 1, 5–92, March 1989.

Hashemi, R., and Leach, R. J., Issues in porting software from C to C++, *Software: Practice & Experience*, vol. 22, no. 7, 599–602, 1992.

Hecht, H., Hecht, M., and Wallace, D., Toward more effective testing for high assurance systems, *Proceedings of the High Assurance Systems Engineering Workshop*, HASE '97, Washington, D.C., 176–181, August 11–12, 1997.

Hennel, M. A., Testing for the achievement of software reliability, in *Software Reliability and Safety*, B. Littlewood and D. R. Miller, eds., Elsevier Science, London, 1991.

Henninger, S., Developing domain knowledge through the reuse of project experience, *Proceedings of the Symposium on Software Reusability*, SSR '95, Seattle, Washington, 186–195, April 28–30, 1995a.

Henninger, S., Information access tools for software reuse, *Journal of Systems and Software*, vol. 30, 231–247, 1995b.

Henricson, M., and Nyquist, E., *Industrial Strength C++: Rules and Recommendations* (Prentice Hall Series in Innovative Technology), Prentice-Hall, Englewood Cliffs, New Jersey, 1996.

Henry, S., and Kafura, D., Software metrics based on information flow, *IEEE Transactions on Software Engineering*, vol. SE-7, no. 5, 510–518, September 1981.

Hoare, C. A. R., Algorithm 64: Quicksort, *Communications of the ACM*, vol. 4, no. 7, 321, 1961.

Hoare, C. A. R., *Communicating Sequential Processes*, Prentice-Hall, Englewood Cliffs, New Jersey, 1985.

Hollingsworth, J. E., Weide, B. W., and Zweben, S. H., Confessions of some used-program clients, *Proceedings of the 4th Annual Workshop on Software Reuse*, Herndon, Virginia, November 1991.

Hooper, J. W., and Chester, R., *Software Reuse: Guidelines and Methods*, Plenum Press, New York, 1991.

Horton, W. K., *Designing and Writing Online Documentation for Self-Supporting Products*, John Wiley & Sons, New York, 1994.

Howden, W. E., *Functional Program Testing and Analysis*, McGraw-Hill, New York, 1987.

Hull, S., 20 obstacles to scalability, *Communications of the ACM*, vol. 56, no. 9, 54–58, September 2013.

Humphrey, W., *Managing the Software Process*, Addison-Wesley, Reading, Massachusetts, 1989.

Humphrey, W., *A Discipline for Software Engineering*, Addison-Wesley, Reading, Massachusetts, 1995.

Ichbiah, J. D., Barnes, J. G. P., Firth, R. J., and Woodger, M., *Rationale for the Design of the Ada Programming Language*, Alsys, France, 1986.

IEEE, *IEEE Standard Dictionary of Measures to Produce Reliable Software*, IEEE STD 982.1-1988.

IEEE, *Standard Glossary of Software Engineering Technology*, IEEE STD 610.12-1990.

IEEE, *Standards for Software Productivity Metrics*, IEEE STD 1045-1992a.

IEEE, *Standard for a Software Engineering Methodology*, IEEE STD 1061-1992b.

International Organization for Standardization, Quality Management and Quality Assurance Standards—Guidelines for Selection and Use, ISO 9000, 1987a.

International Organization for Standardization, Quality Systems—Model for Quality Assurance in Design/Development, Production, Installation, and Servicing, ISO 9001, 1987b.

International Organization for Standardization, Quality Systems—Model for Quality Assurance in Production and Installation, ISO 9002, 1987c.

International Organization for Standardization, Quality Systems—Model for Quality Assurance in Final Inspection and Test, ISO 9003, 1987d.

International Standards Organization, ISO/IEC IS 10040: 1991a, Information Technology—Open Systems Interconnection—Systems Management Overview.

International Standards Organization, ISO/IEC IS 10164: 1991b, Information Technology—Open Systems Interconnection—Systems Management Details.

International Standards Organization, ISO/IEC IS 10165-1: 1991c, Information Technology—Open Systems Interconnection—Structure of Management Information—Part 1: Management Information Model.

International Standards Organization, ISO/IEC IS 10165-4: 1991d, Information Technology—Open Systems Interconnection—Part 4: Guidelines for the Definition of Managed Objects.

Jackson, M. A., *System Development*, Prentice Hall, Englewood Cliffs, New Jersey, 1983.

Jacobson, I., Spence, I., and Ng, P.-W., Agile and SEMAT—Perfect partners, *Communications of the ACM*, vol. 56, no. 11, 53–59, November 2013.

Jacobson, I., Ng, P.-W., Spence, I., and McMahon, P., Major-league SEMAT—Why should an executive care, *Communications of the ACM*, vol. 57, no. 4, 44–50, April 2014a.

Jacobson, I., Ng, P.-W., Spence, I., and Lidman, S., *The Essence of Software Engineering: Applying the SEMAT Kernel*, Addison-Wesley, Upper Saddle River, New Jersey, 2014b.

Jalote, P., *An Integrated Approach to Software Engineering*, Springer, New York, 1991.

Jelinski, F., and Moranda, P. B., Software reliability research, in *Statistical Computer Performance Evaluation*, W. Freiberger, ed., Academic Press, New York, 465–484, 1972.

Jeng, J.-J., and Chen, B. H. C., Specification matching for software reuse: A foundation, *Proceedings of the ACM-SIGSOFT Symposium on Software Reusability*, Seattle, Washington, April 28–30, 1995.

Johnson, L. F., and Cooper, R. H., *File Techniques for Data Base Organization in COBOL*, 2nd ed., Prentice-Hall, Englewood Cliffs, New Jersey, 1986.

Johnson, R. E., and Foote, B., Designing reusable classes, *Journal of Object-Oriented Programming*, vol. 1, no. 2, 22–35, 1988.

Johnson, J. A., Nardi, B. A., Zarmer, C. L., and Miller, J. R., ACE: An application construction environment, *Proceedings of the Fifth Workshop on Software Reuse*, WISR-5, 1992.

Johnson, B., Ornburn, S., and Rugaber, S., A quick tools strategy for program analysis and software maintenance, *Proceedings of the Conference on Software Maintenance*, 206–213, Orlando, Florida, November 9–12, 1992.

Jones, T. Capers, Reusability in programming: A survey of the state of the art, *IEEE Transactions on Software Engineering*, vol. SE 10, no. 5, 488–494, September 1984.

Jones, T. Capers, *Assessment and Control of Software Risks*, Yourdon Press, Prentice-Hall, Englewood Cliffs, New Jersey, 1994.

Jones, T. Capers, Bad days for software, *IEEE Spectrum*, 47–52, September 1998.

Joos, R., Software reuse at Motorola, *IEEE Software*, vol. 11, no. 5, 42–47, September 1994.

Jorgenson, P. C., and Erikson, C., Object-oriented integration testing, *Communications of the ACM*, vol. 37, no. 9, 30–38, September 1994.

Kaczmarczyk, L., Questioning our assumptions about introverts and computing, *ACM Inroads*, vol. 4, no. 4, December, 2013, 24–25.

Kafura, D., and Henry, S., Software quality metrics based on interconnectivity, *Journal of Systems and Software*, vol. 2, no. 2, 121–131, 1981.

Kaiser, G. E., and Garland, D., Melding software systems from reusable building blocks, *IEEE Software*, vol. 4, no. 4, 17–24, July 1987.

Kamath, Y. H., Smilan, R. E., and Smith, J. G., Reaping benefits with object-oriented technology, *AT&T Technical Journal*, vol. 72, no. 5, 14–24, September/October 1993.

Kaner, C., and Fiedler, R., *Foundations of Software Testing*, BBST, 2013.

Karlsonn, E.-A., *Software Reuse: A Holistic Approach*, John Wiley & Sons, New York, 1995.

Karolak, D., Shifting paradigm in software engineering management, keynote address at *International Association of Science and Technology for Development (IASTED) Conference on Software Engineering*, ACTA Press, Las Vegas, Nevada, October 28–31, 1998.

Keiller, P. A., Littlewood, B., Miller, D. R., and Sofer, A., Comparison of software reliability predictions, *Proceedings of the Thirteenth International Symposium on Fault-Tolerant Computing*, IEEE Computer Society Press, 128–134, Washington, D.C., 1983.

Keiller, P. A., and Miller, D. R., Software reliability growth models, in *Software Reliability and Safety*, B. Littlewood and D. R. Miller, eds., Elsevier Science, London, 1991.

Kernighan, B., and Ritchie, D., *The C Programming Language*, Prentice-Hall, Englewood Cliffs, New Jersey, 1982.

Kernighan, B., The UNIX system and software reusability, *IEEE Transactions on Software Engineering*, vol. SE-10, no. 5, 51–3518, September 1984.

Kernighan, B., and Ritchie, D., *The C Programming Language*, 2nd ed., Prentice-Hall, Englewood Cliffs, New Jersey, 1988.

Kitchenham, B. A., Pfleeger, S. L., and Fenton, N. E., Towards a framework for software measurement validation, *IEEE Transactions on Software Engineering*, vol. 21, no. 12, 929–944, 1995.

Knuth, D. E., *The Art of Computer Programming*. Vol. 1, *Fundamental Algorithms*, 2nd ed., Addison-Wesley, Reading, Massachusetts, 1973.

Kontio, J., OTSO: A systematic process for reusable software component selection, University of Maryland, College Park Technical Report CS-TR-3478, UMIACS-TR-95-63, December 1995.

Krishnamurthy, B., *Practical Reusable UNIX Software*, John Wiley & Sons, New York, 1995.

Krueger, C. W., Software reuse, *ACM Computing Surveys*, vol. 24, no. 2, 131–183, June, 1992.

LaMonica, M., Object code is not spurring reuse by IS, *Infoworld*, vol. 17, no. 39, 19–20, August 21, 1995.

Laplante, P. A., Licensing software engineers: Seize the opportunity, *Communications of the ACM*, vol. 57, no. 7, 38–40, July 2014.

Larman, C., and Basili, V. R., Iterative and incremental development: A brief history, *IEEE Computer*, June 2003.

Lea, D., and de Champeaux, D., Object-oriented software reuse technical opportunities, *Proceedings of the Fifth Workshop on Software Reuse*, WISR-5, 1992.

Leach, R. J., *Using C in Software Design*, Academic Press Professional, Boston, 1993.

Leach, R. J., *Advanced Topics in UNIX*, John Wiley & Sons, New York, 1994.

Leach, R. J., *Object-Oriented Design and Programming in C++*, Academic Press Professional, Boston, 1995.

Leach, R. J., *Software Reuse: Methods, Models, Costs*, McGraw-Hill, New York, 1997.

Leach, R. J., Assessment of COTS products from an operating systems perspective, *Pacific Northwest Quality Conference*, Portland, Oregon, 235–243, October 12–15, 1998a.

Leach, R. J., Using reuse for requirements, *IASTED Conference*, Las Vegas, Nevada, October 28–31, 1998b.

Leach, R. J., Measurement of requirements: A case study, *Proceedings of the International Function Point Users Group Conference*, Orlando, Florida, 141–148, September 22–25, 1998c.

Leach, R. J., Software cost estimation, *Encyclopedia of Electrical and Electronics Engineering*, John Wiley, New York, 1999.

Leach, R. J., Using software reuse for requirements, *Fourth Baltic Conference on Databases and Information Systems (IEEE)*, Vilnius, Lithuania, May 1–5, 2000.

Leach, R. J., Using software reuse to drive requirements, in *Databases and Information Systems*, J. Barzdins and A. Caplinskas, eds., Kluwer Academic, Dordrecht, The Netherlands, 2001.

Leach, R. J., Separate money tubs hurt software productivity, *CrossTalk*, vol. 17, no. 12, 19–22, December 2004.

Leach, R. J., Can this COTS-based system be saved?, *PC/104 Embedded Solutions*, vol. 9, no. 4, 38–44, Fall 2005.

Leach, R. J., A model relating software quality and reuse, *CrossTalk*, 27–28, April 2006.

Leach, R. J., Software cost estimation, in *Encyclopedia of Electrical and Electronics Engineering*, John Wiley, New York, 2007a, http://www.mrw.interscience.wiley.com/emrw/9780471346081/home.

Leach, R. J., Reuse perspectives—How much has changed in fifteen years?, *Tenth Annual Joint Conference on Information Systems*, JCIS2007, July 2007b, http://eproceedings.worldscinet.com/9789812709677/978981209677_0106.html.

Leach, R. J., Experiences analyzing faults in a hybrid distributed system with access only to sanitized data, *Journal of Software Engineering and Applications (JSEA)*, vol. 3, no. 5, 446–454, 2010.

Leach, R. J., *Advanced Topics in UNIX*, 2nd ed., AfterMath, 2012.

Leach, R. J., Charles, C., Fagan, K., Kimbrough, T., and Thomas, K. R., A reengineering process using early decomposition and simple tools, *Information and Software Technology*, vol. 40, no. 14, 871–875, 1998.

Leach, R. J., Burge, L. L., and Keeling, H. N., Can students reengineer?, *Proceedings of the 39th Technical Symposium on Computer Science Education*, SIGCSE 2008, Portland, Oregon, March 12–15, 2008.

Ledbetter, L., and Cox, B., Software-ICs: A plan for building reusable software components, *BYTE*, 28–35, June 1985.

Leffingwell, D., *Agile Software Requirements: Lean Requirements Practices for Teams, Programs, and the Enterprise* (Agile Software Development Series), Pearson Education, Boston, 2001.

Leveson, N. G., Software safety: What, why and how, *ACM Computing Surveys*, vol. 18, no. 2, 125–163, 1986.

Leveson, N., *Engineering a Safer World*, MIT Press, Cambridge, Massachusetts, 2012.

Levine, J. R., Mason, T., and Brown, D., *lex & yacc*, 2nd ed., O'Reilly & Associates, Sebastopol, California, 1992.

Lillie, C., Software reuse, *Proceedings of the Second Annual Reuse Education and Training Workshop*, Morgantown, West Virginia, October 1993.

Lim, W. C., Effects of reuse on quality, productivity, and economics, *IEEE Software*, vol. 11, no. 5, 23–30, September 1994.

Lim, W. C., *Managing Software Reuse*, Prentice-Hall, Englewood Cliffs, New Jersey, 1995.

Lindholm, E., Snap-on code, *Datamation*, vol. 40, 63, February 1, 1994.

Littlewood, B., and Miller, D. R., eds., *Software Reliability and Safety*, Elsevier Science, London, 1991. (Also in *Reliability Engineering and System Safety*, vol. 32, nos. 1 and 2, 1991.)

Liu, M., Hansen, S., and Tu, Q., The community source approach to software development and the Kuali experience, *Communications of the ACM*, vol. 57, no. 5, 88–96, May 2014.

Lorenz, M., *Object-Oriented Software Development: A Practical Guide*, Prentice Hall, Englewood Cliffs, New Jersey, 1993.

Luqi, and Ketabchi, M., A computer-aided prototyping system, *IEEE Software*, vol. 5, no. 2, 66–72, March 1988.

Mandl, D., NASA Goddard Space Flight Center Internal Technical Report, 1998.

Marick, B., *The Craft of Software Testing*, Prentice-Hall, Englewood Cliffs, New Jersey, 1997.

Matsumoto, Y., and Ohno, Y., *Japanese Perspectives in Software Engineering*, Prentice-Hall, Englewood Cliffs, New Jersey, 1989.

McCabe, T. J., A complexity measure, *IEEE Transactions on Software Engineering*, vol. SE-2, 308–320, December 1976.

McCarthy, J., *Dynamics of Software Development*, Microsoft Press, Redmond, Washington, 1995.

McClure, C., *The Three R's of Software Automation: Re-Engineering, Repository, Reusability*, Prentice-Hall, Englewood Cliffs, New Jersey, 1992.

McConnell, S., Open source methodology: Ready for prime time?, *IEEE Software*, vol. 16, no. 4, 6–8, July–August 1999.

McGregor, J. D., and Korson, T. D., Integrating object-oriented testing and development processes, *Communications of the ACM*, vol. 37, no. 9, 59–77, September 1994.

McIlroy, M. D., Mass produced software components, in *Software Engineering Concepts and Techniques*, Proceedings of the 1968 NATO Conference on Software Engineering, Petrocelli/Charter, Brussels, Belgium, 88–98, 1968.

Meyer, B., Reusability: The case for object-oriented design, *IEEE Software*, vol. 4, 50–63, March 1987.

Meyer, B., and Arnout, K., Componentization: The visitor example, *IEEE Computer*, vol. 39, no. 7, 23–30, July 2006.

Mili, H., Mili, F., and Mili, A., Reusing software: Issues and research directions, *IEEE Transactions on Software Engineering*, vol. SE-21, no. 6, 528–561, June 1995.

Miller, G., The magic number seven, plus or minus two: Some limits on our capacity for processing information, *The Psychological Review*, vol. 63, 81–97, March 1956.
Miller, R. B., Response time in man-computer conversational transactions, *AFIPS Conference Proceedings*, vol. 22, 267–277, December 1968.
Miller, E., and Howden, W. E., *Software Testing and Validation Techniques*, IEEE Computer Society, Long Beach, California, 1978.
Miller, E., and Howden, W. E., *Software Testing and Validation Techniques*, 2nd ed., IEEE Computer Society, Long Beach, California, 1983.
Mills, H. D., Dyer, M., and Linger, R. C., Cleanroom software engineering, *IEEE Software*, vol. 4, 19–24, September 1987.
Milner, R., *A Calculus of Communicating Systems*, Springer-Verlag, New York, 1980.
Montgomery, D. C., *Introduction to Statistical Quality Control*, John Wiley & Sons, New York, 1991.
Moore, J. M., and Bailin, S. C., Domain analysis: Framework for reuse, in *Domain Analysis and Software Systems Modeling*, R. Prieto-Diaz and G. Arango, eds., IEEE Press, Los Alamitos, California, 1991.
Murphy, G. C., Townsend, P., and Wong, P. S., Experiences with cluster and class testing, *Communications of the ACM*, vol. 37, no. 9, 39–47, September 1994.
Musa, J. D., Iannino, A., and Okumoto, K., *Software Reliability: Measurement, Prediction, Application*, McGraw-Hill, New York, 1987.
Musa, J. D., Operational profiles in software reliability engineering, *IEEE Software*, vol. 10, no. 2, 14–32, March 1993.
Musen, M. A., Dimensions of knowledge sharing and reuse, *Computers and Biomedical Research*, vol. 25, 435–467, 1992.
Myers, G. J., *Software Reliability: Principles and Practices*, Wiley, New York, 1976.
Myers, G. J., *The Art of Software Testing*, John Wiley & Sons, New York, 1979.
Myers, W., Taligent's CommonPoint: The promise of objects, *IEEE Computer*, vol. 28, no. 3, 77–83, March 1995.
NASA, Software reuse issues, *Proceedings of the 1988 Workshop on Software Reuse*, sponsored by NASA Langley Research Center, Melbourne, Florida, November 17–18, 1988.
NASA, *Proceedings of the Second NASA Workshop on Software Reuse*, Research Triangle Park, North Carolina, May 5–6, 1992a.
NASA, Software engineering program: Profile of software within Code 500 at Goddard Space Flight Center, *NASA SEP Report R01-92*, December 1992b.
National Institute for Standards and Technology, *Management Information Catalog*, Issue 1.0, NIST, OIW, and Network Management Forum, Gaithersburg, Maryland, June 1992. (There are subsequent catalogs.)
National Institute for Standards and Technology, *Glossary of Software Reuse Terms*, S. Katz, C. Dabrowski, K. Miles, and M. Law, eds., Gaithersburg, Maryland, 1995.
Navarro, J. J., Organization design for software reuse, *Proceedings of the Fifth Workshop on Software Reuse*, WISR-5, 1992.
Neville-Neil, G., Kode Vicious: A lesson in resource management, *Communications of the ACM*, vol. 56, no. 12, 32–33, December 2013.
Neville-Neil, G., Kode Vicious: This is the foo field, *Communications of the ACM*, vol. 57, no. 4, 20–21, April 2014.
Nielsen, J., *Usability Engineering*, Academic Press, San Diego, California, 1994.
Nielsen, J., and Budiu, R., *Mobile Usability*, New Riders, Berkeley, California, 2012.
Olsen, D. R., MIKE: The menu interaction kontrol environment, *ACM Transactions on Graphics*, vol. 5, no. 4, 318–344, October 1986.
Parnas, D. L., On the criteria for decomposing systems into modules, *Communications of the ACM*, vol. 15, no. 12, 1052–1058, 1972.
Parnas, D. L., The modular structure of complex systems, *IEEE Transactions on Software Engineering*, 259–266, March 1985.

Perry, D., and Kaiser, G., Adequate testing and object-oriented programming, *Journal of Object-Oriented Programming*, vol. 3., no. 1, January/February 1990.

Petri Nets, *Conference on Petri Nets and Related Methods*, MIT July, 1975.

Pfleeger, C. P., Pfleeger, S. L., and Margulies, J., *Security in Computing*, 5th ed., Prentice-Hall, Englewood Cliffs, New Jersey, 2015.

Pfleeger, S. L., *Software Engineering: The Production of Quality Software*, Macmillan, New York, 1989.

Pfleeger, S. L., Model of software effort and productivity, *Information Software and Technology*, vol. 33, no. 3, 224–231, April 1991.

Pfleeger, S. L., and Bollinger, T. B., Economics of reuse: New approaches to modelling and assessing cost, *Information and Software Technology*, August 1994.

Pfleeger, S. L., Fenton, N., and Page, S., Evaluating software engineering standards, *IEEE Computer*, vol. 27, no. 9, 71–79, September 1994.

Pfleeger, S. L., Measuring reuse: A cautionary tale, *IEEE Software*, 118–127, July 1996.

Pfleeger, S. L., and Atlee, J. M., *Software Engineering: Theory and Practice*, 4th ed., Prentice Hall, Englewood Cliffs, New Jersey, 2010.

Pigoski, T. M., *Practical Software Maintenance*, IEEE Press, Los Alamitos, California, 1997.

Plauger, P. J., *The Standard C Library*, Prentice Hall, Englewood Cliffs, New Jersey, 1992.

Plauger, P. J., *The Standard C++ Template Library*, Prentice Hall, Englewood Cliffs, New Jersey, 1994.

Pleszkoch, M. G., Linger, R. C., and Hevner, A. R., Eliminating non-transferable paths from structured programs, *Proceedings of the IEEE Conference on Software Maintenance*, 156–164, Orlando, Florida, November 9–12, 1992.

Pohl, I., *C++ for C Programmers*, Benjamin Cummings, Redwood City, California, 1989.

Poston, R. M., Automated testing from object models, *Communications of the ACM*, vol. 37, no. 9, 48–58, September 1994.

Poulin, J. S., Measuring software reusability, *Proceedings of the Third International Workshop on Software Reuse*, Rio de Janeiro, Brazil, November 1–4, 1994.

Poulin, J. S., and Werkman, K. J., Melding structured abstracts and the world wide web for retrieval of reusable components, *Proceedings of the Symposium on Software Reusability*, SSR '95, Seattle, Washington, 160–168, April 28–30, 1995.

Pressman, R. S., and Maxim, B. R., *Software Engineering: A Practitioner's Approach*, 8th ed., McGraw-Hill, New York, 2015.

Prieto-Diaz, R., Implementing faceted classification for software reuse, *Communications of the ACM*, vol. 34, no. 5, 88–97, May 1991.

Prieto-Diaz, R., and Arango, G., eds., *Domain Analysis and Software Systems Modeling*, IEEE Press, Los Alamitos, California, 1991.

Prieto-Diaz, R., Some experiences in domain analysis, *Proceedings of the Sixth Workshop on Software Reuse*, WISR-6, 1993a.

Prieto-Diaz, R., Status report: Software reusability, *IEEE Software*, 61–66, May 1993b.

Putnam, L. H., A general empirical solution to the macro software size and estimation problem, *IEEE Transactions on Software Engineering*, vol. SE-4, no. 4, 345–361, 1978.

Rada, R., *Software Reuse*, Ablex, 1995.

Ramamoorthy, C. V., and Bastiani, F. B., Software reliability—Status and perspectives, *IEEE Transactions on Software Engineering*, vol. SE-8, no. 4, 354–371, July 1982.

Ramamoorthy, C. V., Garg, V., and Prakash, A., Support for reusability: Genesis, *IEEE Transactions on Software Engineering*, vol. SE-14, no. 8, 1145–1154, 1988.

Randell, B., System structure for software fault tolerance, *Transactions on Software Engineering*, vol. SE-11, no. 2, June 1975.

Ratcliffe, B., and Rollo, A. L., Adapting function point analysis to the Jackson system development, *Software Engineering Journal*, vol. 5, no. 1, 1990.

Ray, G., Software reuse not a panacea: Some firms pursue it as a development goal; others question its viability, *Computerworld*, vol. 26, 47, December 21, 1992.

Rayl, A. J. S., NASA engineers and scientists—Transforming dreams into reality, *NASA 50th Anniversary Magazine*, 2008, http://www.nasa.gov/50th/50th_magazine/scientists.html.

Reifer, D. J., *Managing Software Reuse*, John Wiley & Sons, New York, 1995.

Reifer, D. J., Quantifying the debate: Ada vs. C++, *Crosstalk: The Journal of Defense Software Engineering*, vol. 9, no. 7, 28–30, July 1996.

Renshaw, L., Eliminating GOTOs while preserving program structure, *Journal of the ACM*, vol. 35, no. 4, October 1988.

Riehle, D., A Comparison of the value systems of adaptive software development and extreme programming: How methodologies may learn from each other, in *Extreme Programming Explained*, G. Succi and M. Marchesi, eds., Addison-Wesley, Boston, 2001.

Ritchie, D. M., and Thompson, K., The UNIX time-sharing system, *Bell System Technical Journal*, vol. 57, no. 6, 1905–1929, 1978.

Rix, M., Case study of a successful firmware reuse program, *Proceedings of the Fifth Workshop on Software Reuse*, WISR-5, 1992.

Rombach, H. D., Software reuse: A key to the maintenance problem, *Information and Software Technology*, vol. 33, no. 1, 86–92, January/February 1991.

Rubin, H., *The Rubin Review*, vol. 3, no. 3, July 1990.

Samuelson, P., Self-plagiarism or fair use?, *Communications of the ACM*, vol. 37, no. 8, 21–25, August 1994.

Samuelson, P., Is software patentable?, *Communications of the ACM*, vol. 56. no. 11, 23–25, November 2013.

Saunders, G., COTS controversy: Build-in vs. test-in component reliability—The end of uprating approaches, *COTS Journal*, vol. 5, no. 3, 40–47, March 2003.

Schach, S. R., *Object-Oriented and Classical Software Engineering*, McGraw-Hill, 2011.

Schaefer, W., Prieto-Diaz, R., and Matsumoto, M., eds., *Software Reusability*, Ellis-Horwood, New York, 1994.

Schneidewind, N. F., The state of software maintenance, *IEEE Transactions on Software Engineering*, vol. SE-13, no. 3, 303–310, 1987.

Seaton, B. L., Improving software project estimation within the missions operation and systems development division, Management project for course CSMN 690, University of Maryland, University College Graduate School, College Park, 1995.

Shafer, W., Preito-Diaz, R., and Matsumoto, M., eds., *Software Reusability*, Ellis Horwood, Chichester, United Kingdom, 1994.

Sharma, M., and Padmanaban, R., *Leveraging the Wisdom of the Crowd in Software Testing*, Auerbach/CRC Press, Boca Raton, Florida, 2014.

Shaw, M., and Garlan, D., *Software Architecture: Perspectives on an Emerging Discipline*, IEEE Press, Los Alamitos, California, 1992.

Shneiderman, B., *Software Psychology: Human Factors in Computer and Information Systems*, Scott, Foresman, New York, 1980.

Shneiderman, B., and Plaisant, C., *Designing the User Interface: Strategies for Effective Human-Computer Interaction*, 5th ed., Pearson Addison-Wesley, Boston 2010.

Shooman, M. L., *Software Engineering: Design, Reliability, and Management*, McGraw-Hill, New York, 1983.

Shore, J., and Warden, S., *The Art of Agile Development*, O'Reilly Media, Sebastopol, California, 2008.

Shriver, B., and Wegner, P., *Research Directions in Object-Oriented Programming*, MIT Press, Cambridge, Massachusetts, March 1989.

Sims, C., and Johnson, H. L., *Scrum: A Breathtakingly Brief and Agile Introduction*, Dymaxicon, 2014.

Sinclair, G. C., and Jeletic, K. F., Profile of software engineering within the National Aeronautics and Space Administration (NASA), *Proceedings of the Nineteenth Annual Software Engineering Workshop*, Greenbelt, Maryland, November 30–December 1, 1994.

Sitaraman, M., Welch, L. R., and Harms, D. E., On specification of reusable software components, *International Journal of Software Engineering and Knowledge Engineering*, vol. 3, 207–229, June 1993.

Smith, M. D., and Robson, D. J., A framework for testing object-oriented programs, *Journal of Object-Oriented Programming*, vol. 5 no. 3, 45–53, June 1992.

Software Engineering Laboratory, *Proceedings of the Sixteenth Annual NASA/Goddard Software Engineering Workshop: Experiments in Software Engineering*, NASA/Goddard Space Flight Center, Greenbelt, Maryland, December 1991.

Software Engineering Laboratory, *Software Measurement Handbook*, SEL-94-002, NASA/Goddard Space Flight Center, Greenbelt, Maryland, 1994.

Software Engineering Laboratory, *Impact of Ada and Object-Oriented Design in the Flight Dynamics Division at Goddard Space Flight Center*, NASA/Goddard Space Flight Center, Greenbelt, Maryland, March 1995.

Sommerville, I., *Software Engineering*, 9th ed., Addison-Wesley, Boston, 2012.

Sordillo, D. A., *The Programmer's ANSI COBOL Reference Manual*, Prentice-Hall, Englewood Cliffs, New Jersey, 1978.

Spivey, J. M., *The Z Notation: A Reference Manua*l, Prentice-Hall International, London, 1988.

Stallman, R., Who does that server really serve?, GNU, *Boston Review*, 2010.

Stephens, M., and Rosenberg, D., *Extreme Programming Refactored: The Case against XP*, Apress, Berkeley, California, 2003.

Stone, D., Jarrett, C., Woodroffe, M., and Minocha, S., *User Interface Design and Evaluation*, Morgan Kauffman, San Francisco, 2005.

Stroustrup, B., Classes: An abstract data type faculty for the C language, *ACM SIGPLAN Notices*, vol. 17, no. 1, January 1982.

Stroustrup, B., Data abstraction in C, *AT&T Bell Laboratories Technical Journal*, vol. 63, no. 8, October 1984.

Stroustrup, B., *The C++ Programming Language*, 2nd ed., Addison-Wesley, Reading, Massachusetts, 1991.

Stroustrup, B., *The Design and Evolution of C++*, Addison-Wesley, Reading, Massachusetts, 1994.

SWEBOK, "Software Engineering Body of Knowledge," swebok.pdf, available from http://www.computer.org/web/swebok/v3, IEEE, 2013. The website includes the fifteen individual PDF files listed in the order they appear on the website: "Software Engineering Economics.pdf," "Software Requirements.pdf," "Software Testing.pdf," "Software Construction.pdf," "Software Configuration Management.pdf," "Computing Foundations.pdf," "Software Engineering Models and Methods.pdf," "Software Maintenance.pdf," "Mathematical Foundations.pdf," "Software Design.pdf," "Software Engineering Management.pdf," "Software Engineering Professional Practice.pdf," "Engineering Foundations.pdf," "Software Engineering Process.pdf," "Software Quality.pdf," and "Appendix on Standards.pdf."

Tanenbaum, A., *Modern Operating Systems*, 4th ed., Prentice-Hall, Englewood Cliffs, New Jersey, 2014.

Tate, G., and Verner, J. M., Approaches to measuring size of application products with CASE tools, *Information and Software Technology*, vol. 33, no. 9, 622–628, November 1991.

Tecuci, G., and Keeling, H., Developing intelligent educational agents with the disciple learning agent shell, in *Intelligent Tutoring Systems*, B. P. Goettl, H. M. Halff, C. L. Redfield, and V. J. Shute, eds., pp. 454–463, Springer Verlag, Berlin, 1998. (Best Paper Award.)

Thayer, R. H., and Dorfman, M., *Software Requirements Engineering*, 2nd ed., IEEE Press, Los Alamitos, California, 1992.

Tichy, W. F., Lukowicz, P., Prechelt, L., and Heinz, E. A., Experimental evaluation in computer science: A quantitative study, *Journal of Systems and Software*, vol. 28, no. 1, 9–18, 1995.

Tracz, W., *Software Reuse: Emerging Technology*, IEEE Press, Washington, D.C., 1989.

Tracz, W., *Confessions of a Used Program Salesman: Institutionalizing Software Reuse*, Addison-Wesley, Reading, Massachusetts, 1995.

Tucker, A., Morelli, R., and de Silva, C., *Software Development: An Open Source Approach*, Chapman & Hall/CRC, Boca Raton, Florida, 2011.

Udell, J., Componentware, *BYTE*, vol. 19, no. 5, 46–56, May 1994.

Verner, J. M., Tate, G., Jackson, B., and Haywood, R. G., Technology dependence in function point analysis: A case study and critical review, *Proceedings of the Twelfth International Conference on Software Engineering*, 375–382, 1989.

Voas, J., Payne, J., Mills, J. R., and McManus, J., Software testability: An experiment in measuring simulation reusability, *Proceedings of the ACM-SIGSOFT Symposium on Software Reusability*, Seattle, Washington, April 28–30, 1995.

Voas, J. M., and Miller, K. W., Software testability: The new verification, *IEEE Software*, vol. 12, no. 3, 17–28, May 1995.

Voas, J., Software malleability: We're losing it!, 1999, http://www.cigital.com/papers/download/ndia99.pdf.

Walker, H. M., Homework assignments and Internet sources, *ACM Inroads*, vol. 4, no. 4, 16–17, December 2013.

Walton, G. H., Poore, J. H., and Trammell, C. J., Statistical testing of a usage model, *Software–Practice and Experience*, vol. 25, no. 1, 97–108, January 1995.

Wang, Y., and King, G., *Software Engineering Processes: Principles and Applications*, CRC Press, Boca Raton, Florida, 2000.

Ward, M., Calliss, F. W., and Munro, M., The maintainer's assistant, *Proceedings of the Conference on Software Maintenance 1989*, IEEE Computer Society Press, Miami, Florida, 307–315, October 1989.

Waund, C., COTS integration and support model, in *Systems Engineering in the Global Marketplace: NCOSE International Symposium*, St. Louis, Missouri, July 24–26, 1995.

Webb, N. J., and Thoen, C., *The Innovation Playbook: A Revolution in Business Excellence*, John Wiley & Sons, New York, 2010.

Weiss, D. M., and Lai, C., *Software Product Line Engineering*, Addison-Wesley Longman, New York, 1998.

Wentzel, K., Software reuse—It's a business, *Proceedings of the Fifth Workshop on Software Reuse*, WISR-5, 1992.

Weyuker, E., Axiomatizing software test data adequacy, *IEEE Transactions on Software Engineering*, vol. SE-12, no. 12, 1986.

Weyuker, E., Evaluating software complexity measures, *IEEE Transactions on Software Engineering*, vol. SE-14, 1357–1365, 1988.

Whittaker, J., Arbon, J., and Carollo, J., *How Google Tests Software*, Pearson Education, Upper Saddle River, New Jersey, 2012.

Wilkening, D. E., Loyall, J. P., Pitarys, M. J., and Littlejohn, K., A reuse approach to software reengineering, *Journal of Systems and Software*, vol. 30, no. 1–2, 117–125, July–August, 1995.

Willison, B., *Iterative Milestone Engineering Model*, Parsons Institute for Information Mapping, New York, 2008.

Wohlin, C., and Runeson, P., Certification of software components, *IEEE Transactions on Software Engineering*, vol. 20, no. 6, 494–499, June 1994.

Woody, C., Agile Security—Review of Current Research and Pilot Usage, Software Engineering Institute, April 2013, https://resources.sei.cmu.edu/asset_files/WhitePaper/2013_019_001_70236.pdf.

Wright, P., Mosser-Wooley, D., and Wooley, B., Using color in computer interface design, *ACM Crossroads*, vol. 3, no. 3, 3–6, Spring 1993.

Yourdon, E., and Constantine, L., *Structured Design: Fundamentals of a Discipline of Computer Program and System Design*, Prentice-Hall, Englewood Cliffs, New Jersey, 1979.

Yourdon, E., *Modern Structured Analysis*, Prentice Hall, Englewood Cliffs, New Jersey, 1988.

Zaremski, A. M., and Wing, J. M., Signature matching: A tool for using software libraries, *ACM Transactions on Software Engineering and Methodology*, vol. 4, no. 2, 146–170, April, 1995.

Zweben, S. H., and Heyn, W. D., Systematic testing of data abstractions based on software specifications, *Journal of Software Testing, Verification, and Reliability*, vol. 1, no. 4, 39–55, 1992.

Trademarks and Service Marks

The following are either trademarks or service marks of their respective companies or organizations.

Agilent

Apple, Apple Macintosh, Apple iOS, Pages, Numbers, Keynote

AT&T

Bell Communications Research

Bellcore

Chromebook

Eclipse

Google

Hewlett-Packard

HP

IBM

Java

Microsoft, Microsoft Windows, Microsoft Office, Microsoft Word, Microsoft Excel, Microsoft PowerPoint, Microsoft Project

Oracle

Razor

Sun Microsystems

System Architect

Telcordia

Telelogic

Terremark

Verizon

Visible Systems Corp.

Index

C

D

E

I

J

K

L

M

N

O

P

Q

R

S

X

Y

Z